Flora of the Hudson Bay Lowland and its Postglacial Origins

Inquiries: Monograph Publishing Program, NRC Press, National Research Council of Canada, Ottawa, Ontario K1A 0R6, Canada. Web site: www.monographs.nrc.ca

Correct citation for this publication: Riley, J.L. 2003. *Flora of the Hudson Bay Lowland and its Postglacial Origins*. NRC Press, Ottawa, Ontario, Canada. 236 pp.

A Publication of the
National Research Council of Canada
Monograph Publishing Program

Flora of the Hudson Bay Lowland and its Postglacial Origins

John L. Riley

The Nature Conservancy of Canada
110 Eglinton Avenue West, Suite 400
Toronto, Ontario M4R 1A3

NRC·CNRC
NRC Research Press
Ottawa 2003

Printed in Canada on acid-free paper. ♾

ISBN 0-660-18941-0
NRC No. 44464

Canada cataloguing in publication data

Riley, J.L. (John L.), 1950–
Flora of the Hudson Bay Lowland and its Postglacial Origins

Issued by the National Research Council of Canada.
Includes bibliographical references.
ISBN-0-660-18941-0

1. Botany — Hudson Bay Region.
I. National Research Council Canada.
II. Title.

QK203.H83R54 2003 581.97'14111 C2003-980116-0

CONTENTS

ABSTRACT

The Hudson Bay Lowland is the 325 000 km^2 Paleozoic and Mesozoic geologic basin located south and west of Hudson and James bays, Canada. It occupied a central position in the full-glacial limits of the late-Wisconsinan Laurentide Ice Sheet and is the region of central North America most recently and rapidly emerging from the ocean. During deglaciation, the Lowland was more remote from extraglacial refugia of plant species than any other region of North America, and its patterns of re-colonization provide a youthful analogue for the study of older postglacial floras.

The Lowland flora includes 816 native vascular plant species (857 geographically and ecologically distinct taxa) and 98 non-native species, here catalogued for 11 individual data collection areas. Phytogeographic variability was analysed by calculating the degree of change in geographic affinities of the floras of different areas of the Lowland, resulting in a mapping of floristic zones: maritime tundra (low arctic); peat plateau and woodland (high subarctic); peatland and woodland (low subarctic); southwest James Bay (low subarctic); and boreal peatland (high temperate).

This flora illustrates both conservative (and relict) patterns of gradual species migration (and subsequent range disruption), as well as saltatory or vanguard patterns of long-distance dispersal. The species that are widespread in the Lowland (35% of taxa) are overwhelmingly transcontinental and circumboreal in their continental ranges; 80% of them occur in extraglacial areas east, south, southwest, and northwest of the Wisconsinan late-glacial maximum and likely derived from multiple sources. Fifteen percent of taxa are eastern North American, deriving from eastern refugia and restricted primarily to the south. Eight percent are western North American taxa, mostly subarctic and cordilleran in their ranges and still occurring northwest and southwest of the glacial maximum. More than half of the Lowland's arctic taxa (10%) are restricted to within 20 to 40 km of Hudson Bay, but the others (7%) also occur farther south. Many Lowland taxa (17%) are restricted to its maritime coasts, and more than 30 of them occur on southern James Bay and Hudson Bay as disjuncts, from nearest populations more than 550 to 2100 km away. Other Lowland taxa are similarly disjunct from their core ranges beyond the Precambrian Shield or the ocean that surrounds the Lowland, including 20 western North American taxa; 8 species only at Cape Henrietta Maria, the coldest part of the Lowland; and 8 occurring only on isolated Precambrian outcrops.

The dominant Lowland flora and the immigration of nearly 70% of Lowland species are the product of a gradual migration of species into the Lowland from extraglacial areas east, south, southwest, and northwest of the Lowland. This frontal migration and the consolidation of the dominant vegetation disrupted some previously more continuous distributions of subarctic and western species. However, the occurrence of so many geographically disjunct species likely reflects a pattern of recent long-distance dispersal; their frequency within such a youthful regional flora suggests that this may have been typical of the early development of floras in older glaciated regions.

Twenty-three percent of the native Lowland taxa are presently known from three or fewer sites, and may be considered as regionally rare species. The Lowland's flora of non-native species, although reflecting more than 300 years of European settlement, is still remarkably localized around small human settlements and railheads.

RÉSUMÉ

Les basses terres de la baie d'Hudson sont en fait un bassin géologique de 325 000 km^2, datant du Paléozoïque et du Mésozoïque, situé au sud et à l'ouest de la baie d'Hudson et de la baie James (Canada). Outre le fait que ce bassin occupait une position centrale dans les limites glaciales de l'Inlandsis laurentidien du Wisconsinien tardif, il représente aujourd'hui la dernière région du centre de l'Amérique du Nord qui émerge le plus rapidement de l'océan. Lors de la glaciation, les basses terres étaient l'un des refuges floristiques extraglaciaires les plus éloignés de l'Amérique du Nord et l'analyse des patrons de recolonisation offre un modèle pour l'étude de plantes postglaciaires plus anciennes.

La flore des basses terres comprend 816 espèces de plantes vasculaires indigènes (857 taxons géographiquement et écologiquement distincts) et 98 espèces de plantes non vasculaires exotiques. Cette flore est cataloguée ici pour 11 zones d'échantillonnage. La variabilité phytogéographique a été analysée en calculant le niveau de changement dans les affinités géographiques des plantes de différentes régions des basses terres. Cette procédure a également permis d'établir une cartographie de zones floristiques : toundra maritime (bas-arctique); plateau tourbeux et terrain boisé (haut-subarctique); tourbière et terrain boisé (bas-subarctique); sud-ouest de la baie James (bas-subarctique); tourbière boréale (haut-tempérée).

Cette flore démontre des patrons discrets (et reliques) de migration graduelle d'espèces (suivi ultérieurement par une perturbation de leur distribution) ainsi que des patrons saltatoires ou « pionniers » de dispersion sur de grandes distances. Les espèces très répandues dans les basses terres (35 % des taxons) sont en très grande majorité transcontinentale et circumboréale dans leur répartition continentale; quatre-vingts pour cent d'entre elles ont été répertoriées dans des régions extraglaciaires situées à l'est, au sud, au sud-ouest et au nord-ouest du maximum tardiglaciaire Wisconsinien en provenance, probablement, de sources multiples. Quinze pour cent des taxons sont des espèces de l'est de l'Amérique du Nord et sont originaires des refuges de l'est. Huit pour cent sont des taxons de l'ouest de l'Amérique du Nord, essentiellement caractérisés par des distributions de type subarctique et cordillérien, qui se retrouvent encore aujourd'hui au nord-ouest et au sud-ouest du maximum glaciaire. Plus de la moitié des taxons arctiques (10 %) des basses terres se retrouvent exclusivement dans un rayon de 20 à 40 km de la baie d'Hudson, tandis que les autres taxons (7 %) ont été également répertoriés dans les régions plus au sud. De nombreux taxons (17 %) des basses terres sont limités aux côtes marines et plus de 30 d'entre eux ont été répertoriés dans le sud de la baie James et de la baie d'Hudson en tant qu'individus isolés, leurs plus proches populations étant plus de 550 à 2100 km plus loin. D'autres taxons se retrouvent aussi isolés de leurs domaines centraux au-delà du bouclier précambrien ou de l'océan qui entoure les basses terres, dont 20 taxons de l'ouest de l'Amérique du Nord, huit espèces exclusives au cap Henrietta Maria et huit autres espèces exclusives aux affleurements précambriens isolés.

La flore dominante des basses terres et l'immigration de près de 70 % des espèces des basses terres sont le résultat d'une migration graduelle d'espèces boréales vers les basses terres en provenance des zones extraglaciaires à l'est, au sud, au sud-ouest et au nord-ouest des basses terres. Cette lente migration de front et la consolidation de la végétation dominante ont perturbé une répartition plus ancienne et régulière des espèces subarctiques de l'ouest. Cependant, la présence de nombreuses espèces géographiquement distinctes illustre un patron probable de dispersion récente sur de longues distances; leur présence au sein d'une flore régionale aussi récente semble indiquer qu'il s'agirait d'un phénomène typique dans le développement précoce de la flore des plus anciennes régions glacières.

Les données actuelles sur 23 % des taxons indigènes des basses terres proviennent de trois sites ou moins. Ces espèces pourraient même être considérées comme rares à l'échelle régionale. Même si la flore exotique des basses terres se caractérise par 300 années de colonisation européenne, elle demeure tout de même remarquablement localisée autour de petits établissements humains et de terminaux ferroviaires.

INTRODUCTION

The quintessential landscapes of central Canada are its rich southern lowlands, its immense boreal expanses of forest, rock, and lake, and its uninterrupted arctic terrain to the north. However, at its centre is the Hudson Bay Lowland, one of Canada's most remote landscapes, long recognized as an ecological region clearly distinct from its neighbours (Wicken 1986; ESWG 1995). It occupies 3.5% of the surface area of Canada, a quarter of Ontario, a tenth of Manitoba, and a small portion of Quebec.

The Hudson Bay Lowland is one of the Earth's largest more or less continuous wetland landscapes. By comparison, it is twice the size of South America's Pantanal wetland region and perhaps five times larger than the floodplain forests along the Amazon River. It is also one of the least populated regions in the Western Hemisphere and one of the last regions of North America to have its flora and vegetation documented.

Geology and glacial history

The Hudson Bay Lowland is a broad coastal plain that lies south and west of both James Bay and Hudson Bay. It is a compact geological province of Paleozoic and Mesozoic limestones, shales, and sandstones almost universally sheathed to depth by calcareous marine clays of postglacial age (325 000 km^2; Figs. 1 and 2; Norris et al. 1967; Sanford et al. 1968; Johnson et al. 1992; Dredge 1992).

The Canadian Shield surrounding the Lowland covers half of Canada's land area. It is noted for its harsh landscapes of exposed granites and other predominantly acidic rocks of Precambrian age, locally covered by glacial deposits. This is of particular ecological significance as the extent of Canadian Shield surrounding the Lowland is never less than 325 km in width and isolates the Lowland from plant and animal populations that are not adapted to life on the Shield.

The Hudson Bay Lowland was also the geological province located closest to the centre of the Laurentide Ice Sheet, which dominated the last glaciation of North America. The vast weight of the ice depressed the Earth's crust so that at the time of deglaciation a massive marine incursion flooded almost all of the Lowland area. The extent of this incursion generally coincides with the extent of the geological region.

The weight of this water was, however, much less than the weight of the earlier ice sheet, and in consequence, the Lowland experiences the continent's maximum rates of isostatic rebound, and hence, emergence of land from the ocean. As a result, the Lowland is almost universally covered by marine silts and clays, occasionally reworked by river or sea, and terraced by marine beach ridges deposited along the ocean coast as the postglacial Tyrrell Sea shrank into the modern Hudson Bay and James Bay. At projected rates, Hudson Bay will have decanted entirely into the Atlantic Ocean at about the time of the end of the present interglacial period.

Terrestrial flora and fauna have had less than 7000 years to occupy the Lowland's margins and uplands, which presently are at elevations of approximately 60 m above sea level, and have had much less time to colonize lower elevations. Cape Henrietta Maria has

Fig. 1. Hudson Bay Lowland in relation to the maximum extent of Wisconsinan glaciation.

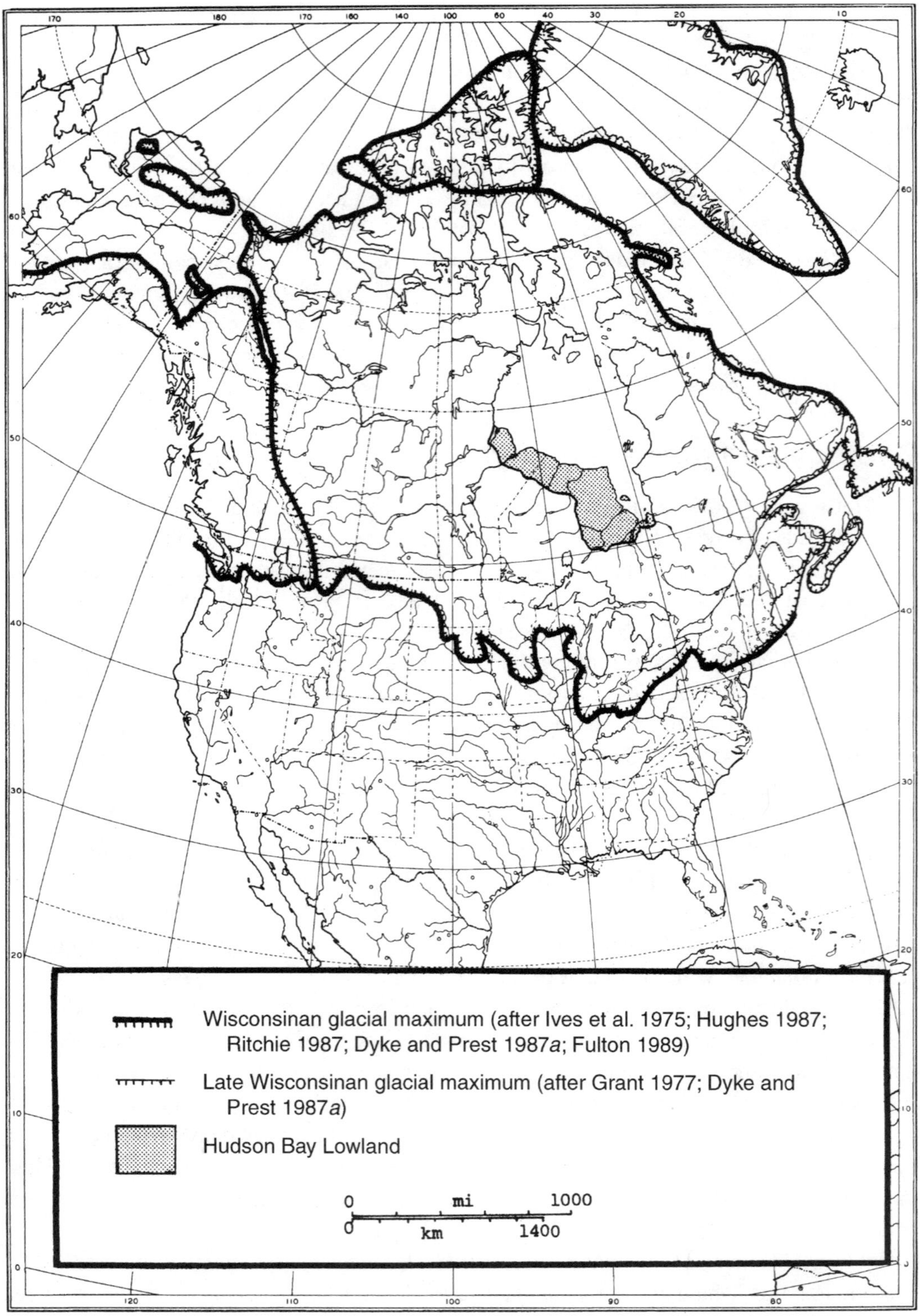

Fig. 2. Hudson Bay Lowland location names.

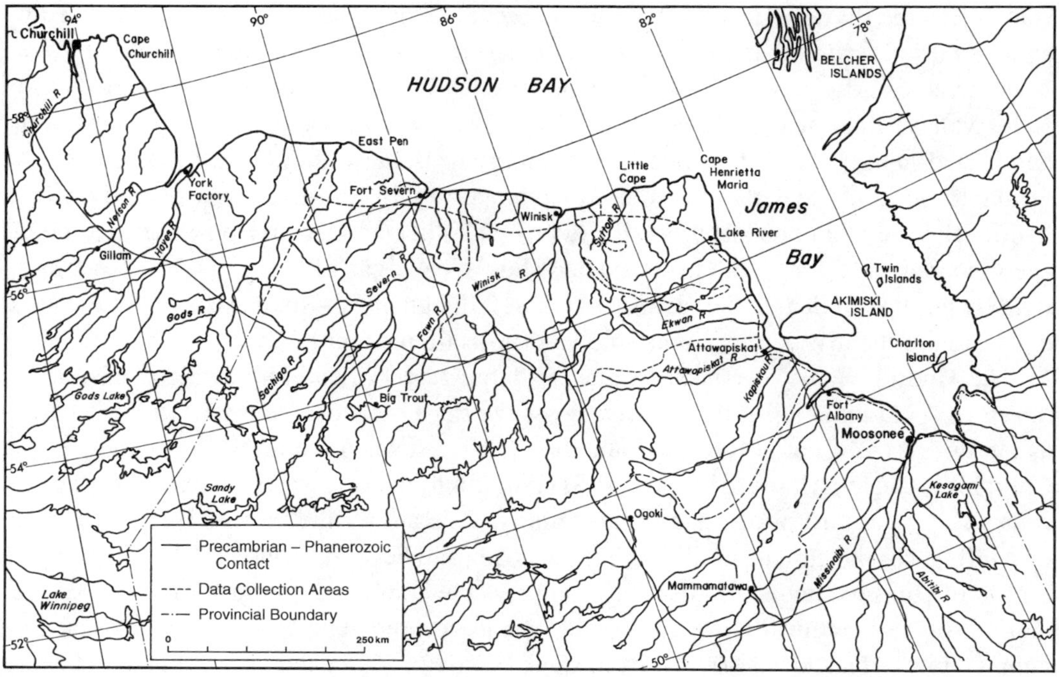

emerged from the ocean at a rate of 1.2 m/century over the last 1000 years, revealing a width of almost 30 km of newly colonizable landscape in that time (Webber and Richardson 1970; Andrews 1968). This emergence rate is an order of magnitude less than during the immediate postglacial period (8000 years BP), at which time the Cape was more than 100 m below the surface of the Tyrrell Sea. The rates of rebound decrease away from the Cape but are still significant everywhere in the Lowland, even to the extent of stranding historic posts like Fort Albany high above their original water access (Hunter 1970). As far away from the Cape as Churchill, rates of rebound are estimated at 40 cm/century, thus the shore is moving northward at a rate of 400 m/century (Dredge and Nixon 1992).

Thousands of beach ridges parallel the coast, thrown up by the sea during major storms. The frequency, separation, and heights of these beach ridges vary along the length of the Hudson Bay and James Bay coasts, largely reflecting the particular exposure and slope of the coast, and the fetch and severity of storm events. Another sea-related landform feature that occurs in the southernmost Lowland, in the Albany interior, is a landscape of iceberg groundings. Here icebergs were pushed about in shallow Tyrrell Sea waters leaving linear scar trenches crisscrossing the marine clays.

The underlying Paleozoic and Mesozoic bedrock is occasionally exposed along major rivers, but exposures of the basement Precambrian bedrock are very rare. In the vicinity of Churchill, Manitoba, protoquartzites intrude through the Paleozoic bedrock to form a low coastal ridge. In Ontario, massive Precambrian inliers, many in the form of dramatic cuestas, define the Sutton Ridges — an area of drier upland ecosystems stretching intermittently about 150 km east to west, about 150 km southwest of Cape Henrietta Maria (Sanford et al. 1968; Craig and Mcdonald 1968).

Glacial features also include a variety of till deposits now overlain by marine clays. In some areas the till deposits date back to the previous interglacial period, as evidenced by ancient buried forests that are exposed along riverbanks, such as along the Moose River. Accumulated surficial sediments are deeper at inland sites (e.g., up to 75 m in the Nelson River valley) than near the coast (e.g., 5 m at Churchill) (Dredge 1992). The origins, depth, and nature of local sediments are important because they, in part, determine the available nutrients and the water holding properties of substrates, particularly where marine clays are thin or absent. Finally, eskers and other subglacial deposits such as deltas occur sporadically across the Lowland and show clear signs of being reworked and modified by the Tyrrell Sea. In the vicinity of Churchill, both north–south eskers and reworked deltas add to the diversity of accessible landforms in that area.

The Hudson Bay watershed is Canada's largest by far, and half of Canada's largest dozen rivers drain in leisurely majesty across the flat Lowland, some carrying water from as far away as the central Rocky Mountains. The fresh waters of these rivers, the Nelson, Churchill, Albany, Moose, Hayes, and Severn, dilute the salt waters of Hudson Bay and James Bay to a third of average ocean salinity. This allows the bays to freeze, which in turn influences the climate of all of central Canada and beyond (Rouse 1991). The climatic effects of Hudson Bay and James Bay compress the continental temperature and precipitation isolines southward, so that arctic, subarctic, and temperate ecosystems are all jammed into a distance of 500 km from Cape Henrietta Maria and Churchill south toward the Great Lakes and Lake Winnipeg.

Vegetation, soils, permafrost, and the tree line

Peatland and wetland vegetation dominates the Hudson Bay Lowland, and the scale and diversity of these vegetation types establish it as a distinct hydrological province (Canada 1974; NWWG 1988). Site conditions shift rapidly away from narrow, dry coastal and riparian vegetation to vast expanses of saturated peatlands. More than 90% of the Hudson Bay Lowland is a saturated peatland plain, with subtle slopes of 65–100 cm/km (Riley 1982; Sims et al. 1979). "The percentage of peatland in relation to total land area in the middle and northern parts of the Hudson Bay Lowland is so close to 100% that it could scarcely be higher anywhere in the world" (Sjörs 1963).

The modern display of vegetation that grows on this thick mat of peat (up to 4 m deep) is remarkably diverse in its patterns of species, available nutrients, and slow lateral water flow. This variation is based on subtle differences in accumulated peat depth, local climate, substrate microtopography, and water flow within the peatlands themselves. The overall trends in vegetation succession, as the slope of the land has flattened over millennia, are toward deeper peat accumulations and toward vegetation types adapted to more acidic, ombrotrophic surface waters.

Permafrost dominates the Hudson Bay coast, decreases southward toward the interior, and is largely absent, except as discontinuous lenses, near the Shield contact in northeastern Ontario. Permafrost occurs wherever land temperatures are less than 0°C in consecutive years. Near Hudson Bay, permafrost underlies almost all landforms at depth (up to 80 m thick at Churchill; Dredge 1992) and extends northward as a wedge under the

Hudson Bay shore for some distance. Peat is a superior thermal insulator and conserves permafrost so effectively that the southern limits of discontinuous permafrost occur, continentally, in the deep peatlands in the southern Hudson Bay Lowland (Brown 1967, 1973).

Within 80 km of the Hudson Bay coast, permafrost dictates the character of the landscape. Near the coast, where organic accumulation is less, permafrost beach ridges support lichen and tundra woodlands, and untreed lowland tundra heath predominates. In mineral soils, freeze–thaw cycles in the active layer can give rise to "frost boils" or "stone circles", and in bedrock exposures, to conspicuous vertical rock heaving.

In shallow peat terrain, multi-sided outlines appear as shallow troughs in the tundra vegetation, each underlain by an ice wedge and clearly visible from the air. These ice-wedge polygons occur at Cape Henrietta Maria, the lower Brant River, and in the Churchill area. Farther toward the interior, in deeper peat, permafrost occupies the entire peat column and extends below it. The insulating properties of peat and the sparse tree cover maintain frost penetration and result in peat expansion that raises the surface of the peatland as individual permafrost "palsen" or as coalesced permafrost landscapes called "palsa fields" or "peat plateaus". These raised and effectively drier active layers support a distinctive vegetation very similar in species and nutrient availability to bogs, but often with dominant lichen cover over broad expanses. Such areas can be elevated sufficiently to support sparse shrub and tree cover as well. These systems are prone to periodic natural fires, especially in the northwestern part of the Lowland.

Extensive thermokarst lake systems dominate portions of the northern peatland interior, such as in the vicinity of the Winisk River and Shagamu Lake. These shallow lakes gradually move downwind across their peat plateaus through the erosion of frozen peat edges downwind and the infilling of peat upwind. Typical measurements taken from five large, lichen-dominated peat plateaus between Cape Henrietta Maria and the Manitoba border (averaging 95 km from the coast) indicated surficial active layers averaging only 35 cm deep in late summer. One of these sites, 73 km from the coast, had a 25 cm active layer over 3.1 m of frozen peat, underlain in turn by frozen marine clay.

Farther south, scattered discontinuous permafrost occurs throughout most of the Lowland (Brown 1967, 1973) where it is increasingly restricted to the deeper peatlands, often persisting permanently under the tear-shaped treed bog "islands" that occur in open peatland complexes.

Sjörs (1959, 1961, 1963) provided the classic treatment of Hudson Bay Lowland peatlands, discriminating between ombrotrophic "bogs", many of them raised and all without laterally moving waters, and more minerotrophic "fens", almost all with slowly flowing waters. Both bogs and fens can occur as treeless or as sparsely or even heavily wooded systems, and both grade imperceptibly into distinct and more minerotrophic "swamp" types. The understoreys of bogs, fens, and swamps in the Lowland can range from pure sphagnum or moss systems to pure graminoid (sedge) and shrub-dominated systems (both still on live moss substrates). Each combination and permutation of these physiognomic types occur on slightly different site types, and in areas with shallower or nonexistent peat, each of these types also grades into more nutrient-rich systems, "marsh", "meadow marsh" or swamp systems, particularly near the coast and along major rivers (Riley and Mckay 1980).

In the Ontario Lowland, wetland types have been estimated to occur in the following proportions: bog (36%), fen (24%), permafrost peatland (palsa and peat plateau, 22%), swamp (13%), and marsh (5%); for a total wetland area of 221 000 km^2 (Riley 1982, 1989*a*, 1989*b*; definitions of types from NWWG 1988). In some parts of the Lowland, particular wetland types occur as homogeneous units over large areas, such as in the Moose River basin where treed fen and conifer swamp cover vast tracts. However, complex patterning is the norm, and throughout the Lowland, the diversity of peatland vegetation patterns have presented major challenges to workers describing and mapping them (Bates and Simkin 1969; Sjörs 1959, 1961, 1963; Pala and Boissonneau 1982). Patterning within these complexes, such as by "string" or stepped fens, bog "islands", and myriad open-water ponds and pools further elaborate the particular vegetation patterns of the Lowland.

In the part of the Lowland not dominated by peatlands, the array of terrestrial ecosystems is also impressive. Coastal ecosystems occupy a broad belt, especially on James Bay, where they include off-shore bottom vegetation, tidal salt marsh, and supertidal meadow-marsh systems, as well as diverse beach-ridge complexes supporting tundra, lichen woodlands, conifer forest, swamp, and treed bog, depending on their distance from the coast and their latitude. On the more exposed coasts of Hudson Bay, these same types of vegetation also occur, but are often truncated in both width and diversity.

Away from the coast, the range of interior upland ecosystems includes, again, innumerable beach ridges as well as scattered, large subglacial deposits of coarse materials that were sorted and reworked by the receding Tyrrell Sea. The major rivers are flanked by a variety of features including forested silt and permafrost-silt levees, and limestone cliffs, fossiliferous limestone reefs and flats with karst and erosional features. Some rivers occupy broad down-cut valleys, with long, unconsolidated erosional slopes and shores supporting dry open forb and shrub communities. On rare occasions bottomland swamps can be found along the rivers, such as the elm bottomlands of the Kenogami River and those around the riparian islands of other major rivers. On the Sutton Ridges and the higher interior terrain straddling the Manitoba–Ontario border, extensive conifer woodlands occur. The Sutton Ridges comprise a distinctive landscape of intermittent Precambrian ridges, cuestas, and boulder pavements, with prominently developed cliffs, scarps, and columnar jointing.

Peatlands and organic soils blanket the Lowland. Along the Hudson Bay coast, perennially frozen Cryosols dominate, with variable depths of active layer depending on exposure, insulation, and microtography. On the well-drained sands and fine gravels of beach-ridge systems, perennially frozen Regosols dominate at youthful, coastal sites. Weak Podzolic soils have developed on ridges 6–10 km from the central Hudson Bay coast, even where frozen subsoils occur within 20 cm of the top of the mineral soil. Visible ice occurs at 40–50 cm depths on beach ridges more than 10 km from the coast at sites where treed-bog vegetation dominates low-relief ridges. On the more prominent, better drained ridges, even as much as 40 km from the coast, organic accumulations of less than 15 cm can occur, with spruce–lichen woodlands occurring there on Podzolic soils (Cowell et al. 1982).

Southward, Regosols and weak Podzols are the typical soils of the more isolated, well-drained sands and fine gravels of uplands, but weak Brunisols develop on some of the river levees with deciduous tree cover, and some Gleysols develop on imperfectly drained mineral sites. The rates of podzolization and carbonate leaching of beach-ridge soils in the Lowland are twice as rapid near southern James Bay as near the Hudson Bay coast, reflecting greater precipitation and higher temperatures. Organic matter can accumulate in these soils for about 2000 years in the southern James Bay area and about 3000 years in the central Hudson Bay region before oxidation begins to deplete those levels, probably as an effect of carbonate leaching (Protz et al. 1984).

The major ecological processes influencing the landscape of the Lowland are directly related to the rapid uplift and flattening of the region and to the rapid accumulation of saturated and frozen plant materials across most of the area. These result in a general paludification of the landscape, the development of permafrost systems, and a relatively rigid gradient of differently aged landscapes from the coasts toward the interior of the Lowland.

The tree line running parallel to Hudson Bay is erratic, extending farthest to the north on river levees and beach ridges, where there is better drainage and a deeper active layer. Locally, the tree line is also controlled by the drying and abrading effects of winter winds and wind-driven snow. The location and topography of snowbeds shape the stunted growth of trees and shrubs by exposing or protecting them (Scott et al. 1993). Exposed tree-line spruce are maintained by vegetative layering, and individual krummholz tree islands may reach great ages. Beach-ridge woodlands, treed peat plateaus, and forested permafrost levees are subject to periodic natural fires, on a more frequent rotation in the drier western Lowland than near James Bay. East of the Lowland, tree-line fires have been shown to reduce the amount of snow accumulating on the landscape and thus to result in the long-term conversion of treed systems to more open lichen–tundra communities (Arseneault and Payette 1992; Gajewski et al. 1993).

Climate, climate change, and other recent stresses

The major gradient of landscape change from the coasts toward the interior is paralleled in the climate of the Lowland. The climate changes rapidly from temperate, continental boreal conditions in the south and southwest to arctic oceanic conditions along Hudson Bay. Hudson Bay freezes over completely in winter and remains frozen or dominated by ice throughout much of the high-sun season, resulting in the winterization of summer. Even in late July, 50% of offshore waters in some areas from Fort Severn to Cape Henrietta Maria can remain covered in ice (Rouse 1991).

The mean summer position of the Arctic front is forced southward by the cold air mass over Hudson Bay (Bryson 1966). Locally, Hudson Bay generates persistent summer onshore winds and land–sea breezes that reduce temperatures and increase fog. Evapotranspiration rates are at their highest 10–20 km inland, decreasing rapidly to levels 25–40% less at the coast (Rouse 1991). Similar maritime-arctic climates occur southward along the coast of Labrador and the Aleutian Islands, also because of cold offshore currents, specifically the Labrador and Kamchatka currents.

Climatic conditions in the Lowland become progressively drier toward the northwest, influencing the frequency of natural fires and the dryness of upland habitats. The Lowland's overall climatic trends are illustrated by data from three coastal Lowland stations, which are compared with the interior stations nearest to them, none of which are in the Lowland itself (Table 1).

In recent years, two major systemic ecological stresses have begun to be documented in the Lowland. Firstly, the regional climate has warmed over the course of the 20th century, with an earlier occurrence of spring melt (Gagnon and Gough 2002). Northern ecosystems, especially at the southern limits of permafrost, are especially sensitive to global climate change because small changes in temperature can have significant effects on snow cover and thaw depth (Maxwell 1992). Palsa peatlands, tundra, ice-aggraded river levees, and other permafrost landforms occur farthest south globally in peat-dominated systems and marine-cooled regions. Both collapse and erosion features and aggrading features are visible in the Lowland's permafrost tension zone. Personal observations in the 1990's compared with those in the mid-1970's, suggest that collapse features are now more widespread, for example, in the Ekwan-to-Lake-River area of the northern James Bay coast. The Hudson Bay Lowland is a major carbon reservoir, and the reduction of the capacity to store carbon, as a result of climate warming, may trigger yet unknown atmospheric impacts, particularly through the release of methane gas. At present, the Hudson Bay Lowland's wetlands are only a moderate source of methane emissions (Roulet et al. 1994).

Secondly, the foraging of rapidly expanding Snow Geese populations in coastal wetlands has led to very extensive, seriously degraded coastal marshes, both intertidal and supertidal. The expanding populations are attributable in large part to the improved health of nesting birds returning from the south. This unnatural increase in their fecundity is largely the result of artificial feeding on their way north through the central North American flyway. The result has been severe vegetation damage, increased erosion, and aggravated halophytic conditions (Abraham and Jeffries 1997; Kotanen and Jeffries 1997).

Table 1. Climatic data from stations in the Hudson Bay Lowland and elsewhere.

	Mean daily temperature (°C)[a]	Annual growing degree days above 5°C[b]	Mean annual precipitation (mm)[a]	Annual days with precipitation[a]
Winisk (55°14´N, 85°07´W)	–5.5	625	608	164
Big Trout Lake (53°50´N, 89°52´W)	–3.0	1025	581	161
Churchill (58°45´N, 94°04´W)	–7.2	625	403	148
Gillam (56°21´N, 94°42´W)	–4.6	830	422	138
Moosonee (51°16´N, 80°39´W)	–1.1	830	728	173
Cochrane (49°04´N, 81°02´W)	–0.6	1300	885	149
Aklavik, N.W.T. (68°13´N, 135°00´W)	–9.1	600	208	107
Toronto (43°40´N, 79°38´W)	7.3	2300	762	137

[a]Canada 1981.
[b]Canada 1974.

Objectives

Over the past century, the Hudson Bay Lowland has been the subject of many studies of its geology, wildlife, and peatlands, the results of which are widely scattered in the literature (Sims et al. 1979). Despite various botanical studies in the region, there is no comprehensive catalogue of its vascular plants. The primary objective of this study was to integrate new floristic data collected across the region with existing published and unpublished data. To this end, original fieldwork was undertaken from the broadest possible range of unique and representative Lowland habitat types within 11 predetermined data collection areas (Fig. 2). These data collection areas were also used to organize the existing body of herbarium collections and literature reports.

A secondary objective was to analyse the flora in terms of its coincident distribution patterns within and beyond the Lowland. This analysis included a quantitative assessment of the change in species occurrence and geographic affinities across the Lowland, resulting in the characterization of distinct floristic zones.

Thirdly, the groups of Lowland species with coincident distributions in their North American ranges were discussed as suites of taxa that share similar inferred postglacial origins, based on their probable co-migration into the Lowlands from particular extraglacial refugia beyond the maximum limits of Wisconsinan ice sheets.

These objectives addressed the practical need for a catalogue of vascular plant species and their geographic affinities and abundance, which is central to appropriate resource use and conservation planning in the Hudson Bay Lowland and to understanding the biological diversity of a unique and relatively inaccessible region of central Canada. Because the Lowland occurs close to the centre of the limits of the Laurentide Ice Sheet and is the region of central North America most recently and rapidly emerging from the ocean, the analysis of its postglacial recolonization provides a unique and youthful analogue for the floras of older glaciated landscapes. Postglacial recolonization and the potential for floras to adjust to climatic and landscape changes is interpreted by contrasting the conservative and relict patterns of gradual species migration (and subsequent range disruption) with the more salutatory or vanguard patterns of long-distance dispersal.

METHODS

Data assembly and field surveys

Systematic research of the flora of the Hudson Bay Lowland did not begin until the last two decades of the 19th century. Earlier published notes consisted of general observations on the natural history and uses of plants in the region, and only incidental collections of plants were made. In the late 1800's, the Geological Survey of Canada began exploring the major rivers and coastal areas, collecting plants and publishing records. Among the most prominent explorers were R. Bell, P. Low, W. Spreadborough, and J. Macoun (1903).

From the turn of the 20th century until the 1950's, other researchers expanded our knowledge of the vegetation and flora of the Lowland, again focusing on the coasts and river systems; for example, E. Beckett (1959), G. Gardner (1937, 1946), W.C. Gussow (1933), I. Hustich (1955, 1957), L. Fagerstrom (1948), F. Johansen (Stormer 1933), D.R. Moir (1958), A.E. Porsild (1932), D. Potter (1934), J.C. Ritchie (1956, 1962), W.B. Schofield (1958), and H.G. Scoggan (1957, 1969).

From 1948 to 1966, E. Lepage, A. Dutilly, and M. Duman published a series of botanical studies of the southern Lowland, the James Bay area, and the Ontario Hudson Bay coast (Dutilly and Lepage 1948, 1952, 1963; Dutilly et al. 1954, 1959; Lepage 1966; Lepage et al. 1962). During the same time period, W.K.W. Baldwin collected widely in the Lowland, particularly in the south (Baldwin 1948, 1953; Baldwin et al. 1959) and worked with both H. Sjors, in his seminal peatland studies (1959, 1961, 1963), and A.E. Porsild. A number of other field studies were not published, but the resulting collections were added to various herbaria. Notable examples are studies by F. Cowell (1968, 1969), A.T. Cringan, J.M. Gillett, J.K. Jeglum (1971), M. Kirk, H.G. Lumsden (Raveling and Lumsden 1977), P.F. Maycock (1968, 1974), J.K. Morton, A.A. Reznicek (1979), G. Ringius, W.B. Scott, R.A. Sims (Sims et al. 1987*a*, 1987*b*), J. Sparling, and G.M. Stirrett. These past studies were reviewed by the author concurrent with new field studies in the region (Sims et al. 1979).

The current study is based on field surveys carried out by the author since 1970, primarily in 1972, 1976–1980, and 1990 (Fig. 3), along with additional recent research by others. In 1972, surveys were taken of the vegetation of the Shipsands Island Waterfowl Sanctuary at the mouth of the Moose River (Riley and Moore 1973) and in the Onakawana area of the Abitibi River (Stanfield et al. 1972). In 1976, field studies concentrated on the Moosonee, Moose River, Moose Factory, southwest James Bay, and Kinoje Lake areas (Riley and McKay 1980). Fixed-wing aircraft and helicopter reconnaissance of the coast was made to Attawapiskat and Cape Henrietta Maria, and westward to the coast northwest of Winisk. In 1977, studies were made of the karst- and limestone-associated vegetation of the lower 100 km of the Attawapiskat River (Cowell and Riley 1979), as well as the Hudson Bay coast near the mouth of the Shagamu River.

In 1978, 1979, and 1990, helicopter surveys were made of the Ontario Lowland with the Ontario Provincial Remote Sensing Office of the Ministry of Natural Resources. More than 310 sites representing wetland, forest, upland, riparian, and coastal systems were

Fig. 3. Hudson Bay Lowland data collection areas and collection sites (names indicated in Table 2).

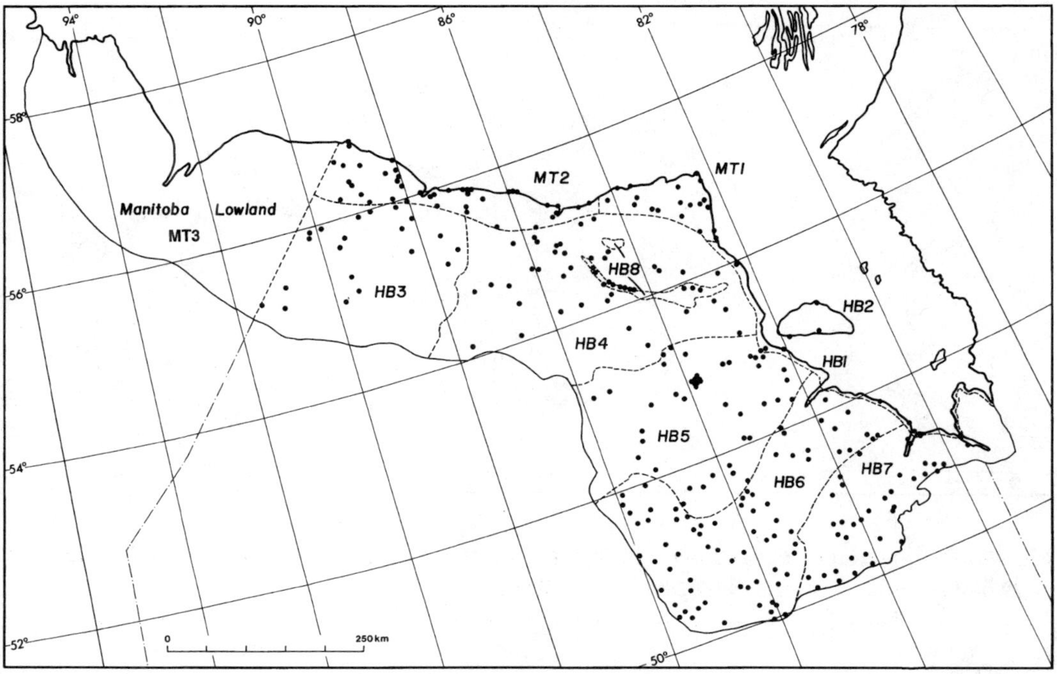

visited for varying lengths of time. In 1979 and 1980, field surveys at the Sutton Ridges were conducted, particularly around Aquatuk Lake (Riley 1979*b*; Riley and Walshe 1985; Riley 1982; McAndrews et al. 1982).

The focus of the field surveys taken from 1976 to 1980 was on historically under-sampled areas such as the Attawapiskat River basin (42 sites), the Winisk-Ekwan River basin (36 sites), Sutton Ridges (more than 20 sites), and the Hudson Bay coast (19 sites in the Cape Henrietta Maria zone, 43 sites west to the Manitoba border). This resulted in relatively well-stratified sampling of the Ontario Lowland (Fig. 3), with the exception of Akimiski Island (Riley 1981). Fortuitously, C.S. Blaney, P.M. Kotanen, and others have since undertaken additional surveys of Akimiski Island (Blaney and Kotanen 2001).

In other collection areas, the new data were added to more significant historic studies, such as in the Severn basin (Moir 1958), the Moose and Albany River basins (Sims et al. 1979, table 1), and the Manitoba Lowland (Sims et al. 1979; Scoggan 1957; Ritchie 1956, 1962; Beckett 1959; Johnson 1987). Recently, M.J. Oldham and D. Sutherland collected in the Ontario Hudson Bay area in 2000 and 2001 (MJO), and C.E. Punter undertook significant collecting in Manitoba's coastal areas in 1999 and 2000 (WIN).

A gap in the field studies remains in the area between the Manitoba and Ontario border and the Nelson River, even at coastal sites in most of that region. The northern half of the Manitoba Lowland, however, is well represented in herbaria collections and in literature reports. The overall evenness of survey coverage across the Lowland can be illustrated by mapping widespread species, such as *Scirpus cespitosus* (Fig. 4*a*) and *Carex limosa* (Fig. 4*b*), and restricted species, such as *Pedicularis* spp. (Fig. 4*d*) and *Rhododendron lapponicum* and *Ledum decumbens* (Fig. 4*c*) in the northernmost Lowland. Overall,

Fig. 4*a*. Distribution of *Scirpus cespitosus* in the Hudson Bay Lowland. (Author's Ontario collections — ● and records — O; and WIN Manitoba collections — ϴ.)

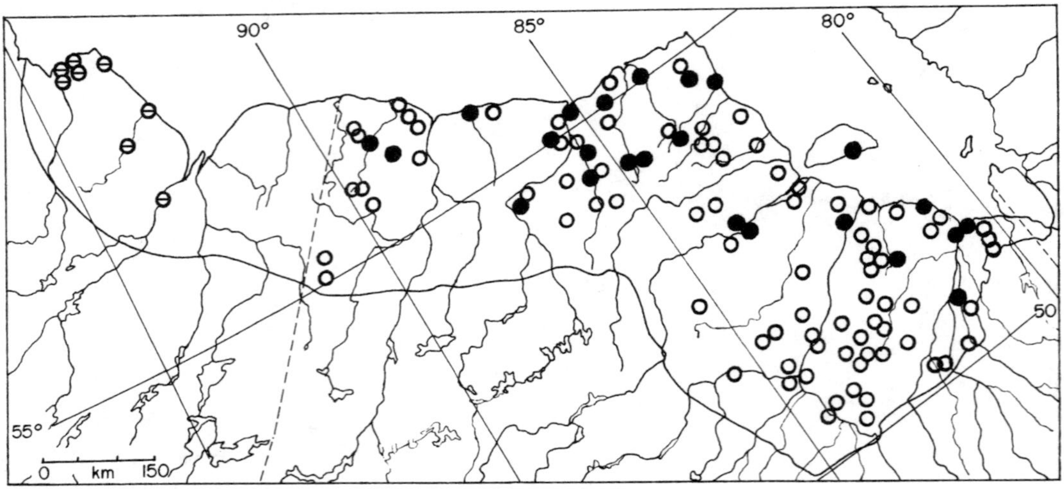

Fig. 4*b*. Distribution of *Carex limosa* in the Hudson Bay Lowland. (Author's Ontario collections — ● and records — O; and WIN Manitoba collections — ϴ.)

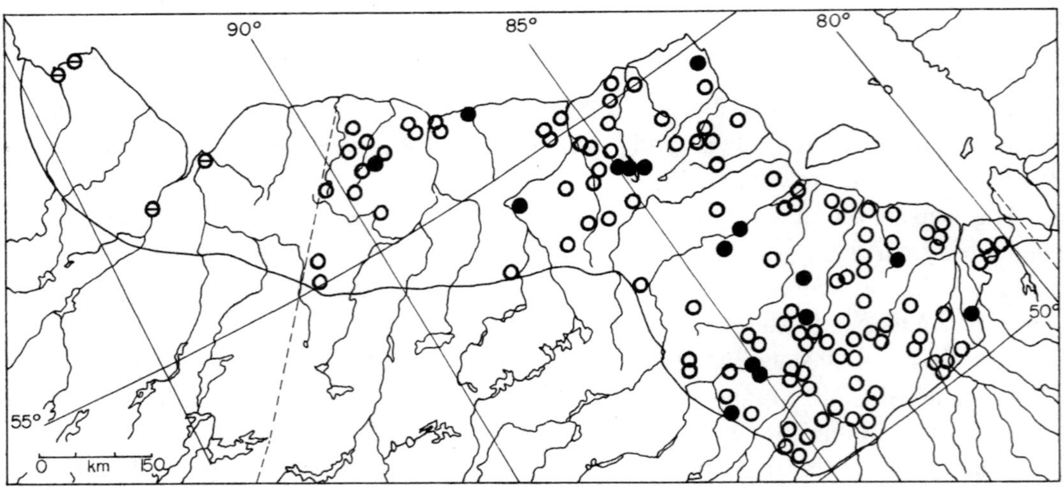

combining the results of recent field surveys with data from botanical studies taken over the past 50 years, makes it possible to accurately characterize the distribution of species within the Lowland and to assess whether the occurrences of individual plant species are common, occasional, or rare (Appendix B).

More than 7000 specimens were collected during the author's field studies, and additional occurrence records were made at more than 310 vegetation-sampling sites. More than 3000 additional specimens were received for identification from the Canadian Forest Service, Canadian Wildlife Service, Ontario Ministry of Natural Resources, and Canadian Centre for Inland Waters. These collections are stored at the herbarium of the Royal Ontario Museum (TRT), with partial duplicate collections at the National Herbarium of

Fig. 4*c*. Distribution of *Rhododendron lapponicum* (●) and *Ledum decumbens* (○) in the Hudson Bay Lowland.

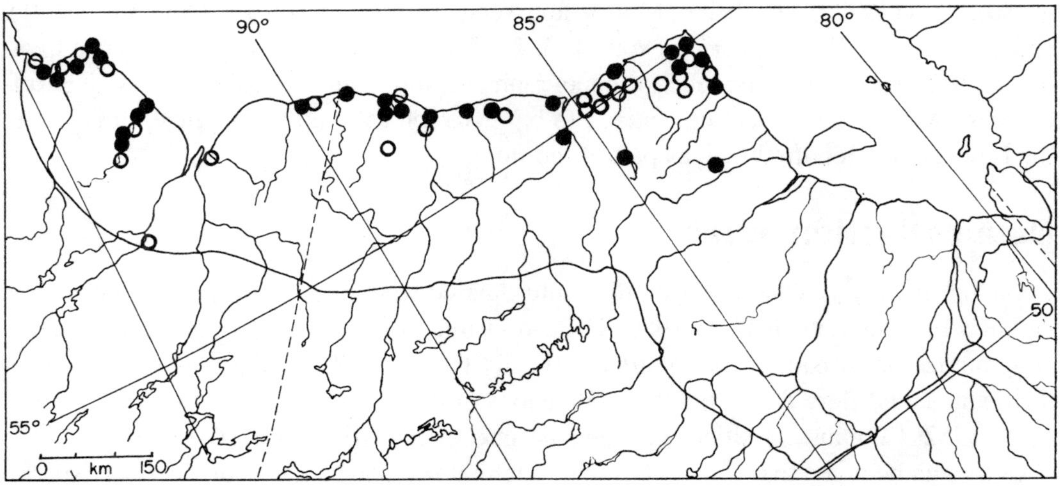

Fig. 4*d*. Distribution of *Pedicularis lapponica* (●), *P. labradorica* (◓), *P. flammea* (◒), and *P. sudetica* (⊖) in the Hudson Bay Lowland.

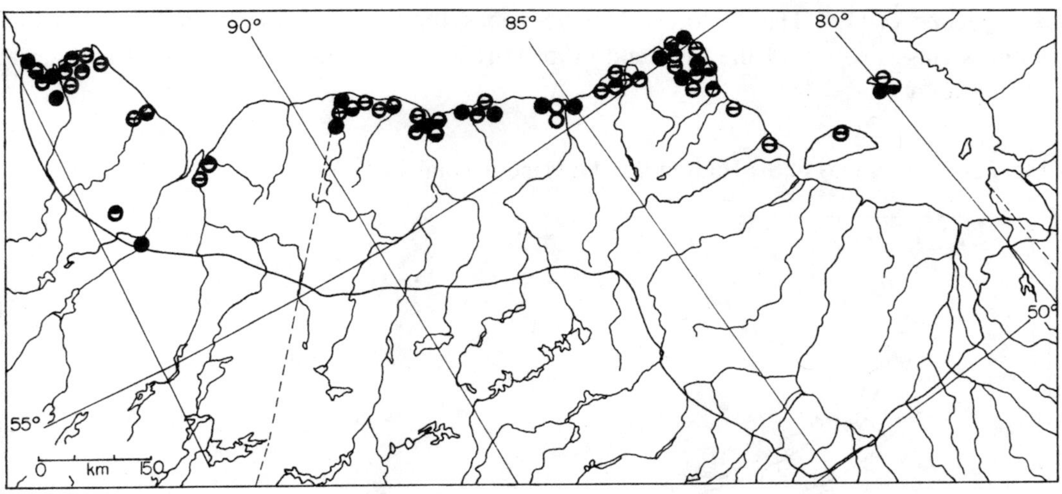

the Canadian Museum of Nature in Ottawa (CAN), the Great Lakes Forest Research Centre at Sault Ste. Marie (SSMF), and the local herbarium of the Moosonee District of the Ontario Ministry of Natural Resources. Collections were examined and verified from herbaria in Toronto (TRT, TRTE, Ritchie, Lumsden, and Maycock), Ottawa (CAN, DAO), Quebec (QFA-Laval), Maple (now at Ontario Natural Heritage International Centre, Peterborough), Sault Ste. Marie (SSMF), Winnipeg (MMMN, WIN), and Helsinki (HEL). Collections held in Peterborough (MJO, at OMNR) have also been catalogued and incorporated. Bruce Ford of the University of Manitoba (WIN) provided access to Manitoba distribution maps based on WIN herbarium collections. Specimen data and literature sources are maintained by the author; many of them summarized elsewhere (Riley 1980).

The botanical nomenclature and synonymy applied here are based on Fernald (1950), Boivin (1966–1967, 1967–1981), Hultén (1968), Scoggan (1978–1979), Porsild and Cody (1980), Voss (1972, 1985, 1996), Morton and Venn (1990), Gleason and Cronquist (1991), and the *Flora of North America* (Vols. 1–3, 22–23, 26), where taxa occur in those treatments. Reference was also made to monographs dealing with particular groups (see References), and taxa that were considered to be geographically and ecologically distinct in the Lowland are reported here as part of the flora.

Data collection areas

The Hudson Bay Lowland was divided into data collection areas in preparation for the author's field surveys in 1976–1980. These divisions were based on a review of existing geographic and climatic zones in order to stratify the field effort and provide checklists of the documented flora for particular subdivisions of the Lowland.

In 1952, Coombs classified the Lowland into three distinct "forest" regions (Fig. 5), even though he noted that the tree cover throughout the Lowland was distinctively sparse. Coombs did not consider the northern coast to be part of a forest region at all, noting that it was a "narrow treeless coastal fringe" (Coombs 1952). These forest zones were generally based on Halliday's forest classification of Canada (1937), Hustich's forest regions for Quebec (1949), Hare's forest–climate zones for Quebec (1950), and Hanson and Smith's classification of the Lowland (1950), as well as on field observations and aerial photography.

Fig. 5. Forest regions of the Hudson Bay Lowland (Coombs 1952).

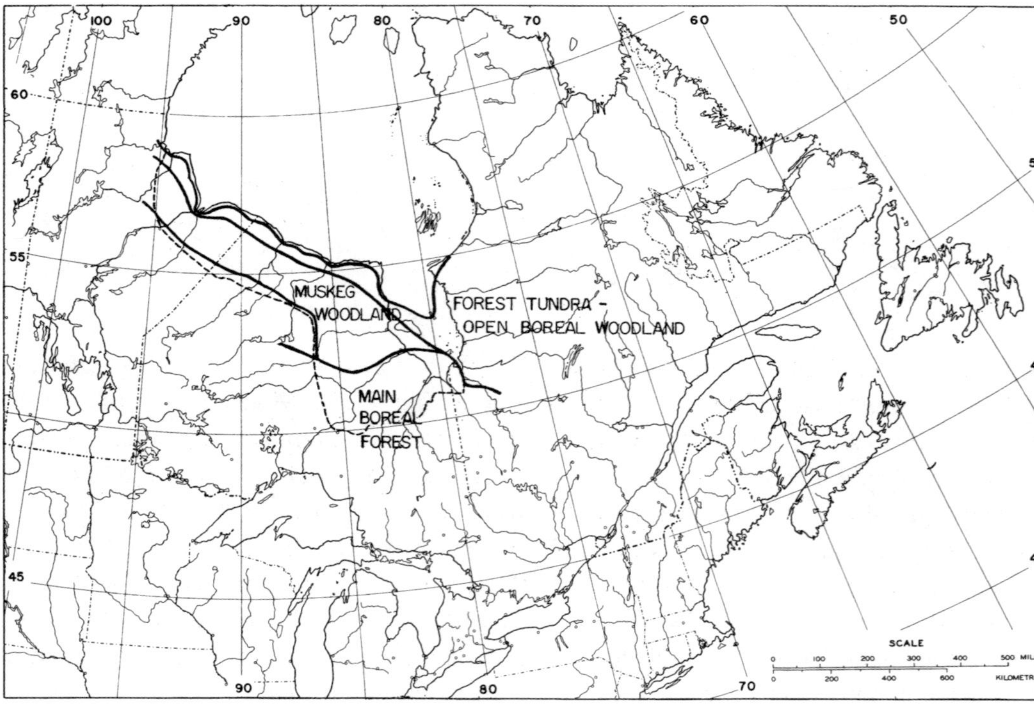

Fig. 6. Vegetation zones of northern Ontario (Ahti 1964).

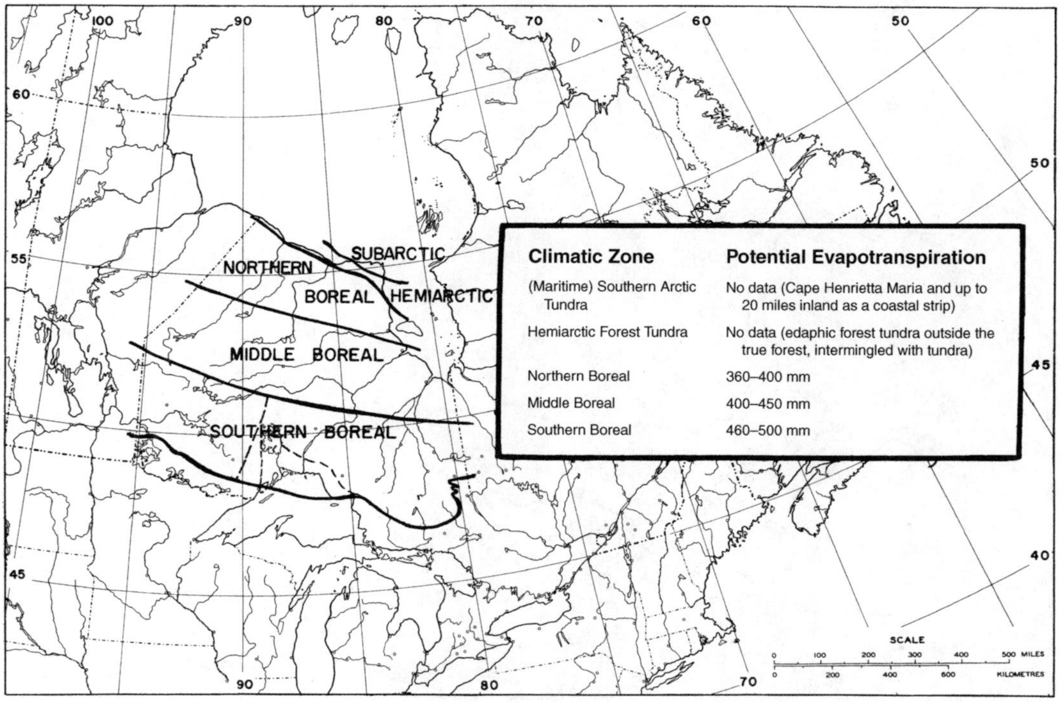

In 1964, Ahti mapped northern Ontario's vegetation into latitudinal climatic zones based on then-current data on potential evapotranspiration (Fig. 6). This division was similar to the grouping of boreal climatic zones by Hare (1954), which was also based on available (but meager) potential evapotranspiration data. Ahti used these zones in his assessment of caribou habitat, which in turn motivated Brokx (1965) to define complex, finer-scale subdivisions of the Lowland, later mapped by Bates and Simkin (1969).

In 1972, Canada's forest regions were mapped by Rowe (Fig. 7). He treated the Hudson Bay Lowland as a distinctive section of the Boreal (Forest and Barren) Region with a subarctic character based on its "open woodland" and its immense areas of swamp, bog, and muskeg. The section along Hudson Bay was treated as part of the transcontinental Forest–Tundra ecotone.

Hills (1959, 1966) developed a system defining ecological site regions for Ontario, characterizing them as regions within which relatively similar plant associations occur on similar landforms because of similar regional climates and rates of biological productivity. In Hills' later work (1976), the Hudson Bay Lowland included a coastal region (OE), a northern region (1E), and a southern region (2E) (Fig. 8). This approach was adopted by Maycock (1979) and later recast as "ecoclimatic regions" (Canada 1989). It is also the approach used by Ontario in its ecoregional land-use planning (Burger 1993; Crins 2000).

In northern Manitoba, Ritchie proposed classifications based on broad-scale vegetation zonation (1960; Fig. 9) and geobotanical zonation (1962; Fig. 10). Both classifications generally reflect the geological boundary of the Lowland, and Ritchie stressed the unity of the Lowland area and the fact that the zones generally paralleled the coast. Finer-scale

Fig. 7. Forest regions and sections of northern Ontario and Manitoba (Rowe 1972).

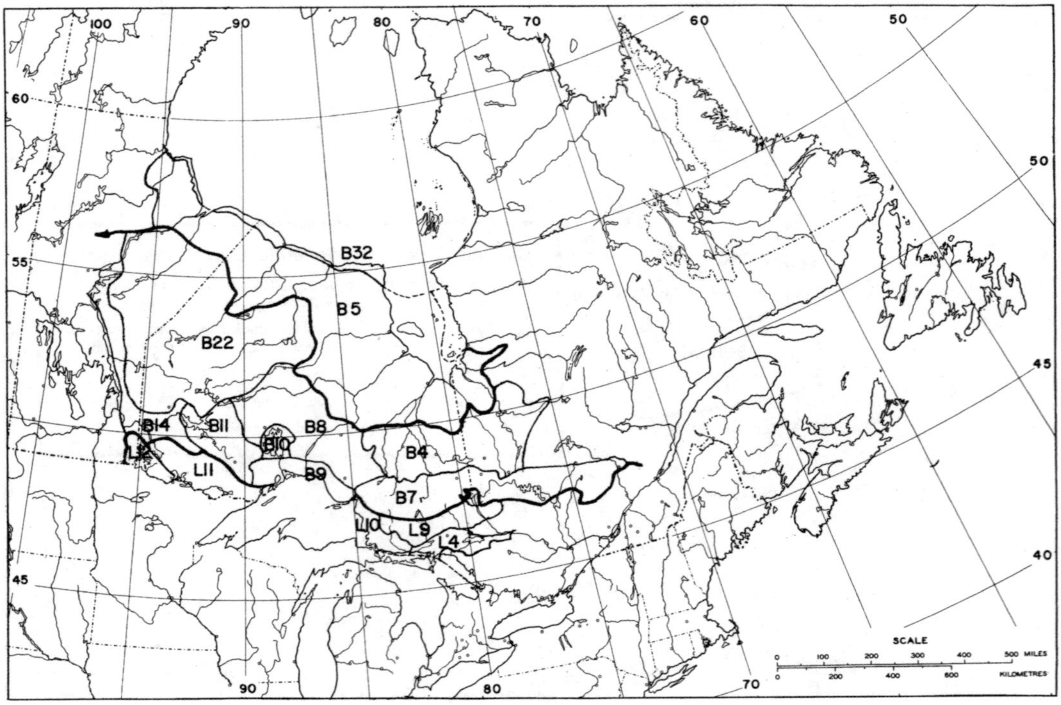

BOREAL (FOREST & BARREN) REGION

B	5	Hudson Bay Lowland	Marine/paleozoic plain, open wetlands.
B	32	Forest-Tundra	Ecotone between subarctic forest and tundra, expressive of climatic influences.

BOREAL (PREDOMINANTLY FOREST) REGION

B	4	Northern Clay	±Lacustrine clays; black spruce.
B	7	Missinaibi-Cabonga	Height of land, rolling & varied; mixed forest limited to the north by decline of white pine, red pine, yellow birch, sugar maple, hemlock.
B	8	Central Plateau	±Sand/gravel plateau; jack pine/black spruce.
B	9	Superior	Variable terrain and forest, with boreal climate.
B	10	Nipigon	Lacustrine sands; black spruce.
B	11	Upper English River	Transitional forest, relatively thin till mantle.
B	14	Lower English River	Glaciolacustrine substrates, low relief; mixed forests.
B	22	Northern Coniferous	Till/bedrock terrain; black spruce/jack pine/tamarack.

GREAT LAKES-ST. LAWRENCE REGION

L	4	Sudbury-North Bay	Lowlands and wide bedrock expanses; aspen and birch.
L	8	Haileybury Clay	Lacustrine substrates; outlier of southern hardwoods.
L	9	Timagami	Uplands; white pine, white birch, white spruce.
L	10	Algoma	Varied terrain; tolerant hardwoods, spruce, fir, cedar.
L	11	Quetico	Originally pine communities, now jack pine, aspen, birch, spruce, balsam fir.
L	12	Rainy River	Waterworked and outwash till, low relief; extensive swamp.

Fig. 8. Ecoregions of northern Ontario (Hills 1976; Burger 1993; Crins 2000).

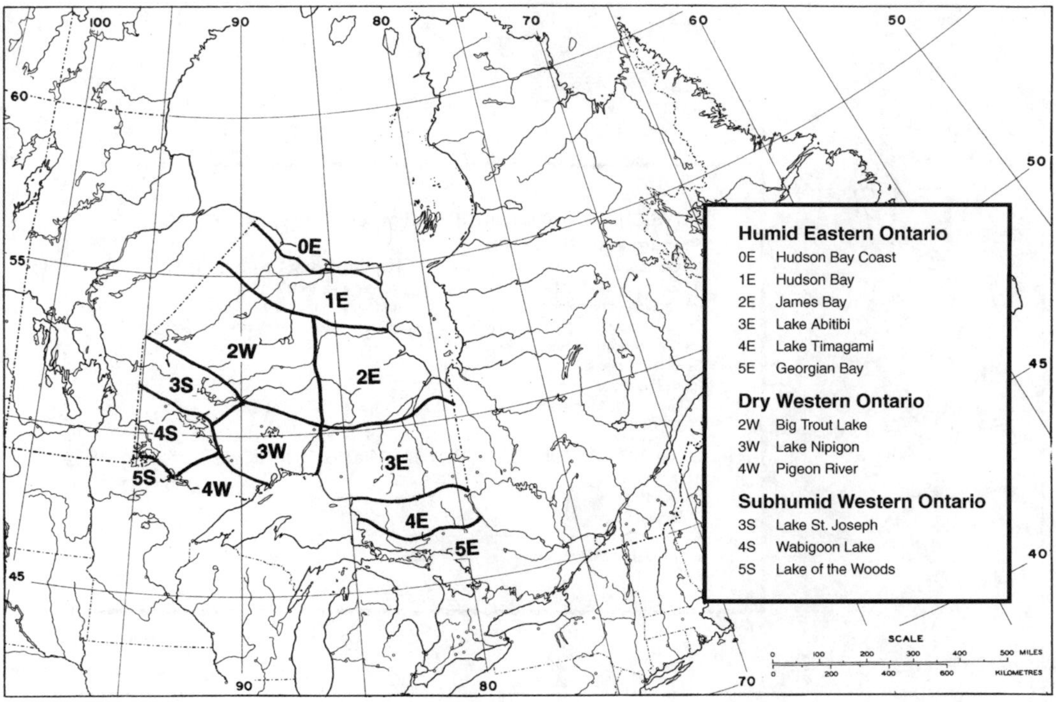

Fig. 9. Vegetation zones of northern Manitoba (Ritchie 1960).

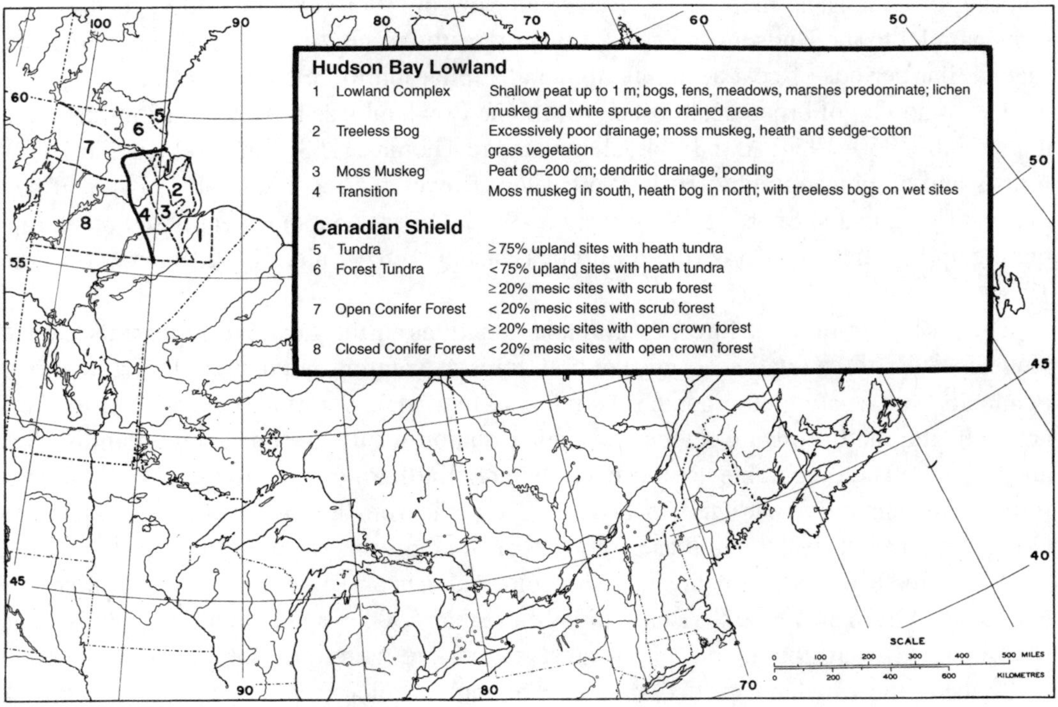

Fig. 10. Physiographic regions and vegetation zones of northern Manitoba (Ritchie 1962).

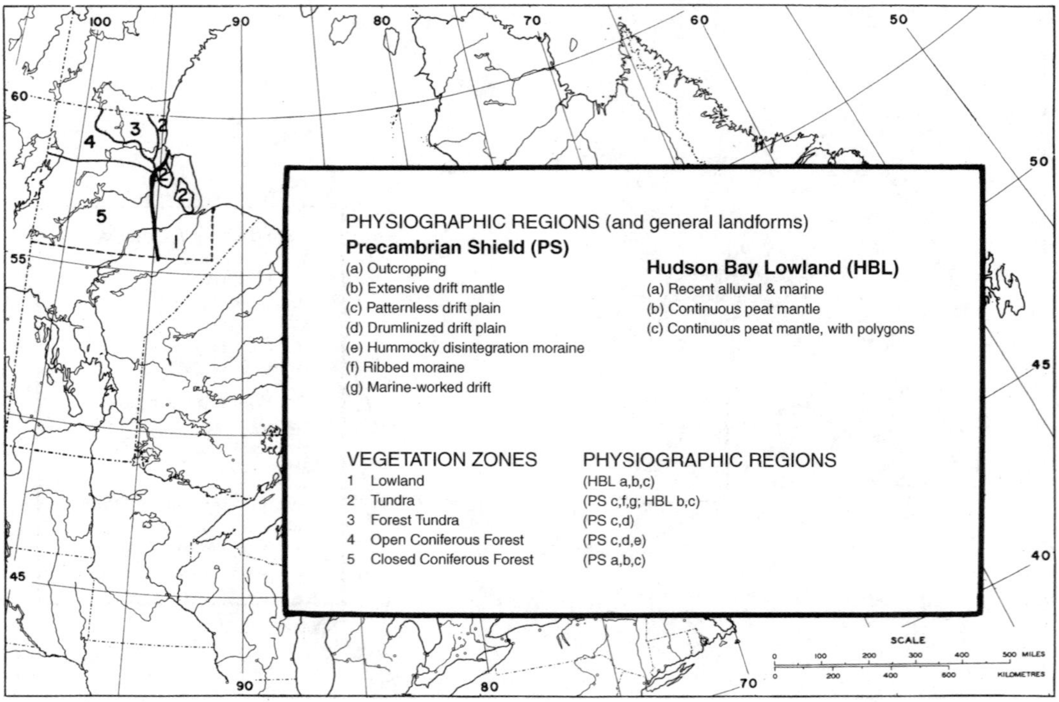

mapping for Manitoba has been undertaken by Mills et al. (1976) and Tarnocai (1974). In Ontario, the fine-scale mapping by Bates and Simkin (1969) also plotted vegetation patterns parallel to the Hudson Bay coast, reflecting similar durations of emergence from the sea, similar periods of peat accumulation, and similar maritime climatic effects.

This mapping of broad-scale zones within the Lowland was based on various data: climate (Hare 1950, 1954; Ahti 1964; Chapman and Thomas 1968; Hare and Ritchie 1972); forests and biophysiography (Hills 1959, 1966; Rowe 1972; Zoltai et al. 1974); and permafrost (Railton and Sparling 1973; Brown 1967). Other zonations were based on extrapolating approaches used in northern Europe (Hustich 1957; Kalela 1962; Sjors 1963; Ahti 1964).

This use of traditional forest or climatic zonations in the Lowland represents a challenge because of the relative absence of both forest vegetation and meteorological stations. Peatlands with a saturated organic substrate 1–4 m thick cover more than 85% of the land area. Upland landforms are scarce and their isolation results in forests communities that are not strictly comparable to those elsewhere. Furthermore, in the Lowland, upland forests are generally restricted to beach ridges, chevron beach complexes, and levees along streams and rivers, and these are not typical landforms in adjacent "forest" regions.

Across the Lowland, climatic and meteorological data are imprecise because isolines are extrapolated from only three coastal stations more than 550 km apart (Churchill, Winisk, and Moosonee). Toward the interior, the nearest stations are more than 150 km from the coast and outside the Lowland itself (Table 1). The data are not accurate enough to establish either a strict climatic zonation or to define practical data collection areas for the Lowland.

On this basis, it was concluded that the existing biogeoclimatic zonations were too imprecise and mapped at scales too coarse to be used to stratify the collection of botanical data. The data collection areas adopted for this study (Fig. 3) do reflect these zones, but with additional consideration being given to size, scale, and mapability.

A bioclimatic boundary parallel to the Hudson Bay coast was used to define the maritime tundra near and beyond the tree line. This area is interrupted occasionally by woodlands on permafrost levees and by lichen woodlands on raised beaches closer to the coast. As noted above, climatic data for this region are not precise, but its boundary generally appears to correspond with the 10-year mean annual lake evaporation isopleth of 28 cm (Canada 1970), the –5.3°C isotherm of mean daily temperature for the year (Chapman and Thomas 1968), and the area with a growing season of less than 65 days (Canada 1974). This coastal area (MT; Fig. 3) is wider than that proposed by Ahti (1964), Rowe (1972), and Hills (1976) and is closer to that of Coombs (1952), reflecting the actual extent of coincident permafrost, treeless tundra, and arctic species. The addition of this wider coastal zone to Ontario site regions (Burger 1993) and ecoregions (Crins 2000) was based on this floristic analysis and other corroborative data. This in turn is reflected in the mapping of a coastal Hudson Bay ecoregion by the Ecological Stratification Working Group (ESWG 1995); this study of Canada's terrestrial ecozones and ecoregions mapped the Lowland as a unique Canadian ecozone.

Hustich (1949), Ahti (1964), Hills (1976), and Maycock (1974) suggested that this coastal area be separated into two units in Ontario; here divided into one unit west of about 84°W longitude (MT2; Fig. 3) and another eastward to Cape Henrietta Maria and to Lake River on James Bay (MT1; Fig. 3). In Ontario, the Cape is the widest area of open treeless tundra, experiencing the most rapid rates of isostatic rebound and the greatest exposure to the cooling effects of Hudson Bay.

Two other data collection areas were used to reflect the distinctive character of the James Bay coast: (1) the coastal area of James Bay up to 15 km from the coast (HB1, up to 25 km in estuaries, and including Charlton Island; Fig. 3); and (2) Akimiski Island (HB2, including nearby Gasket Shoal and Ile Manawanan; Fig. 3).

To divide the remainder of the Lowland, its interior, into data collection areas, Halliday (1937), Ahti (1964), and Hills (1976) suggested using the southern boundary of a northern boreal zone that crosses the Lowland just north of the Attawapiskat River, close to the watershed boundary (HB4, HB5; Fig. 3). Hills (1976) also suggested a boundary segregating the Moose River drainage basin because of its unique dominance by treed fen and treed bog (HB7; Fig. 3); this was corroborated by field studies and LANDSAT2 imagery. The division by Coombs (1952) between areas of boreal forest and muskeg woodland approximated the northern boundary of the Albany River drainage basin (HB5, HB6; Fig. 3). To complete this approach, a watershed boundary between the Severn–Fawn basin and the Winisk–Ekwan basin was also used (HB3, HB4; Fig. 3). Thus, in the Lowland interior, the major watersheds were used to approximate bioclimatic zones proposed by earlier workers. The watersheds are also useful tools for dividing collecting areas as they are of more or less equal size and can be identified more precisely on any scale of mapping.

The geological boundaries of the Precambrian Sutton Ridges were used to define a separate data collection area within the Winisk–Ekwan region (Sandford et al. 1968; HB8, see Fig. 3). This area is primarily an upland zone, with shallower peat deposits and less extensive wetlands. It is largely an area of glacial till and marine deposits, generally at higher elevations rising around exposed Precambrian ridges. The high elevation also indicates a longer period of emergence from the Tyrrell Sea.

The Manitoba Lowland was treated as a single unit (MT3; Fig. 3); the comparable maritime arctic coast was not separated from the Manitoba interior. Published studies and herbarium collections from the Manitoba Lowland remain geographically uneven, with the majority of the studies and collecting areas centred around Churchill, along the Churchill railroad and northwest coast, and at York Factory (Sims et al. 1979). It was concluded that the floristic data for an interior Manitoba Lowland area would not have been, at this time, at a level of sampling comparable to other areas.

Floristic analysis

The geographic affinities of a regional flora are conventionally analysed in relation to the broader continental ranges of constituent species. A complementary challenge is to then assess those continental ranges in relation to the likely refugia of those species during the glacial maximum, in order to assess the origins of that flora on the evolving postglacial landscape.

The former of these tasks is a statistical analysis of coincident distributional patterns within the Lowland itself. Elsewhere for example, Ritchie (1962) characterized the species of northern Manitoba as continentally centred on the arctic, subarctic, boreal, or temperate regions (or more than one), many also with extra-continental circumpolar or amphi-Atlantic distributions. In a study of the flora of Great Whale River and Manitounuk Island, Maycock (1968) identified each species as high arctic, low arctic, high subarctic, low subarctic or high temperate, and used these geographic affinities to characterize the floristic transition across Hudson Bay's maritime tree line. Schofield (1958) used a similar approach in his study of the phytogeography of the salt marsh flora of Churchill, showing that the maritime flora of Churchill had stronger eastern affinities and weaker western affinities than the flora of Churchill as a whole. Similar general approaches have been used to analyze the floras of Manitoba (Scoggan 1957), the Canadian arctic archipelago (Porsild 1957), and the northern Clay Belt of Ontario and Quebec (Baldwin 1958).

Scoggan (1978–1979) categorized all Canadian species by whether their species ranges coincided with broad latitudinal and longitudinal zones: high or low arctic, high or low subarctic, or high and low temperate; east, west, or transcontinental. For example, the code "aST/X" indicated a transcontinental species extending south from the low arctic to high temperate latitudes. These codes portray the total and more precise distribution of a species rather than only the latitudinal zone of which a species is most characteristic.

A major east–west division of the Canadian flora crosses southern Canada from the south shore of Hudson Bay to the general vicinity of Lake Winnipeg and Lake of the Woods. This east–west division is characteristic of temperate and subarctic biota but not arctic biota, most of which are transcontinental or circumpolar. This division can be illustrated, for example, within species complexes of willow (*Salix calcicola* and *S. richardsonii*, *S. glauca* (Argus 1965), *Salix lucida* (Argus 1986)); varieties of *Goodyera repens*

(Kallunki 1976); ploidy levels of *Epilobium latifolium* (Small 1968); monoterpene (and genetic) variability of *Picea glauca* (Wilkinson et al. 1971); Red-wing Blackbird (Snyder and Lapworth 1953); and Canadian orchid species (Riley 1980).

The Hudson Bay Lowland is a distributional bottleneck in the North American subarctic (Fig. 11). Its continentally anomalous habitats constrain the migration of many plant species eastward and westward, and the Lowland flora contains a disproportionate number of species at their range limits. The Precambrian bedrock of the Canadian Shield to the south, between the Lowland and Lake Superior, for example, is largely dominated by thin boreal forest and rock barrens, and probably also served to restrict the migration of species east and west, as well as from the south.

Disjunct plant distributions or the occurrences of plant species at their range limits usually represent (1) the remnants of previously more widespread distributions of species, (2) the initial colonizers of future widespread distributions in an area, or (3) long-range dispersals into restricted but receptive areas. For interpreting the development of a flora, these species are of value disproportionate to their frequency. "The area of a species can only mean the total area, and nothing else. It is just the stations found outside the 'compact area' that are likely to be the most valuable ones, which can give a clue as to how the development has taken place" (Hultén 1937).

Fig. 11. Generalized floristic zonation of Canada (based on Polunin 1951; Hare 1954; Ahti 1964; Rousseau 1968; Young 1971; Scoggan 1978–1979; and the mean annual temperature isotherms of –3.9°C, –1.1°C, and 7.2°C).

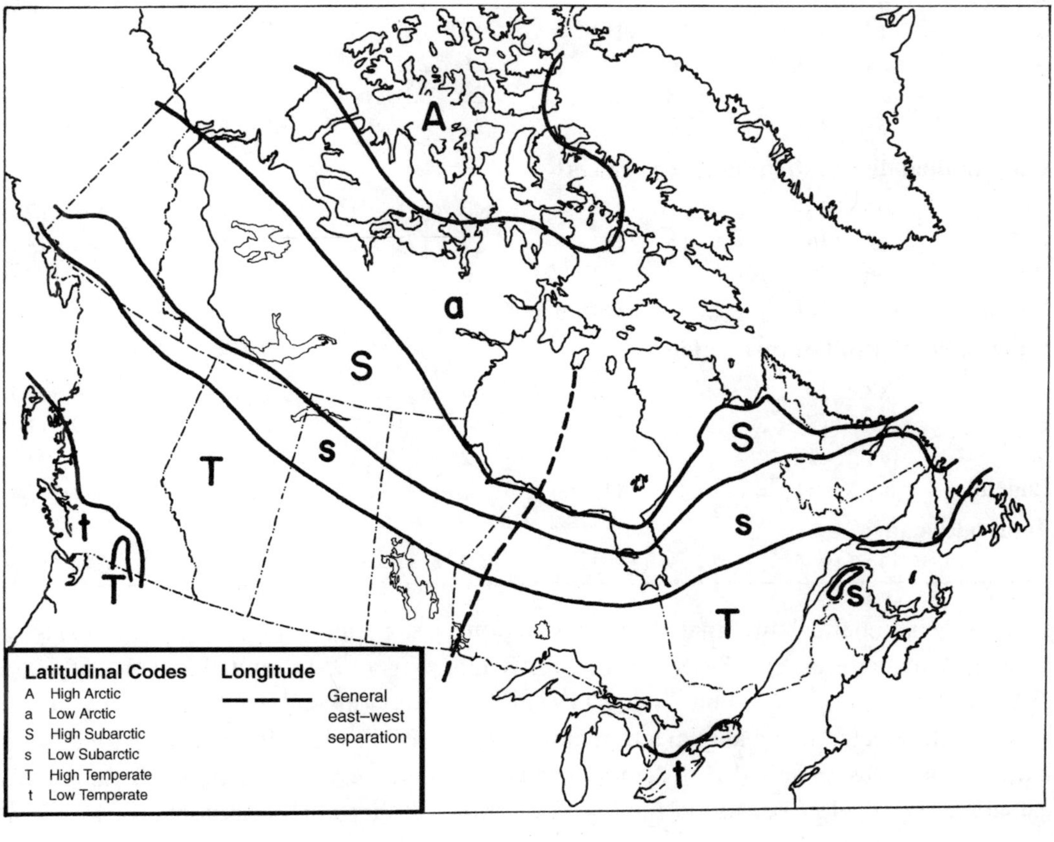

Table 2. Geographic affinities of native taxa occurring in individual data collection areas. (Numbers and relative percentages of taxa; numbers in parentheses are data for the same groups excluding taxa that are widespread across the Lowland.)

	James Bay Coast (HB1)	Akimiski Island (HB2)	Severn Basin (HB3)	Winisk-Ekwan Basin (HB4)	Attwapiskat Basin (HB5)
Latitudinal distribution (number of taxa)					
AST	30 (12)	35 (19)	23 (7)	31 (12)	22 (2)
aST	151 (46)	123 (39)	114 (16)	140 (27)	143 (27)
ST	158 (52)	90 (22)	124 (23)	146 (36)	181 (59)
sT	71 (51)	21 (7)	33 (16)	43 (24)	65 (45)
T	29 (26)	1 (1)	8 (5)	16 (13)	28 (25)
ASs	2 (2)	5 (5)	4 (4)	2 (2)	
aSs	5 (4)	3 (3)	3 (2)	4 (3)	
S		1 (1)			
AS	1 (1)	1 (1)			
aS	2 (2)	2 (2)			
S					
Latitudinal distribution (relative percentages)					
A	2 (3)	4 (7)	3 (5)	3 (4)	2 (1)
a	13 (12)	17 (19)	14 (13)	14 (12)	12 (7)
S	24 (21)	26 (25)	26 (23)	26 (23)	25 (22)
s	29 (30)	28 (26)	29 (30)	29 (29)	30 (32)
T	31 (34)	27 (24)	29 (30)	30 (32)	32 (38)
Longitudinal distribution (number of taxa)					
X	397 (154)	261 (83)	290 (62)	352 (94)	395 (124)
E	45 (36)	15 (14)	13 (7)	24 (18)	39 (30)
W	6 (5)	2 (2)	6 (4)	5 (5)	5 (4)
Endemic	1 (1)	1 (1)			
Longitudinal affinities (percentages)					
X	88 (79)	93 (83)	94 (85)	92 (81)	90 (79)
E	10 (18)	6 (14)	4 (10)	6 (15)	9 (19)
W	1 (<1)	1 (2)	2 (5)	1 (4)	1 (3)
Endemic	<1 (<1)	<1 (1)			
Total native taxa					
	446 (196)	279 (100)	309 (73)	382 (117)	440 (159)

Canada's general latitudinal floristic zonation was mapped (Fig. 11) on the basis of Scoggan's categories (1978–1979), Polunin's mapping (1951), and the work of Hare (1954), Rousseau (1968), and Young (1971). These continental zones generally correspond with selected mean annual temperature isotherms (Canada 1953). Each native Lowland species was assigned a distributional code based on this zonation, on Scoggan's categories (1978–1979), and on published distribution maps. The catalogue of Lowland

Albany Basin (HB6)	Moose Basin (HB7)	Sutton Ridges (HB8)	Cape Henrietta-Maria (MT1)	Ont. Maritime Tundra (MT2)	Manitoba Low-land (MT3)
16 (1)	18 (1)	31 (12)	66 (44)	60 (43)	72 (51)
141 (29)	134 (27)	121 (29)	150 (55)	160 (52)	187 (73)
187 (69)	188 (64)	101 (20)	94 (8)	142 (25)	159 (39)
96 (76)	94 (74)	28 (15)	11 (2)	18 (2)	25 (12)
82 (78)	80 (78)	7 (4)	2 (0)	1 (0)	4 (3)
		1 (1)	13 (13)	12 (12)	13 (13)
1 (0)	1 (0)	3 (2)	11 (10)	10 (10)	11 (11)
				1 (1)	
		1 (1)	3 (2)	3 (3)	6 (6)
			2 (2)	2 (2)	2 (2)
			1 (1)		1 (1)
1 (–)	1 (–)	3 (5)	6 (10)	5 (10)	5 (9)
11 (5)	11 (5)	15 (16)	18 (22)	17 (21)	17 (20)
23 (18)	23 (17)	25 (23)	26 (24)	26 (25)	26 (25)
30 (31)	30 (31)	28 (28)	26 (24)	27 (24)	27 (25)
35 (46)	35 (46)	28 (28)	24 (20)	26 (20)	26 (22)
432 (171)	434 (173)	275 (69)	333 (122)	379 (127)	441 (179)
82 (74)	78 (69)	15 (13)	18 (13)	17 (12)	21 (15)
9 (8)	3 (2)	3 (2)	1 (1)	12 (10)	17 (15)
			1 (1)	1 (1)	1 (1)
83 (68)	84 (71)	94 (82)	94 (89)	93 (85)	92 (85)
16 (29)	15 (28)	5 (16)	5 (10)	4 (8)	4 (7)
2 (3)	<1 (1)	1 (2)	<1 (<1)	3 (6)	4 (7)
			<1 (<1)	<1 (<1)	<1 (<1)
523 (253)	515 (244)	293 (84)	353 (137)	409 (150)	480 (211)

species (Appendix B) includes these distribution codes, and these codes were then tabulated for the native flora of the Lowland as a whole and for individual data collection areas. The relative percentages of each of these distribution codes (latitudinal and longitudinal) were calculated as a measure of the floristic differences between data collection areas within the Lowland (Table 2). Some of these data are also mapped to make their variability within the Lowland easier to interpret (Fig. 12).

Fig. 12*a*. Relative percentage of taxa with high arctic occurrence, by data collection area (total Hudson Bay Lowland 4%).

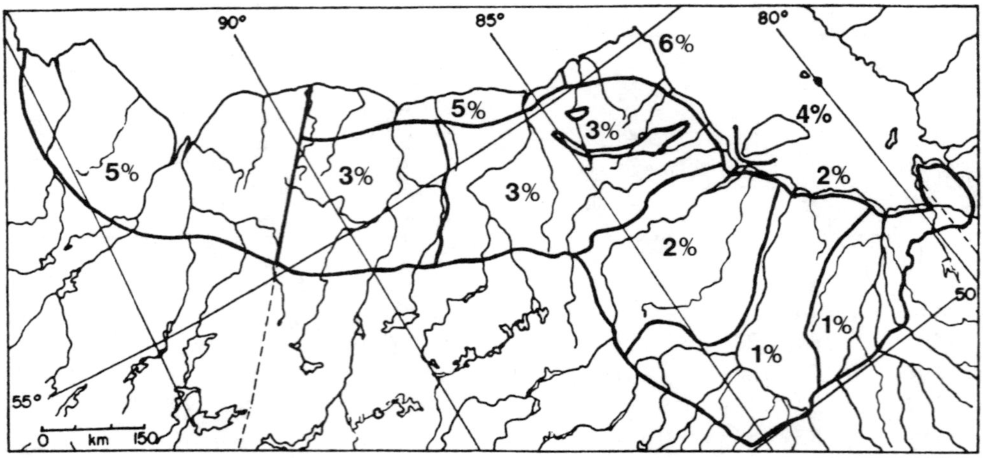

Fig. 12*b*. Relative percentage of taxa with high and low arctic occurrence, by data collection area (total Hudson Bay Lowland 17%).

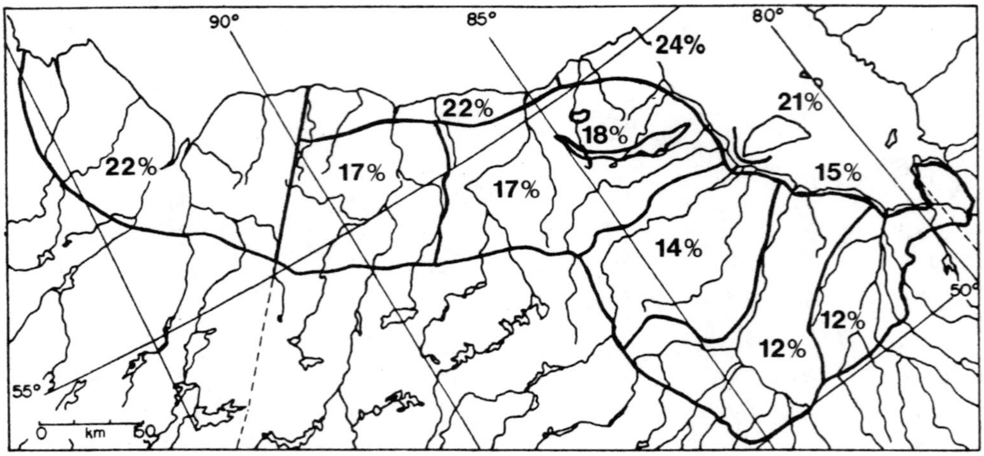

Fig. 12*c*. Relative percentage of taxa with high temperate occurrence, by data collection area (total Hudson Bay Lowland 31%).

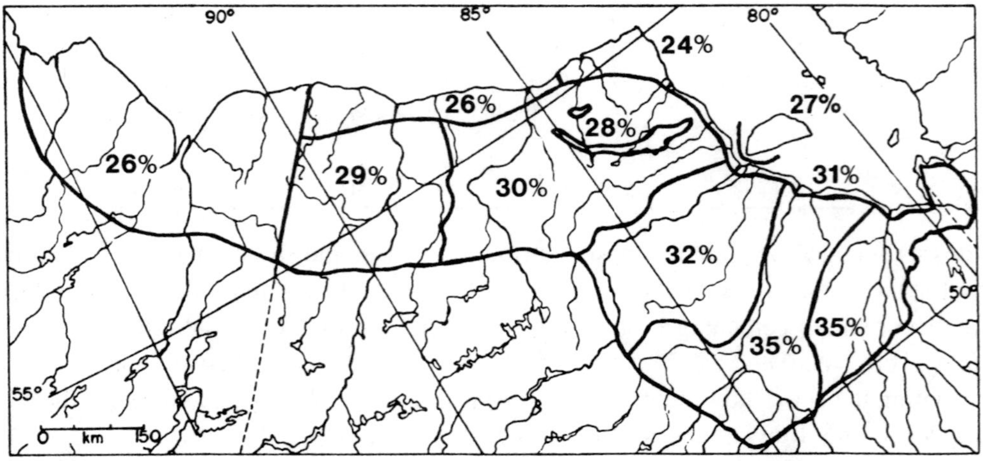

Fig. 12*d*. Relative percentage of taxa with high and low arctic occurrence, by data collection area, excluding widespread taxa.

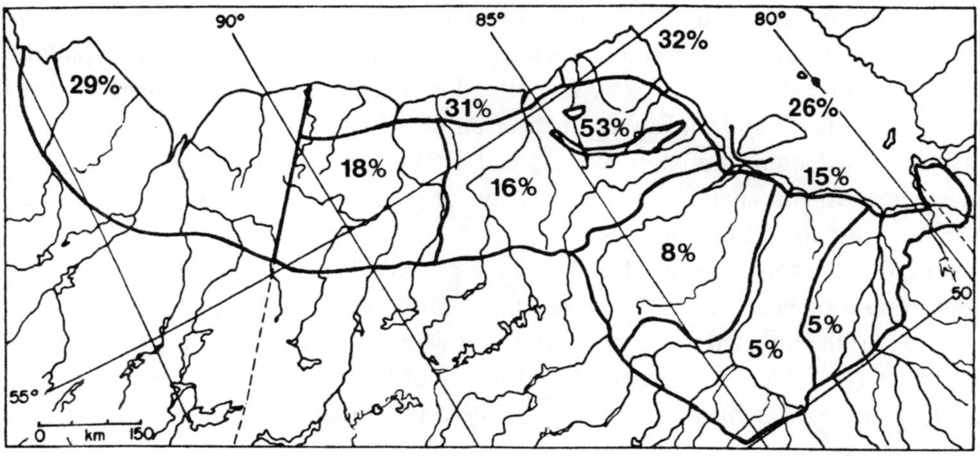

Fig. 12*e*. Percentage of western taxa, by data collection area (total Hudson Bay Lowland 4%).

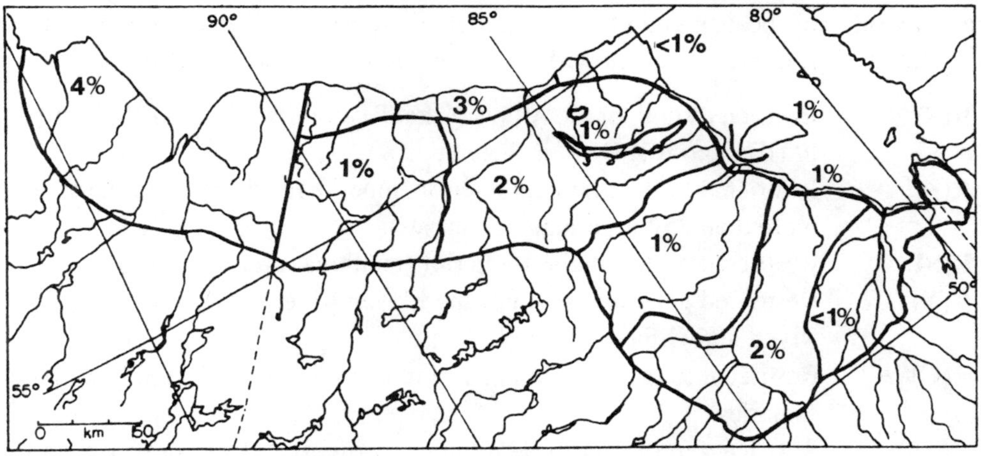

Fig. 12*f*. Percentage of eastern taxa, by data collection area (total Hudson Bay Lowland 15%).

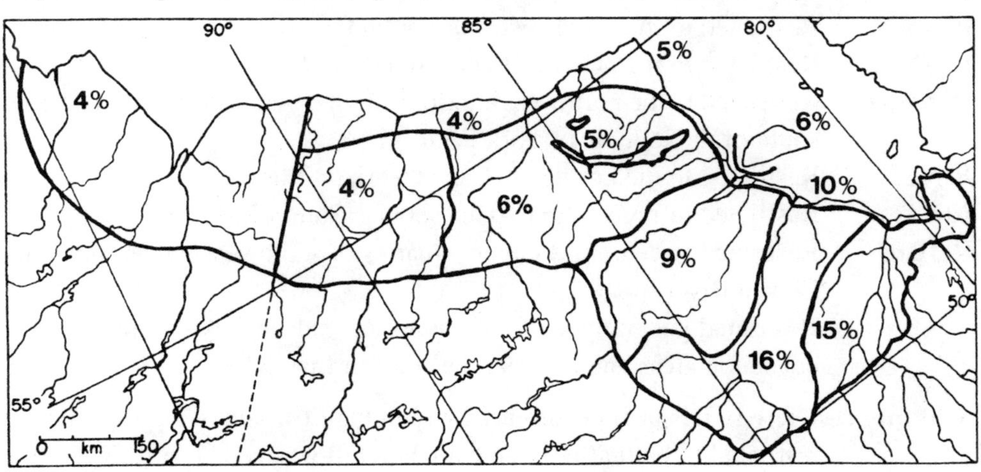

In addition to this analysis of the flora of each data collection area, a similar analysis of geographic affinities was completed for groups of taxa with particular coincident distribution patterns within the Lowland as a whole. Fourteen of these general Lowland distribution patterns were recognized and coded. The largest single distributional pattern was that of "widespread" Lowland taxa, W, defined in this study as those known from at least three of five coastal and four of six interior data collection areas. A large group of other taxa were considered "probably widespread", (W), but as yet may be under-sampled. These were defined as occurring in at most two areas less than the above criteria, with both coastal and interior areas represented.

Using these criteria, a total of 303 taxa are classified as widespread or probably widespread in the Lowland (Table 3). These taxa form the dominant vegetation of the region. For assessing the vegetation of the region as a whole, this ambient flora is of paramount importance. (In Table 2, the geographic affinities of the flora of each data collection area, excluding the widespread taxa, are tabulated to emphasize the floristic differences between areas.)

Each native taxon was also coded for its particular distributional pattern within the Lowland (Appendix B). Widespread taxa, W and (W), which comprise 35% of the native taxa, have already been mentioned. The other, more restricted distribution patterns of native Lowland taxa are listed below. These abbreviations are used in Table 3.

R(S) Restricted south of the Attawapiskat River (HB6 and (or) HB7, infrequent in HB5 and (or) HB1, if there at all).

R(I) Restricted to the Lowland interior (more than 15 km from the coast, and more than 25 km in major estuaries).

R(SI) Restricted to the southern interior (combination of 1 and 2).

R(N) Restricted generally north of the Ekwan River, Akimiski Island, Fawn River, and Gillam (Manitoba).

R(C) Restricted to the coast (up to 15 km from the coast, 25 km in major estuaries).

R(NC) Restricted to the maritime north coast (up to 20 km from the Hudson Bay coast, from the Manitoba border to the Lake River area and Akimiski Island, up to 40 km from the coast in the Cape Henrietta Maria area).

R(NW) Restricted to the northwest (interior and coastal areas in Manitoba).

R(NW) Restricted to the northwest coast (coastal Manitoba only).

R(NWI) Restricted to the northwest interior (interior Manitoba Lowland, generally south of the latitude of York Factory).

R(NE) Restricted to the northeast (Cape Henrietta Maria).

R(LS) Restricted to Precambrian outcropping (Sutton Ridges).

R(JB) Restricted to coast and major estuaries of southwest James Bay (south of Ekwan River mouth).

t Restricted generally south of Ontario's Hudson Bay coastal data collection areas and not near the coast in Manitoba.

Multiply restricted Disjunct occurrences within the Lowland (e.g., a species restricted to both at HB1 and at Churchill was coded R(JB,NWC)).

Taxa conforming to these distribution patterns were treated as specific groups for analysis of floristic affinities in the same way as those of the data collection areas (Table 3). The boundaries that define those restricted distribution patterns were then used to split the Lowland flora into separate floras on each side of that particular boundary, and the floristic affinities of the floras on each side of the boundary were calculated, as was the amount of change in those floristic affinities across the different boundaries (Riley 1980). The calculations of the rapidity or steepness of floristic change across those boundaries were the basis for mapping the floristic zonation of the Hudson Bay Lowland (see Results, p. 32).

Table 3. Geographic affinities of groups of native taxa with coincident Lowland distribution patterns. (Numbers and relative percentages of taxa.)

	W and (W)	R(S)	R(I)	R(SI)	R(N)	R(C)	R(NC)	R(NW)	R(NWC)	R(NWI)	R(NE)	R(LS)	R(JB)	t	Multiple restrictions	Other	Totals (native taxa)
Latitudinal distribution (number of taxa)																	
AST	21				21	11	15		6		1	3			1		80
aST	122	8	8	8	22	21	17	2	6	1	2	2	2	8	2	1	234
ST	133	19	21	18	5	10	5	4	2	3		2	4	15	8	3	256
sT	21	35	14	27		2			1	5	1	1	10	13	3	7	135
T	3	28	7	56				1	2	1			5	8		2	111
ASs					4	1	9		1						1		16
aSs	1				5	1	4		1								12
Ss							1										1
AS						1	2		3		2						8
aS							2										2
S									1		1						2
Latitudinal affinities (relative percentages)																	
A	2				10	7	12		12		14	10			4	2	4
a	14	4	6	4	22	19	22	10	21	4	23	16	4	7	9	10	13
S	27	14	22	13	24	25	25	29	25	17	27	23	13	21	26	26	23
s	29	33	33	27	24	25	23	29	21	38	18	26	36	32	32	31	28
T	29	48	38	56	20	24	17	33	21	42	18	26	47	40	30	31	31
Longitudinal distribution (number of taxa)																	
X	289	63	40	57	46	37	50	3	18	7	6	7	17	38	9	11	698
E	11	27	8	50	6	9	1		1		1	1	3	6	1		125
W	1		2	2	5	1	3	4	4	3			1		5	2	33
Endemic							1										1
Longitudinal affinities (percentages)																	
X	96	70	80	52	81	79	91	43	78	70	86	88	81	86	60	85	81
E	4	30	16	46	11	19	2		4		14	13	14	14	7		15
W	<1		4	2	9	2	5	57	17	30			5		33	15	4
Endemic							2										<1
Total native taxa																	
	303	90	50	109	57	47	55	7	23	10	7	8	21	44	15	13	857

RESULTS

The vascular plant flora of the Hudson Bay Lowland includes 816 native vascular plant species, documented here as 857 geographically and ecologically distinct taxa. In addition, 98 non-native species have also been identified, almost all of which are still confined to the region's towns, villages, docks, and railheads.

Data assembly and field surveys were organized around 11 distinct data collection areas to maximize the even distribution of sampling and deliberately target gaps in the data. As a result, the specimens collected across the Lowland, by the author and by others, are more evenly distributed than those of most studies from remote parts of northern Canada. Gaps in collection coverage remain in the northern interior of the Lowland, closest to the boundary with the Canadian Shield, in both Ontario and Manitoba, and in the Manitoba Lowland interior in general, especially between the Hayes River and the Ontario border (Fig. 4; Appendix A).

The Lowland, covering 3.5% of Canada's land area, supports 25% of the vascular plant species occurring in Canada. Canada's arctic and subarctic plant species are overwhelmingly transcontinental in their ranges (70 and 51%, respectively; Scoggan 1978–1979), and the majority of the arctic species are circumpolar or amphi-Atlantic beyond North America (61%). The Lowland flora as a whole is similarly overwhelmingly transcontinental (81% of taxa), reflecting its low-arctic and high-subarctic latitudes and its relative location farthest from the glacial refugia of these wide-ranging and quick-migrating species (Fig. 1). The latitudinal affinities of the flora also reinforce the categorization of the Lowland as part of the low arctic and high subarctic.

The relative percentages of geographic affinities of the flora are given below (after Table 3; and Scoggan 1978–1979):

	Canada	Hudson Bay Lowland
High Arctic	3	4
Low Arctic	10	13
High Subarctic	17	23
Low Subarctic	23	28
Temperate	47	31
Transcontinental	33	81
Western	38	4
Eastern	29	15

Within the Lowland itself, the flora varies significantly across its 1000 km of latitude and 1100 km of longitude. This variability is evident in the geographic affinities of the total flora of each data collection area (Table 2) and in the geographic distinctiveness of the suites of taxa that have coincident distribution patterns within the Lowland (Table 3). These two measures both contribute to mapping the floristic zonation of the Hudson Bay Lowland (Fig. 13).

Data collection areas

The number of native taxa decreases northward, from 515 and 523 taxa in the Moose and Albany basins to 382 taxa in the Winisk – Ekwan River basins and 309 in the Severn basin. Along the Hudson Bay coast, the number of species rises again to 409 in the Ontario maritime tundra area (MT2) and 353 at Cape Henrietta Maria (MT1), where the number of arctic and coastal species increases significantly.

Relatively few high-arctic species are found in the Lowland interior south of the Ekwan River. The exceptions are approximately 20 wide-ranging species that are found everywhere in Canada from high arctic to temperate regions (AST), such as *Equisetum arvense, Deschampsia cespitosa, Carex aquatilis, Carex capillaris,* and *Hippuris vulgaris*, and are equally widespread across the Lowland. Along the Hudson Bay coast, however, the relative percentage of high-arctic species reaches 5–6%, and the relative percentage of all arctic species exceeds 22% (Figs. 12*a* and 12*b*). The relative percentage of non-widespread species with arctic occurrences rises from about 5% in the south and 15% on the southern James Bay coast to 32% at Cape Henrietta Maria (Fig. 12*d*).

The relative percentage of taxa that occur in high-temperate areas elsewhere in their ranges decreases from the southern interior towards the coast and the north (Fig. 12*c*). Excluding widespread species, this percentage decreases from 46% in the southern interior to 34% on the southern James Bay coast and to about 20% at Cape Henrietta Maria (Table 2). The representation of species with subarctic occurrences varies very little across the Lowland (between 52 and 55% overall); these constitute the stable majority of species in all data collection areas. Relatively few western species occur in the Lowland (33 taxa, 4% of native Lowland taxa), and only one of these, *Elymus innovatus*, is widespread. The percentage of western species reaches 4% (17 taxa) in the Manitoba data collection area and 3% (12 taxa) in the Ontario maritime tundra area (MT2). Farther east, this decreases to less than 1% at Cape Henrietta Maria and the Moose River basin (1 and 3 taxa, respectively).

Eastern species are much more strongly represented in the Lowland flora (120 taxa, 14% of native taxa). The percentage of eastern species increases from 4% in the Manitoba Lowland, the Hudson Bay area, and the Severn River basin, to 16% and 15% in the Albany and Moose River basins, respectively (Fig. 12*f*). Only 11 of these eastern taxa are widespread in the Lowland, including *Trisetum melicoides, Bromus ciliatus, Iris versicolor, Salix cordata*, and *Salix pellita.* Excluding these widespread species, the percentage of eastern taxa rises from 7% in the Manitoba Lowland to more than 28% in the Albany and Moose River basins. Eastern taxa are effectively southern elements in the Lowland flora, and their numbers reflect the distance from the southwest Lowland to extraglacial areas south of the Laurentide Ice Sheet, which is comparatively a much shorter distance than to other glacial refugia.

Transcontinental species dominate the Lowland flora (698 native taxa), from 94% of native species in the Hudson Bay area and Sutton Ridges to about 83% in the southern interior areas where there are more eastern species. Excluding widespread species, the percentage of these transcontinental species varies from 89% (Cape Henrietta Maria and Sutton Ridges) to about 68% in the Albany River basin, very much paralleling the frequency of arctic species. Over 96% of the Lowland's widespread taxa are transcontinental, but only 41% of its transcontinental species are widespread in the Lowland; this reflects the increase in arctic species northward and the fact that most arctic species are transcontinental.

Overall, these data illustrate the transcontinental and subarctic dominance of the flora. Subarctic elements comprise a majority of the taxa in all areas, while transcontinental elements are particularly dominant in the data collection areas along Hudson Bay and northward in the interior. There is a very strong arctic influence along the Hudson Bay coast, which is only slightly stronger at Cape Henrietta Maria and the Manitoba coast. There is an even stronger floristic salient of temperate eastern species, strongest in the Lowland interior south of the Albany River, and extending locally as far north as the Attawapiskat River. A comparatively weak western salient of species extends into the Lowland interior east into the Severn River basin and along the Ontario maritime tundra (MT2).

Coincident distribution patterns

A similar geographic analysis of coincident distribution patterns, ranging from those of widespread taxa to those restricted to different Lowland distributions, illustrates even more precisely the character of the Lowland flora (Table 3).

For example, more than 100 species are restricted to the southern Lowland interior, particularly the Albany and Moose River basins inland from the James Bay coast. These species exhibit very strong temperate, eastern floristic affinities, reflecting the more continental climate and southeast location. An additional 90 species are also generally southern species in the Lowland, occurring occasionally as far north as the Attawapiskat basin and not necessarily restricted to interior sites.

The flora restricted solely to coastal southwest James Bay (21 species) is not in itself distinct from that of the adjacent interior. This suite of taxa, most of them disjunct east or west, occupy halophytic habitats in similar subarctic–temperate latitudes elsewhere in their continental ranges. However, when combined with the 47 other species restricted to coastal habitats in general, the southwest James Bay coast has a flora very distinct from its adjacent interior.

The 47 species that are restricted to coastal areas throughout the Lowland from Manitoba to southern James Bay define a distinct floristic zone, with clear arctic and high subarctic affinities and an increased proportion of transcontinental taxa. An additional group of 55 species is restricted to the immediate Hudson Bay maritime coast, from Manitoba to Cape Henrietta Maria, south towards Lake River and Akimiski Island. This group, and the additional seven species restricted to Cape Henrietta Maria, have the strongest arctic and high-subarctic affinities in the Lowland flora (Table 3); they define a distinctive maritime tundra zone that extends some distance to the interior, reflecting the extent of predominantly untreed, permafrost ecosystems. Many of these restricted northern coastal species are not halophytic or shoreline species, but are more generally arctic-maritime. "It should be remembered that all arctic plants are to be regarded as more or less oceanic" (Hultén 1937). Likewise, the few species restricted only to the northwest coast in Manitoba (23 taxa) form a somewhat distinct floristic group; the western arctic affinities of this flora account for most of this distinction.

In addition to these northern coastal species, an additional 57 are restricted to the general area north of the Ekwan River, Fawn River, and Gillam area. In total, 150 taxa are restricted to areas north of that general boundary (Table 3; R(N), R(NC), R(NE), R(LS), and R(NWC)). Of these, 118 are only found within about 50–60 km of the Hudson Bay coast, equivalent more or less to the MT1 and MT2 data collection areas in Ontario (Fig. 3). This northern floristic boundary across the Lowland is based both on the presence of species restricted north of it (118 taxa) and on the absence of other species that are more or less widespread south of the Hudson Bay rank of data collection areas (45 taxa). The distributions of plants in this northernmost part of the Lowland reflect a rapid and steady shift in the flora based on changes in both southern and northern range limits of species.

Smaller groups of species are restricted to the northwest, the northwest interior, the general Lowland interior, and the Sutton Ridges. These areas support distinct sets of restricted species, but are more minor themes within the dominant patterns of the Lowland flora (Table 3).

Floristic zonation

Absolute differences between numerical sums of the geographic affinities of the floras restricted on either side of various boundaries drawn across the Lowland were calculated to quantify the degree of floristic change across those boundaries (Riley 1980). The most significant of these boundaries are portrayed in Fig. 13, and the characteristics of the resultant zones are as follows.

The *Maritime Tundra* (low arctic) along the Hudson Bay coast corresponds to zones identified by others as marine arctic (Hustich 1949), tundra (Hare 1954), hemiarctic (Ahti 1964), continuous permafrost (Hills 1976; Burger 1993), and subarctic tundra (Maycock 1974, 1979). It becomes narrower and less floristically distinct along the northern James Bay coast, where arctic species are more localized on the most exposed sites and on offshore islands.

In southern James Bay, most of the floristic dissimilarity with adjacent interior sites is based on the presence of widespread or disjunct coastal taxa rather than on the presence of non-coastal northern elements. Nevertheless, floristically, the narrow *Southwest James Bay* coast (subarctic) is distinct enough from adjacent interior zones and from the coast to the north to warrant its segregation.

These two zones merge along the James Bay coast between Akimiski Island and Lake River. In this part of the coast, few of the taxa characteristic of more arctic areas north of Lake River or of the southwestern James Bay coast are present. Its distinctiveness from the adjacent interior rests primarily on the presence of widespread coastal species. The north coast of Akimiski Island, from Houston to Duncan points, has strong affinities with the James Bay coast north of Lake River. Its south coast is more similar to the southwestern James Bay coast (Riley 1981). The interior of Akimiski Island as a whole is somewhat depauperate of species (Table 2); its under-sampled interior landscapes are most similar to the immature peatland landscapes inland along western James Bay.

A zone segregating the southernmost interior of the Lowland was identified by Coombs (1952, main boreal forest) and Hare (1954, closed forest), and intimated by others (Hustich 1949). Past descriptions inaccurately suggest a predominance of terrestrial ecosystems rather than the bogs and fens (treed and open) and conifer swamps that dominate the area. The term *Boreal Peatland* is more suggestive of its floristic and ecological character and its temperate floristic affinities.

The remaining central Lowland interior has been variously described as northern boreal (Ahti 1964), boreal woodland (Hare 1954), central Canadian taiga (Hustich 1949), and boreal (forest and barren) (Rowe 1972). Coombs (1952) segregated the interior into two zones: a southern muskeg woodland and a northern forest tundra – open boreal woodland, similar to the zones deriving from this floristic analysis.

This change largely reflects the increasing proportion, towards the north, of plant species dependent on permafrost-based peat plateaus, beach-ridge systems, lichen woodlands, and outcrop-associated uplands. These habitats effectively support more terrestrial, open, and subarctic vegetation types than do the complex, open peatlands to the south (Riley 1982). The latitudinal affinities of the flora are high subarctic to the north and low subarctic to the south. The descriptive terms applied to these zones in this study are *Peat Plateau and Woodland* for the north and *Peatland and Woodland* for the south (Fig. 13). These zones are less distinct from each other, and encompass a more gradual shift in floristic affinities between them, compared to the other floristic zones.

Fig. 13. Floristic zonation of the Hudson Bay Lowland.

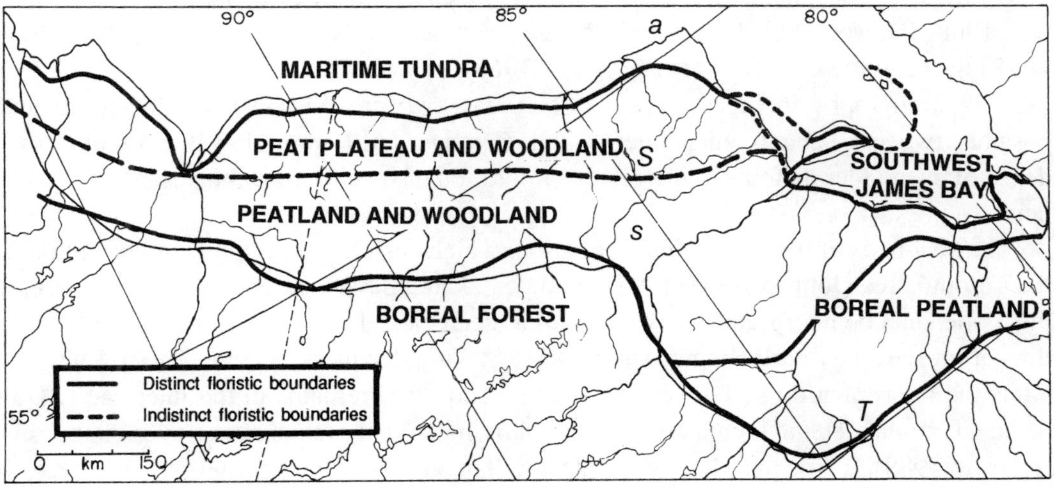

POSTGLACIAL ORIGINS OF THE FLORA

The migration of plants and the development of the vegetation of the Hudson Bay Lowland during the present interglacial period occurred within a framework of continental and regional events that are documented by a variety of geological, stratigraphic, and paleoecological studies. This migration and development has taken place over the past 8000 years, although the length of time available to each part of the region for this evolution varies with its distance from the centre of the late-Wisconsinan glaciation and from the refugia located beyond that glacial maximum (Hultén 1937; Ritchie 1989).

Early vegetation development

There is a general conformity of geological strata across the Lowland, which has been summarized on the basis of exposed sections and boreholes (Terasmae and Hughes 1960; McDonald 1968; Dredge and Cowan 1989):

(1) a generally continuous surface stratum of Holocene postglacial marine clays, silts, and sands, overlying;

(2) several Wisconsinan-age tills separated by sand and silt beds of uncertain origins, overlying;

(3) Sangamonian Interglacial peat beds (Missinaibi Formation), overlying;

(4) three or more tills believed to predate the Sangamonian Interglacial, overlying;

(5) a variety of Paleozoic and Mesozoic sedimentary bedrock types (Norris and Sanford 1968; Johnson et al. 1992), with Precambrian bedrock types outcropping in the Sutton Ridges and the Churchill area (Bostock 1968).

This stratigraphy includes evidence of at least two pre-Wisconsinan glaciations and possibly two pre-Sangamonian interglacials (Dredge and Cowan 1989), providing evidence of even older patterns of interglacial migration of species assemblages into the Lowland.

The peat beds of the Missinaibi Formation, which correlate with the Don Formation in the Toronto area (Johnson et al. 1992), include oak and basswood pollen. The presence of this pollen can be interpreted as evidence of a mean daily July temperature +1.7°C higher than at present (oak pollen is not known to stray long distances; Ritchie 1967). This parallels the interpretation of *Liquidamber* pollen and other remains in the interglacial Don beds, which indicate an increase of +3°C in July mean daily temperature over the present (Karrow 1989). Thus, the climate during the Sangamonian Interglacial is considered to have been as warm or warmer than it is today (Barnett 1992), and the vegetation that developed in the Lowland during that time was much more temperate.

At the beginning of the late-Wisconsinan glaciation, Hudson Bay may have been largely dry land, and the Laurentide Ice Sheet probably spread into it from the surrounding plateaus in the west (the Keewatin Sector) and the east (the Labrador Sector) (Fulton 1989; Dredge and Cowan 1989). The Hudson Strait was (and is) the critical link between Hudson Bay and the ocean, and depending on the relative elevations of Hudson Bay and the Atlantic, the Laurentide Ice Sheet may have been grounded in Hudson Bay (Dredge and Cowan 1989) or may have capped a large and largely frozen subglacial lake (Shoemaker 1991; Kor et al. 1991). In either case, there was probably a structural divide between the sectoral ice sheets, running from the Strait to the southern part of the modern Hudson Bay, and this probably influenced the pattern of deglaciation in the Lowland (Dredge and Cowan 1989).

The Laurentide Ice Sheet contained more than 50% of the difference in interglacial and glacial amounts of water in the global water budget, so the pattern of deglaciation of the Hudson Bay area influenced oceanic volumes and climates far beyond the region (Andrews 1987). The southern margin of Laurentide ice was rimmed by proglacial lakes prior to and during deglaciation. The shifting ice front and the reworking of glacial materials in the shifting lakes resulted in a complex array of moraines and glaciolacustrine features on the Canadian Shield south and west of the Lowland (Barnett 1992; Teller 1987).

Large and linked proglacial lakes Agassiz, Kelvin, Nakina, Barlow, Ojibway, and Opemiska occupied the depressed land area south of the decaying ice front (Fig. 14). The lakes along the ice sheet first drained to the Atlantic Ocean southward through the Mississippi basin, until the ice sheet decayed sufficiently to permit the lakes to began draining to the north Atlantic. This latter input of cold glacial water (11 000 to 10 000 BP) significantly lowered north Atlantic water temperatures, contributing to the cool Younger Dryas event in Europe (Ruddiman 1987).

The fastest rates of glacial melting were probably between 10 500 and 9500 BP (Ruddiman 1987), with drainage eastward through the Great Lakes and Ottawa Valley to the St. Lawrence Valley until about 9000 BP (Fig. 15). Increasingly, the rebound of the land and

Fig. 14. Cumulative proglacial and marine inundation in central Canada following the last continental ice sheet (after Dadswell (1974), Teller (1987), and Barnett (1992)).

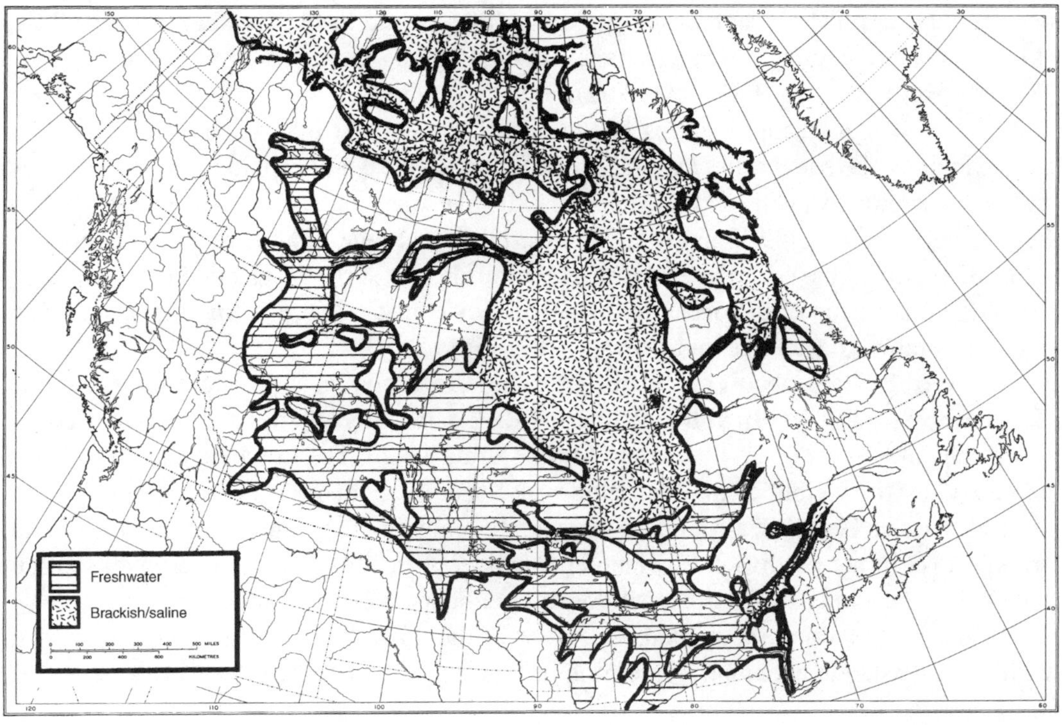

Fig. 15. Approximate position of ice sheet margin, 10 000 years BP (after Dyke and Prest 1987*a* and others).

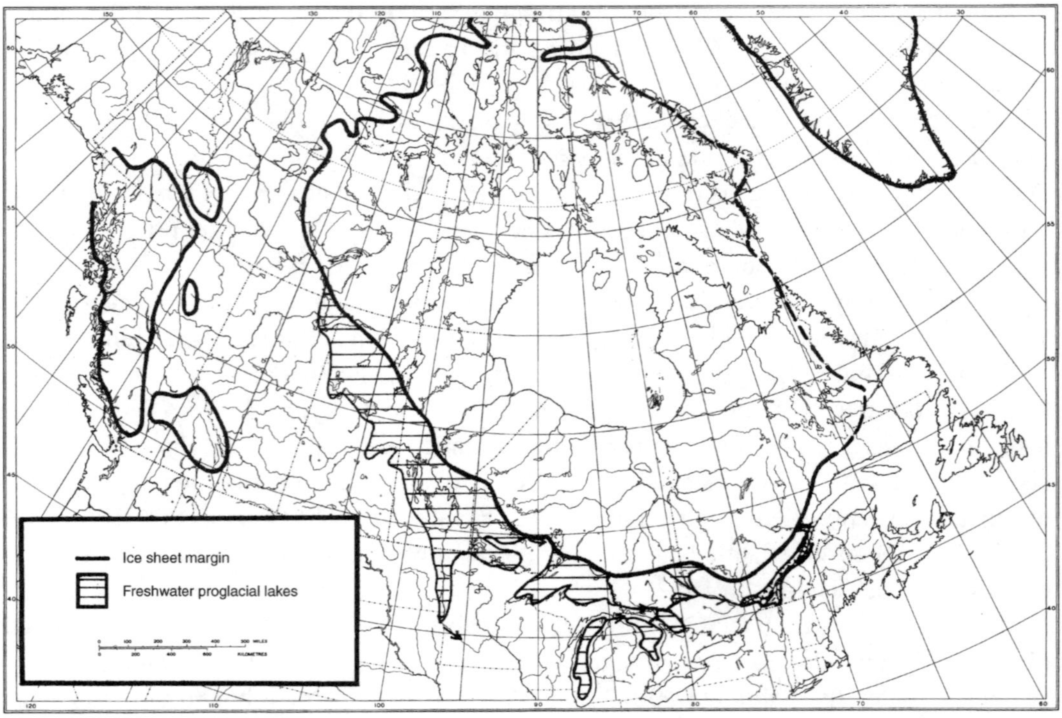

the slowing of meltwater runoff confined the lakes north of the watershed divide, although some drainage continued through the Lake Timiskaming Rift Valley (Fig. 16). It is likely that at about 8000 BP, the Hudson Strait bottleneck opened and a major marine incursion occurred through the Hudson Strait into the still depressed basin, creating the Tyrrell Sea (Fig. 17). The remaining ice-margin lakes drained rapidly into the Tyrrell Sea at that point (Vincent and Hardy 1979; Barnett 1992).

A series of small, rapid surges in the receding ice sheet may have also occurred at approximately the same time, leaving few if any moraines, and capping eskers, moraines, and other sediments with a thin veneer of till-like clay and loam (Boissonneau 1966; Dredge and Cowan 1989; Barnett 1992; called the Cochrane I and II readvances, 8200 to 7800 BP). Others have interpreted this as the result of a surge of glacial slush caused by the sudden incursion of the ocean into Hudson Bay (Prest 1970), floating the basin's remnant ice southward (Dredge and Cowan 1989)(Fig. 17).

The Tyrrell Sea probably inundated the whole Lowland basin at about the same time (Dredge and Cowan 1989); although, marine shell dates indicate that it may have occurred somewhat later in the Churchill area (7300 BP, Craig 1968). The Lowland mantle had been depressed by the weight of the Laurentide Ice Sheet to about 160 m below present elevations. Thus, the Tyrrell Sea occupied lands up to a maximum of about 130 m above modern sea level in the northwest Lowland and about 210 m above modern sea level south of James Bay, generally to about the geological boundary with the Shield.

As the Earth's crust rebounded, the shore of the Tyrrell Sea receded, rapidly at first, with upland islands eventually emerging around the Sutton Ridges and elsewhere (Fig.

Fig. 16. Approximate position of ice sheet margin, 8400 years PB (after Dyke and Prest (1987*a*) and Hughes (1987)).

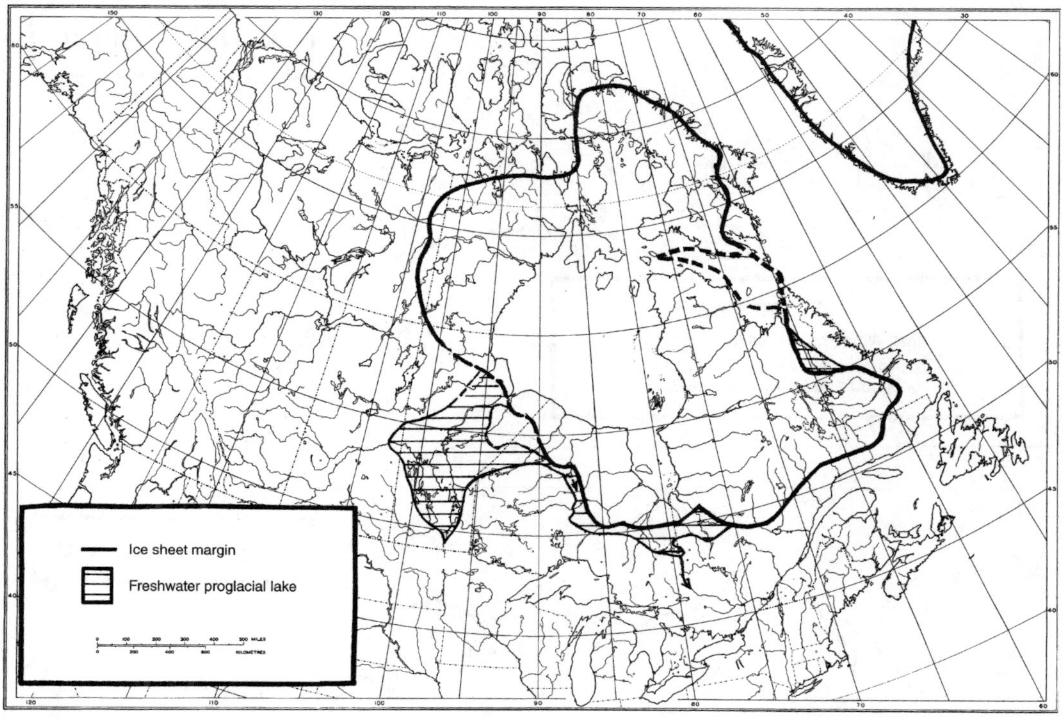

Fig. 17. Approximate position of ice sheet margin, 8000 years BP (after Dyke and Prest 1987*a* and others).

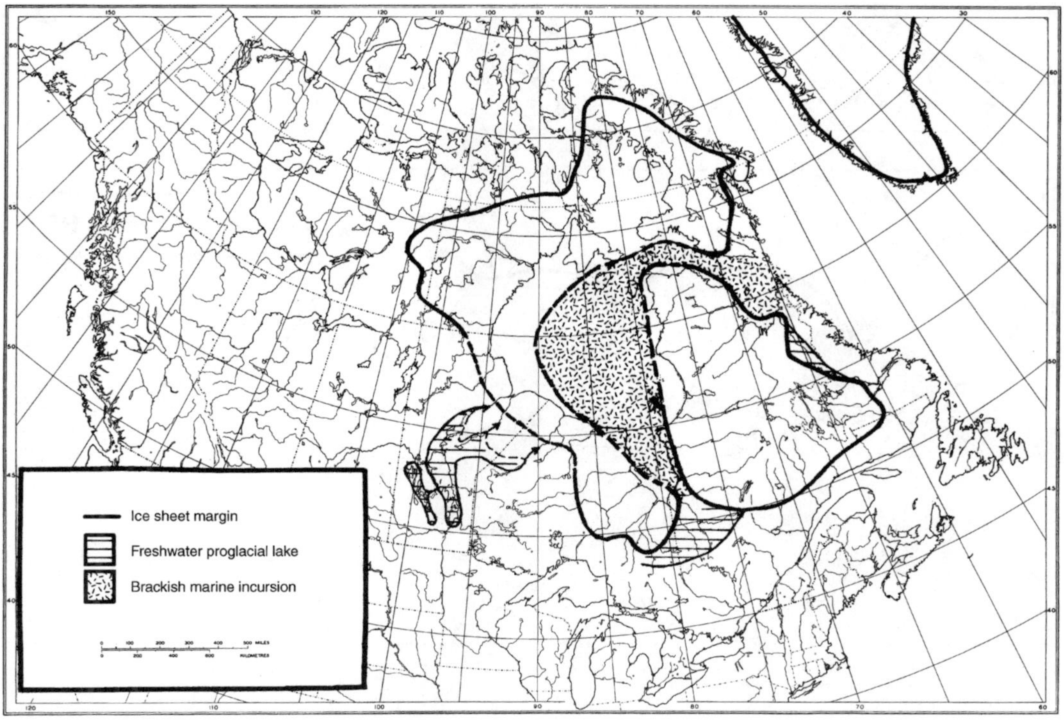

Fig. 18. Pattern of emergence of the Hudson Bay Lowland from the Tyrell Sea.

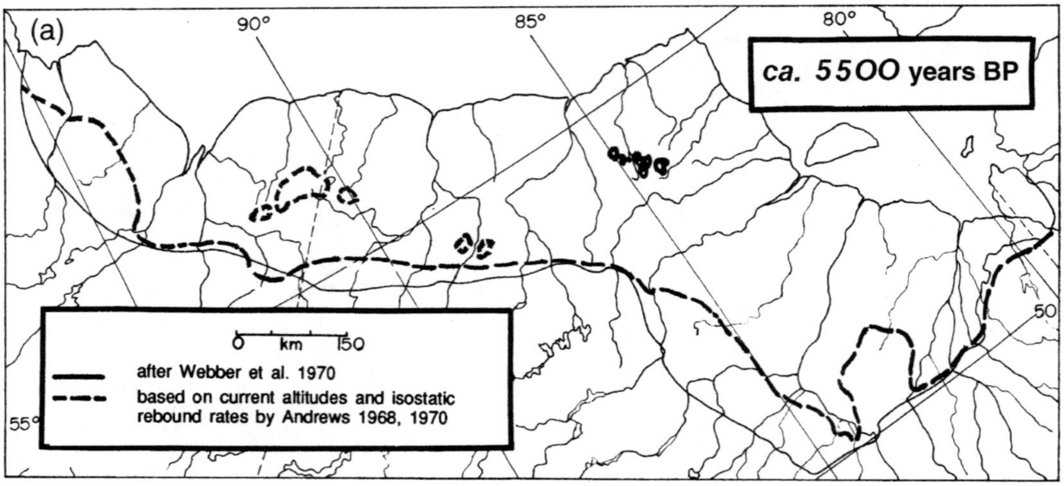

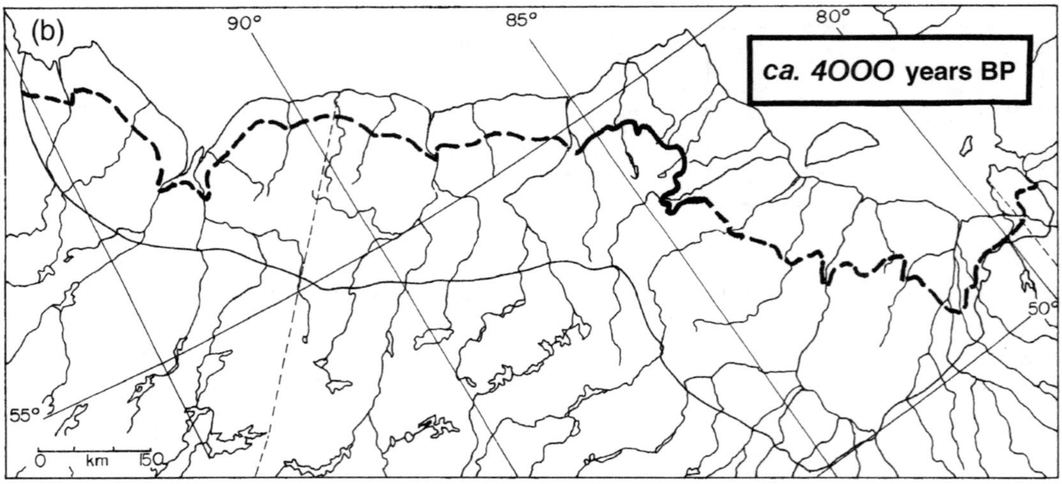

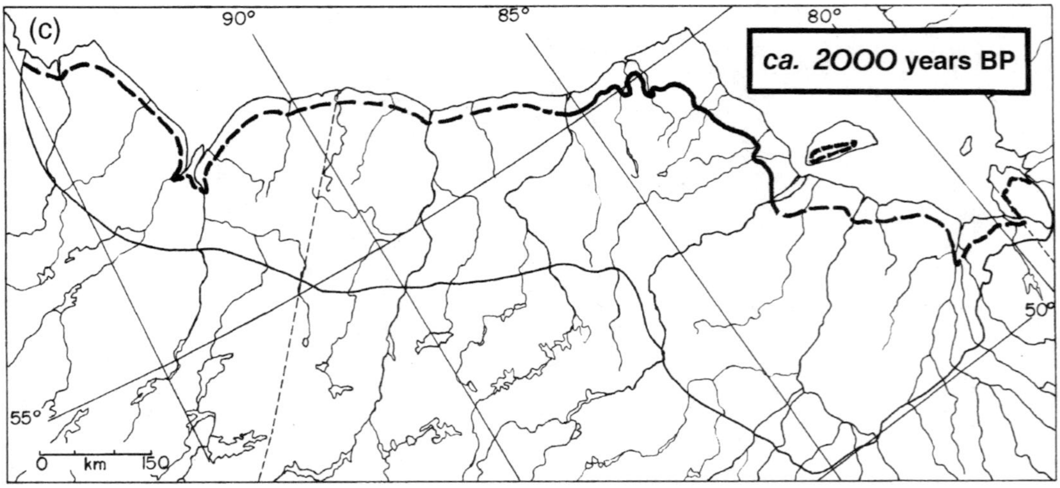

18). Marine clays, silts, and sands were deposited throughout the Lowland, and continue to be deposited as the same coast continues to emerge from Hudson and James bays.

A lake core taken from near Aquatuk Lake in the Sutton Ridges has shown that those uplands, at 120 to 125 m above sea level, emerged from the ocean about 8200 BP (McAndrews et al. 1982). Pollen from the core indicated a distinct basal zone of sparse tundra dominated by *Dryas*, with a high representation of sedges, willow, and grasses, and obligative coastal species such as *Plantago maritima*. This was followed by a shrub tundra, and then by a modern woodland of spruce (*Picea* spp.), tamarack (*Larix laricina*), and alder (*Alnus* spp.) between 8200 and 6500 BP. These zones closely reflect the modern pattern of change in vegetation from the Hudson Bay coast toward the interior Lowland. The subsequent analysis of the July mean temperatures derived from these pollen types showed that temperatures were probably colder at 6000 BP than at present, probably because of the cooling effects of the Tyrrell Sea (McAndrews 1993).

Another coastal paleobotanical study near Churchill interpreted the composition of pollen from a 110-cm deep peat core, with a 33-cm active layer over permafrost. It indicated that herbaceous-ericaceous heath tundra was present since the inception of vegetation in that area (Johnson 1949).

The results of palynological work south of the Lowland in the Ontario Clay Belt suggest that the climate at the time of emergence of the edges of the Lowland was warmer than now, near the beginning of the thermal optimum or hypsithermal (7000 BP; Vincent 1973). Pollen there was dominated by white pine (*Pinus strobus*) and red pine (*Pinus resinosa*), followed by jack pine (*Pinus banksiana*), and this has been interpreted to indicate the existence of a low-density, species-depauperate woodland. White spruce woodlands began to dominate soon after (Liu 1990), with subsequent temperature declines and landscape paludification resulting in closed boreal forests dominated by spruce, with some pine, yellow birch (*Betula alleghaniensis*), trembling aspen (*Populus tremuloides*), maple (*Acer*), and beech (*Fagus grandifolia*). The continuation of these trends has resulted in the present forest types, which have, for example, lost some of their temperate elements, such as beech (Liu 1990).

This warmer hypsithermal period is considered to have occurred south of the Lowland from about 7000 to 3250 years ago (Vincent 1973; Liu 1990). It was a period of rapid expansion and later decline in white cedar (*Thuja occidentalis*) in the Ontario Clay Belt (Liu 1990). Also during this period, the range of white pine extended northward, and fossil wood has been documented 95 km north of its present range (Fig. A116) at just over 5000 BP (Terasmae and Anderson 1970).

Indirect evidence of other early northward range extensions can also be interpreted from the presence of the hybrid *Populus ×jackii* (*Populus balsamifera × deltoides*) in the southern Lowland and the presence of introgressive types of *Populus tremuloides* with *Populus grandidentata* in the Kirkland Lake area (Hills 1962); both of which suggest that two presently southern poplars (*P. deltoides* and *P. grandidentata*) may have once had more northerly distributions. Stands of red pine remain in disjunct northern sites protected from fire, such as at Lake Pierre (Riley 1978), and disjunct white elm (*Ulmus americana*) bottomlands persist along the Kenogami River of the Hudson Bay Lowland, all possibly relicts from the period of warmer postglacial climates.

In northwest Ontario, white pine reached its maximum northward distribution between 6500 and 6000 years ago, during the thermal optimum and about the same time as a peak influx of prairie pollen into northwest Ontario (Ritchie 1983; Bjorck 1985).

The postglacial vegetation of the region west of the Great Lakes and east of the Red River was predominantly boreal spruce forest or woodland (Wright 1964; Bjorck 1985). This spruce forest shifted north to take up its modern position about 11 500 BP, and was replaced by a mixed conifer–deciduous forest that has persisted with relatively little change since. (Further west, the late-glacial pollen assemblage has been interpreted as originating as a youthful spruce-dominated community (Ritchie 1966), with some indications of a treeless, tundra-like episode preceding it (Ritchie 1976).)

By 12 000 BP, an essentially boreal assemblage extended from southern Alberta to the Great Lakes in a band still lying south of the present boreal region (Ritchie 1987). These boreal systems had spread northward since 20 000 BP from mid-continental refugia stretching from Kansas to the Carolinas and, variably, as far south as Texas and Missouri (Ritchie 1987, 1989; Delcourt and Delcourt 1987). After 10 000 BP, the western prairie grasslands spread northeast beyond their modern range, presumably originating from southwest of the glacial maximum, in response to a warmer, drier climate comparable to the hypsithermal. Following 5500 BP, boreal forest spread southward again to assume its modern limits as the climate became cooler and wetter (Ritchie 1976).

Species migration

Even by 7000 BP, the modern Hudson Bay Lowland remained almost entirely under the proglacial Tyrrell Sea (Fig. 18). The regions just to the south were beginning to see significant development of their modern vegetation, and pine and spruce woodlands dominated areas within 150 km of the Tyrrell Sea shore.

The consolidation of these early arrivals into postglacial woodlands and forests south of the Lowland reflected climate changes, soil development, and the variable migration rates of individual tree species. This has been called “a great natural experiment resembling paper chromatography” (Drury 1956). The effect of this “migration lag” (Davis 1975) may have largely waned by 6000 BP, with the the composition of the boreal forests around the southern margins of the Tyrrell Sea approximating present assemblages, albeit in different zlandscape patterns.

Because of the rapid establishment of temperate and boreal forests northward through this warmer period, it is highly unlikely that tundra-like vegetation dominated by arctic species occurred near the south shore of the Tyrrell Sea. Although an earlier presence of forest tundra and open woodland is indicated by pollen records from the southeastern margins of the receding proglacial lakes Barlow and Ojibway (Vincent 1973), they were not present in this area at the time of the Tyrrell Sea incursion.

Far to the south, and much earlier, open tundra and subarctic woodland developed adjacent to the southern margin of the maximum Wisconsinan ice sheet, until 17 000 BP (Martin 1959; Gleason 1922) and until 10 000 BP in coastal Maine (Davis et al. 1975). Relict species from those associations were subsequently left on the Gaspé Peninsula as the ice cap left that area about 13 000 BP (Prest 1970) and were also left on the Appalachian and Adirondack peaks.

In northern Minnesota, Watts (1967) concluded that communities with tundra affinities disappeared before 10 500 BP. Around Lake Superior, tundra-like associations of arctic alpine species probably occupied Lake Superior's north shore and islands from 11 000 to 10 000 BP. The latter date concurs with deglaciation of this shoreline (Prest 1970; Dyke and Prest 1987*a*; Barnett 1992). Saarnisto (1974) considered that there was a static ice front located between the north shore of Lake Huron and the Algonquin Stadial ice margin, at the Cartier Moraine, for long enough (11 000 to 10 100 BP) to explain pollen deposits there that were indicative of tundra that lasted several hundred years.

However, by the time deglaciation occurred in the Wawa area to the north (at the Chapleau Moraine, at 9500 BP), the warming climate and the prevalence of boreal species prevented the development of tundra-like vegetation. A spruce pollen zone occupied the earliest deposits there and was rapidly replaced by more temperate species about 9000 BP (Saarnisto 1974). This rapid forest development north of Superior reflected the presence of terrain that was probably better drained than at present, because of the steeper, prerebound topography, as well as the early-hypsithermal climatic conditions. As a result, it is unlikely that the Lowland's present arctic species and tundra vegetation migrated there from the south.

The origins of the Lowland's arctic tundra assemblage lay in the developing postglacial floras and vegetation of the continental Yukon, Northwest Territories and Nunavut, the arctic archipelago, and the Atlantic maritimes. These were the closest extraglacial refugia from which arctic species could migrate to assume their current, largely circumpolar, ranges. The locations and timing of arctic and east coast ice sheets resulted in broad areas that acted as late-Wisconsinan refugia for arctic and maritime species (Grant 1977; Ives 1978; Boulton 1979; Ritchie 1987; Fulton 1989).

While there is general agreement that there was an overall frontal migration of many species northward as the climate warmed latitudinally (Delcourt and Delcourt 1987, 1991), there is less agreement about the routes that east-coast species may have taken north and west. For example, at the southern edge of the Shield in Quebec, Richard (1977) interpreted pollen assemblages as indicating that tundra occurred to the south of Laurentides Provincial Park as late as 7300 BP. Did some tundra or maritime-tundra species migrate towards southern James Bay via the eastern shores of proglacial Lake Ojibway? At that time, the shortest distance between the Gulf of St. Lawrence and southern James Bay, about 800 km, was along the ice margin of the receding Labrador ice sheet (Dyke and Prest 1987*a*).

The proglacial lakes across northeast Ontario and northwest Quebec were primarily glacier fed, cold and fresh, and their shorelines may have been appropriate sites for the colonization and migration of coastal or arctic species. They would not have been a suitable habitat for strictly halophytic species migrating westward into the continent from the Gulf of St. Lawrence (Boivin 1952), and no relict halophytes are known to persist in the intervening area. Less obligative halophytes do occur in the Lac St. Jean area, where a brackish marine incursion occurred 9000 BP (Prest 1970).

In western Canada, on the prairies, halophytic species were probably more widespread. A warmer, drier period occurred on the prairies from 7000 to as late as 5000 BP, during which time the mix of habitats may have shifted towards that of modern arid prairies, with highly saline sloughs and high rates of evaporation (Walker and Coupland 1968). The more widespread incidence of saline sloughs on the western prairies at that time could have contributed to the occurrence of western halophytes in the Hudson Bay Lowland; however there are no inland relict occurrences of halophytes in the Lowland or on the Canadian Shield between the Lowland and the prairies.

Beringia, northwest of the glacial maximum, supported extensive arctic and subarctic herb and shrub tundra during the full- and late-glacial periods (Ritchie 1984, 1987). After 11 000 BP, the Beringian climate warmed rapidly and spruce forests and woodlands developed on non-montane uplands, with birch and alder increasing in importance later. There is strong evidence from modern plant distributions of the contribution of northwest refugia to the flora of central Canada (e.g., Figs. A45 and A47; Hultén 1937, 1968). On the other hand, many boreal and subarctic species occur in extraglacial areas both northwest and southwest of the glacial limits (e.g., Figs. A26 and A51), and it is difficult to distinguish between southern, southwestern, and northwestern sources for some widespread species. This is the case even for critical species like white spruce, which may or may not have survived the Wisconsinan in Beringia (Raup 1947; Wilkinson et al. 1971; Ritchie 1984).

In terms of the mechanics of plant migration, species are not static in their distributions. Hemlock and white pine may still be able to move westward even though their northern distributions have contracted since the hypsithermal. The migration patterns of major tree species have been mapped, and these maps reflect the varying migration speeds of species over several thousand years, for example jack pine migrates at 350–500 m/year, hemlock at 200–300 m/year, and beech at 250–300 m/year (Davis 1976; Dyke and Prest 1987*b*).

Savile (1956, 1972) stressed the importance of winter wind dispersal and storm events, which are typical dispersal methods for arctic and subarctic species. Adaptations in arctic plants include censer and vibratory mechanisms, and gradual, delayed, or wind-triggered seed release. Shoreline or marine dispersal mechanisms include ice rafting of shoreline communities, production of fleshy edible fruits, hooked or otherwise adhesive propagules, and the dispersal of propagules by waterfowl.

Dispersal by waterfowl was ranked by Savile as second in importance only to winter wind dispersal, particularly in the distribution of coastal and wetland species. The chance of successful colonization based on individual dispersal events is increased by the reproductive specialization of northern plants. Mosquin (1966) reviewed how various mechanisms, such as self-pollination, apomixis, and aneuploid decreases in chromosome number reduce the need for genetic recombination in northern species. Diploid outcrossers are largely absent from the arctic.

In the higher latitudes during the Pleistocene, interglacial stages were the exception and glacial periods were the norm, at a ratio of about 1:9 (Davis 1976). Rapid climatic change, glaciated landscapes, and arctic conditions select species for their advanced abilities to colonize new habitats. "The dispersal at any time of even a single disseminule of a hermaphrodite species to a new spot may lead to the full establishment of the species at that new spot. Such 'saltatory' migrations over great distances may actually be the origin of the supposed 'relict' populations 'depleted in biotypes' which advocates of persistence have called 'conservative'" (Polunin 1955).

Biotic communities did not migrate intact as tightly co-evolved associations, but instead migrated as individual species with actively invading leading edges of distribution and with simultaneous and progressive fragmentation of ranges through local extirpations (Polunin 1955). Modern, northern associations are largely based on the vagaries of individual dispersals and available environments, over short periods of time (Ritchie 1987).

Two general concepts of plant migration must be considered in tandem in the interpretation of a flora such as that of the Lowland. The vanguard or saltatory concept describes the long-distance movement of species, and can be particularly applied to the migration of arctic and coastal species. The conservative concept descibes gradual immigration and emigration, and range consolidation and fragmentation, over long periods, and has a more widespread application to the dominant continental flora.

Because of its gradualism and the longer periods of time involved, the conservative concept of plant migration tends to accumulate corroborative paleoecological evidence, while the vanguard concept is most often applied to the otherwise inexplicable. Disjunct populations of species can either be judged to be indicative of past range fragmentation or of long-distance dispersal, depending on the overall range of the particular species, its biology, and its habitats. Reference to these concepts and to the documented record of species migration will be made in the following discussion of the major geographic themes in the flora of the Lowland.

Widespread species of the Hudson Bay Lowland

Thirty-five percent of the native species of the Hudson Bay Lowland are widespread throughout the region. Volumetrically, they dominate its vegetation. This group of Lowland species has the widest range of latitudinal occurrences and the highest proportion of transcontinental species (Table 3).

These species have been able to rapidly colonize and dominate mature boreal and subarctic communities in postglacial environments. Eighty percent of them still occur in multiple extraglacial areas south, southwest, northwest, *and* east of the continental ice sheet, and may have survived the late-Wisconsinan in all or any of these regions. For example, almost 90% of these widespread species also occur in unglaciated areas of Alaska or the Yukon (Hultén 1937, 1968).

This dominant flora represents a wide range of plant families and genera. In this group are five widespread tree species (*Populus balsamifera*, *Populus tremuloides*, *Larix laricina*, *Picea glauca*, *Picea mariana*) and over 40 shrubs (12 *Salix* spp., 6 *Ribes* spp., *Betula pumila*, *Myrica gale*, *Alnus crispa*, *Alnus rugosa*, *Ledum groenlandicum*, *Kalmia polifolia*, etc.) (Appendix B). These are the dominant species of wetlands (Figs. 4*a* and 4*b*, *Scirpus cespitosus*, *Carex limosa*) and of uplands and tundra (Fig. A64, *Erigeron acris*; Fig. A119, *Poa alpina*). Many of these widespread species occur more frequently in the interior but are also found on the coasts; others prefer coastal sites, but are not halophytes. Among the latter are *Polygonum fowleri* (Fig. A123), *Catabrosa aquatica* (Fig. A41) and *Carex saxatilis* (Fig. A38).

The locations of probable refugia or source areas are deduced from the persistence of those species in specific areas that were unglaciated in the late-Wisconsinan (after Hultén 1937). For example, northwestern areas are almost certainly the source for species not now known south or southwest of the glacial maximum, as in the case of widespread species such as *Cicuta mackenzieana* (Fig. A47) and *Pedicularis parviflora* (Fig. A108). Another widespread species occurring to the northwest of the late-Wisconsinan is *Cypripedium passerinum* (Figs. A50 and A51), but it is also disjunct to Lake Superior and the Gulf of St. Lawrence and occurs in Montana at the edge of the glacial maximum. This species is likely to have had a postglacial refugium south of the ice sheet that has since been extirpated.

Other species are presently known from both southwest and northwest of the former ice sheet and may have migrated from both (or either) northwestern and southwestern populations. These include *Elymus innovatus* (Fig. A61), *Arnica chamissonis* ssp. *foliosa* (Fig. A17), *Antennaria pulcherrima* (Fig. A8; disjunct to the St. Lawrence and perhaps also a late-Wisconsinan survivor in that region or elsewhere in the east), *Astragalus alpinus* (Fig. A23; with a distinct eastern var. *brunetianus* confined to New England and the Gulf of St. Lawrence), and *Elaeagnus commutata* (disjunct to the Gaspé).

Other widespread Lowland species have modern ranges restricted to eastern North America, and almost certainly originated postglacially from south of the maximum ice sheet; for example, *Aster simplex, Trisetum melicoides* (Fig. A145), *Bromus ciliatus, Iris versicolor, Salix cordata, S. lucida, S. pellita, Potentilla tridentata, Oxytropis campestris* var. *johannensis*, and *Campanula aparinoides.*

Another widespread Lowland species, *Pedicularis groenlandica* (Figs. A104 and A105) has a bimodal North American distribution, suggesting that the Lowland population was derived from either the east coast refugia or from the southwest continental range of the species. Some other widespread Lowland species are simply generally disjunct, such as *Astragalus eucosmus* (Fig. A26) and *Tanacetum huronense* (Fig. A46). The restriction of these species to open disturbed littoral or coastal habitats suggests an ecological basis for their widespread but disjunct postglacial distributions; probably indicating a history of range fragmentation by later closed-forest ecosystems.

Two other groups of Lowland species appear to have similar postglacial histories to those of widespread species: interior species (Table 4; 48 species largely widespread but not occurring in coastal habitats) and species restricted to interior areas south of the Hudson Bay coastal data collection areas and away from the Manitoba coast (Table 5; 44 species). Their floristic affinities indicate a prevalence of more temperate species with less geographic amplitude latitudinally. The percentage of transcontinental species remains at 80–90%. Less than 30% of them are circumpolar (as opposed to over 50% for widespread species), and 70% of them are exclusively North American (about 25% for widespread species as a whole).

These two groups are more continental (non-oceanic) in their ranges, and their current distributions suggest that the northwest continent played only a minor role as a late-Wisconsinan refugia for the Lowland populations. Again, species like *Astragalus americanus* (Fig. A24) and *Aquilegia brevistyla* (Figs. A10 and A11) may have redispersed from the northwest, but their presence in the southern cordillera and southwest indicate other possible sources as well. *Arnica lonchophyllus, Carex atratiformis, Aster modestus*, and *Lathyrus ochroleucus* only barely reach the Yukon and Alaska but are widespread in the central part of the continent, and almost certainly radiated from south of the ice sheet. Other species of these groups can have had only eastern postglacial sources: *Pinus banksiana* (Figs. A113 and A114; Yeatman 1967), *Utricularia cornuta, Solidago hispida, Thalictrum venulosum* var. *confine, Kalmia angustifolia, Andromeda glaucophylla, Cirsium muticum*, and many others.

Table 4. Species restricted to the interior of the Hudson Bay Lowland, R(I).

Lycopodium annotinum (typ.)	*Betula papyrifera*
Gymnocarpium dryopteris	*Nuphar variegatum*
Woodsia ilvensis	*Coptis groenlandica*
Pinus banksiana (A113, A114)	*Aquilegia brevistyla* (A10)
Abies balsamea	*Caltha natans*
Sparganium fluctuans	*Corydalis sempervirens*
Scheuchzeria palustris	*Cardamine pensylvanica*
Elodea canadensis	*Sarracenia purpurea*
Danthonia intermedia	*Amelanchier alnifolia* (typ.)
Oryzopsis asperifolia	*A. bartramiana*
Muhlenbergia glomerata	*Prunus pensylvanica*
Rhynchospora alba	*Astragalus americanus* (A24)
Carex arcta	*Lathyrus ochroleucus*
C. atherodes	*Polygala paucifolia*
C. atratiformis	*P. senega*
C. deflexa	*Rhamnus alnifolia*
C. houghtoniana	*Arala nudicaulis*
C. lenticularis	*Kalmia angustifolia*
C. oligosperma	*Vaccinium myrtilloides*
Juncus filiformis	*Trientalis borealis*
J. stygius	*Utricularia cornuta*
J. subtilis	*Arnica lonchophylla*
Salix humilis	*Solidago hispida*
Cypripedium parviflorum var. *planipetalum* (A49)	*S. ptarmicoides*
C. parviflorum var. *pubescens*	

Note: Numbers in parentheses refer to the map number in Appendix A.

Overall, the species that are widespread in the Lowland have relatively broad latitudinal distributions outside the Lowland, as well as within, and are largely transcontinental, either circumboreal or North American. They are very likely the product of multiple or frontal redispersal into the Lowland from northwest, south, *and* east of the late-Wisconsinan ice sheets, and are the dominant species of both the Lowland and the transcontinental boreal vegetation.

Eastern species of the Hudson Bay Lowland

Of the eastern North American species in the Lowland flora, most still occur in extraglacial areas south and east of the Laurentide maximum (126, Table 8). Sixty-two percent of them are restricted in the Lowland to areas south of the Attawapiskat River (Table 8). Overall, these eastern North American species correspond significantly with the species that are restricted to the southeast Lowland (199, Tables 6, 7), and even more so with those restricted to the extreme south in the Boreal Peatland floristic zone (Fig. 13).

This correlation of eastern and southern elements is the product of the progressive migration of species from south of the glacial maximum. The migration routes of eastern

Table 5. Species restricted to south of the Hudson Bay collection areas and away from the Manitoba coast, "t" (excluding species also restricted to the interior, Table 4).

Botrychium virginianum	*A. spicata*
Sparganium angustifolium	*Viola palustris*
S. chlorocarpum	*Sanicula marilandica*
Phalaris arundinacea	*Cornus stolonifera*
Potamogeton natans	*Andromeda glaucophylla*
Sphenopholis intermedia	*Arctostaphylos uva-ursi* (typ.)
Schizachne purpurascens	*Vaccinium oxycoccus* var. *microphyllum*
Poa interior	*Apocynum cannabinum*
Glyceria grandis	*Prunella vulgaris*
Scirpus microcarpus	*Veronica scutellata*
Carex flava	*Galium tinctorium*
C. lasiocarpa	*Lonicera dioica*
C. pauciflora	*L. villosa*
C. praticola	*Antennaria neglecta*
C. sterilis	*Aster modestus*
Calla palustris	*A. umbellatus*
Juncus dudleyi	*Cirsium muticum*
Cypripedium parviflorum var. *makisin*	*Erigeron philadelphicus*
	Eupatorium maculatum
Thalictrum venulosum var. *confine*	*Hieracium kalmii*
	Prenanthes racemosa
Drosera linearis	*Solidago graminifolia*
Amelanchier sanguinea	*S. uliginosa*

North American tree species into, through, and beyond the Great Lakes area have been documented on the basis of first-arrival times from pollen records (Whitehead 1972; Davis 1976, 1983; Kapp 1977; Delcourt and Delcourt 1987; Dyke and Prest 1987*b*). The migration of aquatic vascular plants and freshwater fishes into the Great Lakes region probably followed similar routes, with a major migration northward through the west Mississippi basin (Vesper and Stuckey 1977; Mandrak and Crossman 1992).

For some eastern deciduous trees, specific extraglacial sources have been suggested; for example, hemlock (*Tsuga canadensis*) survived in a small refugium in the southeast United States and white pine recolonized from a small area of coastal Virginia (Delcourt and Delcourt 1987). On the other hand it has also been suggested that future palynological studies will indicate that the full-glacial refugia of eastern temperate species were more widespread than present data suggest (Ritchie 1987).

Soper interpreted the coincident distributions of Carolinian species in southern Ontario as confirmation of "two rather obvious routes from the area to the south…around the two ends of the present Lake Erie", with another route extending up the Ottawa Valley (1962; see also Cody 1982). These routes follow low-lying landforms (less than 330 m above sea level) with warmer climates (Chapman and Thomas 1968).

Table 6. Species restricted to the southern Hudson Bay Lowland, R(S).

Matteuccia struthiopteris	*C. trisperma*	*Epilobium strictum*
Dryopteris carthusiana	*Juncus articulatus*	*Oenothera biennis*
Thuja occidentalis	*J. brevicaudatus*	*O. parviflora*
Typha latifolia	*J. effusus*	*Circaea alpina*
Sparganium eurycarpum	*J. longistylis*	*Myriophyllum alterniflorum*
Alisma triviale	*J. pelocarpus*	*Osmorhiza depauperata*
Sagittaria latifolia	*Lilium philadelphicum*	*Lysimachia ciliata*
Danthonia spicata	*Platanthera psycodes**	*L. terrestris*
Spartina pectinata (A140)	*Arethusa bulbosa* (A15)	*Gentiana linearis* (A67)
Cinna latifolia	*Listera auriculata*	*Gentianella crinita*
Elymus canadensis (A60)	*Goodyera repens*	*Scutellaria lateriflora*
E. virginicus (A63)	*Calypso bulbosa*	*Lycopus americanus*
E. wiegandii	*Salix discolor*	*L. uniflorus*
Scirpus acutus	*S. myriocoides*	*Chelone glabra*
S. pungens	*S. eriocephala*	*Mimulus ringens*
S. cyperinus	*Polygonum anchoreum*	*Veronica americana*
S. validus	*Ranunculus flammula*	*Galium asprellum* (A66)
Eleocharis elliptica	var. *ovalis*	*G. triflorum*
E. erythropoda	*R. hispidus*	*Viburnum trilobum*
Carex bebbii	*Thalictrum dasycarpum*	*Lonicera hirsuta*
*C. crawfordii**	*Spiraea alba*	*L. oblongifolia*
C. deweyana	*Fragaria vesca*	*Artemisia caudata*
C. peckii	*Rosa blanda*	*Solidago altissima*
C. pedunculata	*Prunus virginiana*	*S. juncea*
C. prairea	*Astragalus canadensis* (A25)	*S. serotina*
C. projecta	*Acer spicatum*	*Aster lateriflorus*
C. retrorsa	*Impatiens biflora*	*Erigeron annuus*
C. sartwellii	*Viola adunca*	*Megalodonta beckii*
C. scoparia	*V. incognita*	*Lactuca biennis*
C. stipata	*V. selkirkii*	*Prenanthes alba*
C. stricta		

Note: Numbers in parentheses refer to the map number in Appendix A.

*Species restricted to the extreme south, south of a line that extends from Ogoki to Kinoje to the junction of the Moose and Abitibi rivers, the Boreal Peatland floristic zone (Fig. 13).

This kind of climatic amelioration also extends to the Bruce Peninsula and Manitoulin Island in the central Great Lakes. These limestone areas formed a bridge to northern Ontario from about 10 600 to 9000 BP (Dyke and Prest 1987*a*; Barnett 1992). The Ottawa Valley and the Bruce–Manitoulin bridge would have provided receptive terrain northward past the barren Shield landscapes east of Georgian Bay, as well as the cold Algonquin Precambrian dome and the Precambrian uplands north of the Ottawa River valley. Many species currently found on those routes are largely absent from neighbouring Shield areas: *Celastrus scandens*, *Rhamnus alnifolius*, *Shepherdia canadensis*, *Viburnum lentago*, *Lonicera dioica*, etc. (Soper and Heimburger 1982; Rousseau 1974). However,

Table 7. Species restricted to the southern interior of the Hudson Bay Lowland, R(SI).

*Lycopodium clavatum**	*C. leptonervia**	*Anemone quinquefolia*
L. dendroideum	*C. loliacea*	*A. virginiana*
Isoetes echinospora	*C. ormostachya**	*Clematis occidentalis**
*Equisetum hyemale**	*C. richardsonii**	*Corydalis aurea**
*Thelypteris palustris**	*C. umbellata**	*Armoracia aquatica**
*Pteridium aquilinum**	*C. viridula*	*Physocarpus opulifolius*
*Onoclea sensibilis**	*Cladium mariscoides**	*Sorbus americana**
Athyrium filix-femina	*Acorus calamus**	*S. decora*
*Cystopteris bulbifera**	*Xyris montana** (A146)	*Amelanchier alnifolia*
*Dryopteris cristata**	*Eriocaulon aquaticum**	var. *compacta*
D. expansa	*Juncus tenuis**	*Aronia prunifolia** (A18)
*D. intermedia**	*Luzula acuminata**	*Prunus pumila**
*Taxus canadensis**	*Clintonia borealis**	*Astragalus tenellus* (A27)
*Pinus resinosa** (A115)	*Maianthemum canadense*	*Geranium bicknellii*
*P. strobus** (A116)	*Prosartes trachycarpa**	*Oxalis montana**
*Potamogeton epihydrus**	*Streptopus amplexifolius**	*Rhus radicans**
*P. oaksianus**	*S. lanceolatus**	*Hypericum majus**
*Andropogon gerardii** (A6)	*Trillium cernuum*	*Triadenum fraseri**
Bromus kalmii	*Cypripedium reginae** (A52)	*Viola pubescens**
Muhlenbergia mexicana	*Platanthera lacera** (A118)	*Aralia hispida**
M. richardsonis	*P. orbiculata**	*Osmorhiza claytoni**
*Panicum capillare**	*Pogonia ophioglossoides** (A122)	*O. longistylis**
P. lanuginosum	*Corallorhiza striata*	*Chimaphila umbellata*
var. *implicatum**	*Malaxis paludosa** (A95)	*Monotropa uniflora*
*P. lindheimeri**	*M. unifolia** (A96)	*Fraxinus nigra**
*P. virgatum**	*Populus ×jackii*	*Apocynum androsaemifolium*
*Phragmites australis** (A110)	*Salix petiolaris*	*Calystegia sepia**
*Poa saltuensis**	*S. pyrifolia*	*Monarda fistulosa**
*Eriophorum tenellum**	*Corylus cornuta**	*Melampyrum lineare**
E. virginicum	*Betula ×sandbergii*	*Sambucus pubens**
*Scirpus atrovirens**	*Ulmus americana**	*Symphoricarpos albus*
*Rhynchospora fusca**	*Arceuthobium pusillum*	*Diervilla lonicera**
*Carex arctata**	*Asarum canadense**	*Aster macrophyllus**
C. crawei	*Nymphaea odorata*	*Solidago nemoralis**
C. exilis	*N. leibergii*	*S. rugosa**
*C. granularis**	*Nuphar microphyllum**	
*C. intumescens**	*N. ×rubrodiscum*	

Note: Numbers in parentheses refer to the map number in Appendix A.

*Species restricted to the extreme south, south of a line that extends from Ogoki to Kinoje to the junction of the Moose and Abitibi rivers, the Boreal Peatland floristic zone (Fig. 13).

both these routes were probably less important and shorter-lived than the Michigan peninsula, which was unglaciated and lake-free 3000 years earlier, by 13 000 BP (Dreimanis 1977; Barnett 1992). The lower Michigan peninsula also provides evidence of a modern floristic tension zone, suggesting the potential of that landscape as a source for northward pulses of southern species (Voss 1972).

Table 8. Eastern North American species of the Hudson Bay Lowland.

Thelypteris palustris*	Platanthera lacera* (A118)	Oxalis acetosella*
Onoclea sensibilis*	P. psycodes*	Polygala paucifolia
Cystopteris bulbifera*	Pogonia ophioglossoides* (A122)	Acer spicatum*
Dryopteris intermedica*	Arethusa bulbosa* (A15)	Hypericum majus*
Taxus canadensis*	Listera auriculata*	Triadenum fraseri*
Pinus resinosa* (A115)	Goodyera repens*	Viola incognita*
P. strobus* (A116)	Malaxis unifolia* (A96)	V. pubescens*
Thuja occidentalis*	Salix ballii	Epilobium strictum*
Potamogeton oakesianus*	S. cordata	Aralia hispida*
Bromus ciliatus	S. eriocephala*	Osmorhiza claytoni*
B. kalmii*	S. humilis*	Angelica atropurpurea
Panicum virgatum*	S. lucida	Ligusticum scothicum (A84)
Poa flexuosa	S. myricoides*	Chimaphila umbellata*
P. saltuensis*	S. pellita	Pyrola grandiflora
Puccinellia ambigua	Ulmus americana*	Kalmia angustifolia
P. lucida (A127)	Polygonum caurianum	Andromeda glaucophylla
Trisetum melicoides (A145)	Arceuthobium pusillum*	Primula laurentiana
Eriophorum tenellum*	Asarum canadense*	Lysimachia terrestris*
Scirpus cyperinus*	Atriplex glabriuscula (A28)	Fraxinus nigra*
Rhynchospora fusca*	Minuartia groenlandica (A98)	Gentiana linearis* (A67)
Carex arctata*	Nymphaea odorata*	Gentianopsis nesophila (A70)
C. exilis*	Nuphar microphyllum*	Chelone glabra*
C. granularis*	N. ×rubrodiscum*	Mimulus ringens*
C. intumescens*	Ranunculas hispidus*	Bartsia alpina (A29)
C. leptonervia*	Thalictrium venulosum	Utricularia cornuta
C. norvegica (typ.)	var. confine	Galium asprellum* (A66)
C. ormostachya*	Armoracia aquatica	G. tinctorium
C. paleacea (A37)	Draba incana (A55)	Lonicera hirsuta*
C. projecta*	D. norvegica (A57)	L. oblongifolia*
C. recta	Arabis alpina	Diervilla lonicera*
C. salina	Spiraea alba*	Campanula aparinoides
C. stricta*	Aronia prunifolia* (A18)	Solidago altissima*
Cladium mariscoides*	Sorbus americana*	S. hispida*
Xyris montana* (A146)	S. decora*	S. juncea*
Eriocaulon aquaticum*	Amelanchier bartramiana	S. rugosa*
Juncus pelocarpus*	A. sanguinea	Aster lateriflorus*
J. subtilis	A. spicata	A. macrophyllus*
Clintonia borealis*	Potentilla crantzii	A. lanceolatus
Trillium cernuum*	P. tridentata	Bidens hyperborea
Iris versicolor	Prunus pumila*	Senecio aureus
Cypripedium parviflorum	Oxytropis campestris	Cirsium muticum
var. planipetalum (A49)	var. johannensis (A100)	Prenanthes alba*
C. reginae* (A52)	Lathyrus japonicus var. pellitus (A81)	

Note: Numbers in parentheses refer to the map number in Appendix A.

*Species confined to the southeast Lowland, R(S), R(SI).

Landform analysis has suggested similar routes. Hills suggested that the upper Ottawa Valley and Lake Timiskaming Valley was an area of increased site diversity that encouraged the migration of a variety of species from one receptive, warmer-than-normal site to the next (1962). Also in this area, high-lime landforms frequently occur (Hills 1959). Hills considered the landforms east and west of this route to be hurdles to migration, specifically the Precambrian Gogama-Hornepayne upland axis and the Laurentide Shield east of Lake Timiskaming. The Lake Timiskaming and Little Clay Belt areas north to 48°N have longer frost-free periods than adjacent areas, and are similar in climate to areas farther south, such as North Bay, Sudbury, and Sault Ste. Marie (Chapman and Thomas 1968). Hills also suggested that the east part of the Northern Clay Belt was particularly suitable for migration because of its combination of calcareous clay and silt substrates, more consistently rolling terrain, and broad esker complexes running long distances south to north.

Edaphic controls on plant distributions have long been recognized; calciphiles are absent from the Canadian Shield but are found in Paleozoic areas adjacent to the Shield, and acidophiles have the contrary substrate fidelity (Wynne-Edwards 1937; Scoggan 1978–1979). Southern species near their northern limits tend to become facultative calciphiles (Porsild 1955). "For instance, *Acer nigrum*, *Ulmus racemosa* [*U. thomasii*], *Celtis occidentalis*, not distinctively calcicolous within their main areas, are exclusively so in southeast Quebec where they reach their northern limit" (Marie-Victorin 1938). Calcareous substrates lengthen effective growing seasons (Marie-Victorin 1938), and the Northern Clay Belt and the till deposits in the Geraldton area, for example, are rich in calcium carbonate (more than 30% calcium carbonate clasts; Barnett 1992).

Corroborative evidence of these migration routes lies in modern plant distributions. For example, more than 60 native species reach their northern limits in the Little Clay Belt (Riley 1979*a*), 18 of which are trees or shrubs, and most of which are not found in the Algoma or Laurentide uplands. Further north, Baldwin (1958) noted 139 native species that were unreported north, east, or west of the Northern Clay Belt (now reduced to just over 100 on the basis of this study and other work just south of the Lowland; Gaudreau 1972; Riley 1978). Examples of this distribution pattern into the northeast from the Great Lakes and Ottawa Valley include *Cornus alternifolia*, *Cornus rugosa*, and *Rhus typhina* (Rousseau 1974; Bergeron et al. 1978; Riley 1979*a*).

The Lowland species that have distributions suggesting migration through the Ontario Clay Belt axis are restricted to the southern part of the Lowland. They include *Cypripedium reginae* (Fig. A52), *Spartina pectinata* (Fig. A140), *Elymus canadensis* (Fig. A60), *Elymus virginicus* (Fig. A63), *Scirpus pungens*, *Carex peckii*, *Pogonia ophioglossoides* (Fig. A122), *Astragalus canadensis* (Fig. A25), *Malaxis unifolia* (Fig. A96), *Platanthera orbiculata*, *Ulmus americana*, and *Nuphar microphyllum*.

In the same way that the precursors of lakes Erie, Ontario, and Huron were outflanked by species migrating northward, similar patterns are evident around Lake Superior, for example, northward occurrences of species such as *Sorbus americana*, *Pinus strobus*, *Cornus alternifolia*, and *Viburnum rafinesquianum*. However, most of the eastern North American species in the southern Lowland have ranges that are more or less continuous southward to northeast Ontario and northwest Quebec. None of them are presently known to have ranges that are continuous only to the region west of Lake Superior. Southern species of plants may have had little chance of reaching the Lowland from west of Lake Superior because of the great distances involved in crossing unreceptive Shield terrain, much of it covered by thin or no tills. On the other hand, the more widespread, transcontinental species of boreal systems are certainly likely to have advanced on a front from that direction as well as from the southeast.

A small number of these southern Lowland species are disjunct from their nearest eastward and southern populations. Excluding disjunctions to coastal James Bay, which are discussed elsewhere, these include eastern species such as *Panicum virgatum* and *Xyris montana* (Fig. A146) and transcontinental species such as *Andropogon gerardii* (Fig. A6), *Panicum capillare*, and *Panicum lindheimeri*.

Over 80% of the 78 species restricted to the extreme south in the Lowland are presently known from three or less occurrences. Over 55% of the Lowland species that are considered to be rare by this criteria are restricted to the south. Some of these are enigmatic. For example, *Xyris montana* (Fig. A146), *Eriocaulon aquaticum*, and *Rhynchospora fusca* are known from only single sites in the Kinoje Lake area; an area that emerged from the Tyrrell Sea about 3500 years ago. The presence of these species suggests a relatively recent dispersal into developing open peatlands. This parallels other disjunctions into the Lowland, such as the single records of *Spergularia marina* along southwest James Bay, of *Astragalus tenellus* and *Astragalus agrestis* on prairie-like, juniper-woodland valley slopes along the Albany River, of *Aster alpinus* in a common river bar habitat along Hudson Bay, and of *Juncus ensifolius* in a similar habitat along the lower Harricana River.

The major succession in Lowland habitats over time has been the progressive paludification of the landscape. This can explain the stranding of populations like the dry-site *Astragalus* spp., a pattern similar to the less-advanced state of range disruption of *Astragalus alpinus* (Fig. A23) and *Erigeron acris* (Fig. A64). However, it is not possible to exclude the possibility of long-distance dispersal, particularly for extremely small disjunct populations, as is the case for *Aster alpinus* and *Spergularia marina* along recently emerged coasts.

The extraglacial refugia of eastern North American members of the Lowland flora were probably: (1) Atlantic coastal areas north of Maine, including coastal Gulf of St. Lawrence, Newfoundland, and Labrador; (2) Atlantic coastal areas south of Maine; (3) the Appalachian uplands, including the Gaspé; or (4) the northern Mississippi basin. Sixty-five percent of the eastern species in the Lowland currently occupy all four of these areas and may have redispersed from all or any of them. This group includes *Pogonia ophioglossoides* (Fig. A122), *Arethusa bulbosa* (Fig. A15), *Malaxis unifolia* (Fig. A96), and *Galium asprellum* (Fig. A66), and others that are disjunct into the Lowland, such as *Conioselinum chinense* (Fig. A48) and *Cypripedium reginae* (Fig. A52). *Pinus strobus* (Fig. A116) is also one of these but it is considered from present palynological records to have survived the glacial maximum only in the south Atlantic coast (Delcourt and Delcourt 1987).

Other eastern species currently occupy widespread ranges from the Mississippi Basin through New England to the St. Lawrence, but do not occur on the Atlantic seaboard south of New England. These include *Pinus resinosa* (Fig. A115), *Trisetum melicoides* (Fig. A145), *Listera auriculata*, *Lathyrus japonicus* var. *pellitus* (Fig. A81), *Oxalis acetosella*, *Andromeda glaucophylla*, *Lonicera oblongifolia*, *Malaxis monophyllos* (Fig. A94), and

Saxifraga paniculata (Fig. A136); bringing to over 75% the proportion of the Lowland's eastern species that may have survived the late-Wisconsinan in widespread eastern refugia.

A smaller group of eastern species is currently only found at or near the north Atlantic coast and Appalachian extraglacial areas: *Carex norvegica*, *Minuartia groenlandica* (Fig. A98), *Oxytropis campestris* var. *johannensis* (Fig. A100), *Primula laurentiana*, *Hierochloe alpina* var. *orthantha* (Fig. A74), *Montia lamprosperma*, *Arabis alpina*, and *Draba incana* (Fig. A55).

Seventeen of the eastern species are amphi-Atlantic in their distributions (Hultén 1958), such as *Eriocaulon aquaticum*, *Bartsia alpina* (Fig. A29), and *Draba norvegica* (Fig. A57). These species probably migrated to the Hudson Bay Lowland from the north Atlantic coast, as did the North American *Ranunculus pallasii* (Fig. A130) and *Pedicularis groenlandica* (Figs. A104 and A105), which both also have distant western populations.

In comparison with western species in the Lowland flora — and with the Lowland flora as a whole — the eastern species are distinctly more temperate in their distributional affinities. Some of the temperate and boreal transcontinental species that are distinctive of the southernmost Lowland are *Lilium philadephicum, Potamogeton natans, Sagittaria latifolia, Thalictrum dasycarpum, Rosa blanda, Scirpus atrovirens, Carex viridula, Maianthemum canadense, Corylus cornuta, Anemone quinquefolia, Clematis occidentalis, Corydalis aurea, Physocarpus opulifolius, Monotropa uniflora, and Sambucus pubens.*

Western species of the Hudson Bay Lowland

The western North American species in the Lowland are mostly subarctic and high-temperate species. The 65 western taxa (Table 9) are distinctly more subarctic in their continental distribution than the eastern taxa in the Lowland, and almost half are restricted to northern and interior (more continental) sites in the Lowland. Climatically, the Lowland interior north of the Albany River is much more similar to the dry, low-subarctic continental interior of the northern prairies than it is to the moist, temperate Great Lakes region.

Eight of these native western species are only known as modern introduced weeds in the Lowland: *Chenopodium pratericola, Lepidium densiflorum, Lepidium ramoissimum, Lepidium bourgeauanum, Artemisia ludoviciana, Monolepis nuttalliana, Descurainia sophioides*, and *Thermopsis rhombifolia* (now extirpated, Staniforth and Scott 1991). These taxa are known mainly from Churchill, a railhead for grain shipments from the prairies, and the location of numerous prairie introductions (as is Thunder Bay on Lake Superior).

Eleven of these species presently have distributions that extend northwest of the late-Wisconsinan ice sheet (Yukon Valley or Beringia), and these species almost certainly migrated to the Lowland from the northwest: *Pedicularis parviflora* (Fig. A108), *Gentianopsis detonsa* (Fig. A70), *Cicuta mackenzieana* (Fig. A47), *Castilleja raupii* (Fig. A40), *Arnica lonchophylla*, *Pedicularis sudetica* (Fig. A109), *Rorippa palustris* var. *williamsii*, *Salix alaxensis* (Fig. A134), *Carex rotundata*, *Achillea sibirica*, and *Picea glauca* var. *porsildii*. Within the Lowland, many of these species are generally restricted to the extreme north or occur in the south only in coastal areas.

Table 9. Western North American species of the Hudson Bay Lowland.

Picea glauca var. *porsildii*
Bromus pumpellianus
×*Agroelymus hirtiflorus*
×*Agrohordeum macounii*
Elymus innovatus (A61)
Eleocharis kamtschatica† (A59)
Carex rossii
C. rotundata
Juncus ensifolius† (A78)
Malaxis paludosa (A95)
Salix alaxensis (A134)
S. arbusculoides (A135)
S. athabascensis
S. lutea
Betula neoalaskana
Chenopodium glaucum var. *salinum* (A44)
*C. pratericola**
*Monolepis nuttalliana**
Aquilegia brevistyla (A10, A11)
Caltha natans†
Thalictrum sparsiflorum var. *richardsonii*
*Lepidium bourgeauanum**
*L. densiflorum**
*L. ramoissimum**
Rorippa palustris var. *williamsii*
*Descurainia sophoides**
*Thermopsis rhombifolia**
Astragalus adsurgens†
A. agrestis† (A22)
A. tenellus† (A27)
Oxytropis campestris var. *varians*†
O. deflexa var. *sericea*†
O. splendens† (A101)
Kalmia microphylla
Collomia linearis
Phacelia franklinii
Gentianopsis detonsa† (A70)
Achillea sibirica
Agoseris glauca† (A2)
*Artemisia ludoviciana**
Aster alpinus† (A20)
Solidago decumbens

Other effectively western species

Agropyron violaceum† (A3)
Carex atratiformis ssp. *raymondii*
Prosartes trachycarpa
Cypripedium passerinum (A50, A51)
Lathyrus ochroleucus
Astragalus americanus† (A24)
Hedysarum mackenzii (A73)
Cicuta mackenzieana (A47)
Gentianopsis macouni† (A68, A69)
Gentianella propinqua† (A71)
Lycopus asper† (A93)
Castilleja raupii (A40)
Pedicularis parviflora (A108)
P. sudetica (A109)
Antennaria pulcherrima (A8)
A. parvifolia
Arnica lonchophylla
A. chamissonis (A17)
Artemisia tilesii ssp. *elatior*† (A19)
Aster brachyactis† (A21)
A. modestus
Erigeron lonchophyllus† (A65)
Lactuca pulchella

Note: Numbers in parentheses refer to the map number in Appendix A.

*Species that only have introduced populations in the Lowland, mostly in Manitoba.

†Species disjunct for more than 500 km into the Lowland.

Twelve of these western species presently have distributions extending only to areas southwest of the late-Wisconsinan ice sheet: *Aster modestus*, *Prosartes trachycarpa*, *Salix athabascensis*, *Salix lutea*, *Astragalus adsurgens*, *Phacelia franklinii*, *Collomia linearis*, *Agoseris glauca* (Figs. A1 and A2), *Lycopus asper* (Fig. A93), *Lathyrus ochroleucus*, *Lactuca pulchella*, and *Gentianopsis macounii* (Fig. A69).

The other western species are currently known from extraglacial areas both northwest and southwest of the glacial maximum and may have origins in either or both refugia. A major corridor opened up early separating the Laurentide and Cordilleran ice sheets: "major portions of the valleys were open by 10 000 to 12 000 YBP" (Reeves 1973), and previous to that, a wide corridor at this latitude (50°N) existed throughout most of the mid-Wisconsinan (Westgate et al. 1974). Nunatak refugia occurred on the unglaciated mountain peaks and foothills west of the corridor (Packer and Vitt 1974), and the postglacial reconvergence of western taxa through this widening corridor illustrates the migrational abilities of these taxa.

Most of these species now have generally continuous eastward distributions; such as *Carex rossii*, *Bromus pumpellianus*, *Elymus innovatus*, *Aquilegia brevistyla*, *Thalictrum sparsiflorum*, *Oxytropis deflexa* var. *sericea*, and *Aster modestus*. A few have fragmented ranges eastward: *Juncus ensifolius* (Fig. A78), *Salix arbusculoides* (Fig. A135), *Astragalus tenellus* (Fig. A27), *Astragalus agrestis* (Fig. A22), *Astragalus americanus* (Fig. A24) *Oxytropis campestris* var. *varians*, *Oxytropis splendens* (Fig. A101), *Arnica chamissonis* (Fig. A17), *Aster alpinus* (Fig. A20), *Antennaria pulcherrima* (Fig. A8), *Antennaria parvifolia*, *Artemisia tilesii* (Fig. A19), *Cypripedium passerinum* (Figs. A50 and A51), *Erigeron lonchophyllus* (Fig. A65), *Gentianella propinqua* (Fig. A71), and *Hedysarum mackenzii* (Fig. A73). Twenty of the Lowland's western species (31%) are disjunct for more than 500 km into the Lowland (Table 9), with the disjunction often representing a major leap across the intervening Precambrian Canadian Shield from the Paleozoic prairies.

The study of eastward occurrences of western North American species has focused on "the numerically small but historically significant species of the Pacific slope or the Cordillera" that are isolated around the upper Great Lakes and the Gulf of St. Lawrence, and on the Appalachian peaks (Fernald 1935). The early hypothesis was that these species were survivors of preglacial floras in areas that were not glaciated (Fernald 1925, 1935). This has since proven not to be the case. The hypothesis now is that there was a widespread late-glacial development of pseudo-arctic or alpine periglacial habitats in the east, with subsequent survival of disjunct taxa in edaphically or climatically suitable, low-competition sites following deglaciation (Wynne-Edwards 1937, 1939; Marie-Victorin 1938; Drury 1969; Rousseau 1974). In addition, the confirmation of coastal refugia along the northeastern Atlantic coast added a significant factor to interpreting geomorphological, faunal, and botanical studies on the east coast (Dahl 1946, 1955; Gillett 1960; Ives 1963; Lindroth 1963; Morisset 1971; Packer and Vitt 1974).

The presence of extraglacial refugia for arctic and subarctic species in the immediate vicinity of the glacial limits explains some of the disjunct occurrences of species in eastern North America. For example, the arctic species now known from late-glacial macrofossils east of Minnesota and south of the Wisconsinan ice sheet also have extant distributions that are usually disjunct in the south and suggest more widespread late-glacial or postglacial ranges (Martin 1958; Watts 1967; Miller and Benninghoff 1969; Bertie 1975). Their current distributions suggest that they are relict of more widespread extraglacial populations, probably indicating the presence there of postglacial tundra-like habitats (Whitehead 1972). Examples of macrofossil species are *Carex supina*, *Dryas integrifolia*, *Vaccinium uliginosum* var. *alpinum*, *Salix herbacea*, *Rhododendron lapponicum*, *Sedum rosea*, *Potentilla nivea*, *Alnus crispa*, and *Arctostaphylos alpina*. However, none of these are restricted western North American species.

Most of the western species in the Lowland flora, certainly the members of the Fabaceae and Asteraceae, occur in dry, non-arboreal, mostly calcareous habitats in the Lowland, and are mostly non-arboreal elsewhere in their ranges. In western Canada, they are prairie and parkland species, without strong affinities to either the Great Lakes, the Gulf of St. Lawrence, or the western cordillera. Any southern and eastern ranges they may once have had have been disrupted and restricted by forest development and landscape paludification. The present ranges of the transcontinental *Astragalus alpinus* (Fig. A23) and *Astragalus eucosmus* (Fig. A26) suggest that they had more widespread postglacial distributions that were later fragmented by habitat changes. Some of these western species still exist in prairie-like habitats in the Lowland, such as the juniper-woodland bluffs along the Albany River, where *Astragalus tenellus* occurs (Fig. A27).

The ranges of several of the western taxa that redispersed from the northwest are notable. An important species of the subarctic tree line is *Picea glauca* var. *porsildii*. *Cicuta mackenzieana* (Fig. A47) is a wetland, non-montane Beringian radiant that has also spread eastward in a relatively narrow subarctic band, but without the Cordilleran radiant characteristic of many subarctic-subalpine taxa. *Pedicularis parviflora* (Fig. A108) and *Castilleja raupii* (Fig. A40) have similar distributions. The last three of these occur farthest south in non-arboreal habitats at low elevations in Alberta and Saskatchewan, but their eastward ranges narrow parallel to the northern limit of boreal forest in Manitoba and Ontario. All three may find suitable habitat east of their present ranges, if they can cross or otherwise skirt the migrational hurdle presented by James Bay.

Coastal species of the Hudson Bay Lowland

The Hudson Bay Lowland has a distinctive, low-lying coast that is more than 1900 km long. It is internationally renowned for its wildlife, such as its shorebirds, geese, and polar bear. This area supports over 160 plant species that are restricted to coastal habitats in the Lowland, and many of these dominate coastal vegetation assemblages (Tables 10–15).

The dominant coastal flora is relatively consistent throughout the Lowland. The intertidal marshes of James Bay are dominated by *Scirpus maritimus* (Fig. A138), *Eleocharis* spp., *Carex paleacea* (Fig. A37), *Senecio congestus*, and *Hippuris vulgaris*. Shoreward, *Festuca rubra, Hordeum jubatum, Potentilla anserina, Juncus balticus, Scirpus rufus*, and *Deschampsia cespitosa* dominate the supertidal meadow-marshes.

The progression of marshes and meadows shoreward is subtly terraced by sequences of parallel beach ridges. On the most exposed ridges *Elymus mollis* (Fig. A62), *Artemisia tile-*

sii (Fig. A19), *Lathyrus japonicus, Honkenya peploides* (Fig. A75), *Rhinanthus cristagalli*, and other species frequently occur, but at low cover values. Away from the shore, the ridges support shrub and tree cover, and more diverse, mesic and species-rich associations.

Northward, the coastal marshes of Hudson Bay are narrower, the slopes are steeper, and the beach ridges occur in denser swarms. *Puccinellia phryganodes* (Fig. A128), *Carex glareosa* (Fig. A33), *Carex subspathacea* (Fig. A39), *Carex mackenziei* (Fig. A34), *Hippuris vulgaris*, and *Senecio congestus* are dominant in the lower marshes, and *Dupontia fisheri* (Fig. A58), *Calamagrostis stricta, Juncus balticus, Petasites sagittatus*, and *Salix* spp. are dominant shoreward.

The near-shore Hudson Bay beach ridges are sparsely dominated by *Elymus mollis, Artemisia borealis, Hedysarum mackenzii* (Fig. A73), *Astragalus alpinus* (Fig. A23), *Trisetum spicatum, Salix arctophila, Salix reticulata, Salix brachycarpa, Saxifraga tricuspidata* (Fig. A137), *Honkenya peploides, Euphrasia arctica, Castilleja raupii* (Fig. A40), *Draba aurea, Lomatogonium rotatum, Stellaria longipes, Androsace septentrionalis, Solidago multiradiata, Rhinanthus borealis*, and numerous other species. Permafrost underlies the coast of Hudson Bay and northern James Bay, and the more protected leeward flanks of the ridges begin to support tundra heath not far from the coast, generally with an organic active layer. The strictly coastal species decline in frequency away from the coast, but the arctic species persist locally towards the interior in tundra heath, low shrub thickets, and moss–lichen woodlands.

Along the whole coast, above the normal high-water line, salt can accumulate through evaporation, the removal of surface clays by ice-rafting, and by an upward hydraulic gradient bringing fossil salt to the surface (Riley and McKay 1980; Price et al. 1988). These higher-salinity, panne–sward mosaics are frequent, but their occurrence is relatively local in comparison with the extensive, but less saline intertidal marshes and supertidal meadow-marshes. The pannes support more strictly halophytic species, such as *Salicornia europaea* (Fig. A133), *Spergularia canadensis* (Fig. A141), *Suaeda calceoliformis* (Fig. A143), *Plantago maritima* (Fig. A117), *Puccinellia phryganodes* (Fig. A128), and *Triglochin* spp.

There are strong latitudinal differences between the floras of the northern and southern coasts. Forty-seven species occur in coastal areas throughout the Lowland, although some are relatively rare (Table 10). Another 21 coastal species are restricted to the southwest James Bay coast (Table 11). Many other species again are restricted along the northern coast between Lake River on James Bay and the northwest coast of Manitoba, but these species are largely maritime arctic species, not halophytes (Tables 13, 14).

Widespread coastal species

Of the 47 widespread coastal species (Table 10), almost 80% also occur in the Alaska–Yukon area (Hultén 1968, 1973), 60% in Greenland (Böcher et al. 1968), and 83% on the Labrador – northeast Newfoundland coastline (Hultén 1968; Scoggan 1978–1979). Only 6% are confined to North America, the remainder being circumpolar or amphi-Atlantic species. Coastal species that are even further restricted to just eastern North America are very uncommon and include *Atriplex glabriuscula* (Fig. A28), *Puccinellia ambigua, Puccinellia lucida* (Fig. A127), *Carex paleacea* (Fig. A37), *Carex recta*, and *Carex salina*.

Table 10. Species restricted to the coast in the Hudson Bay Lowland, R(C).

*Zostera marina** (A148)	*Spergularia marina*
Zannichellia palustris (A147)	*Ranunculus cymbalaria*
Deschampsia cespitosa var. *maritima**	*R. pallasii* (A130)
Dupontia fisheri (A58)	*R. pedatifidus*
*Elymus mollis** (A62)	*R. scleratus*
Phleum alpinum	*Cardamine pratensis* var. *angustifolia*
Puccinellia ambigua	*Draba aurea*
P. lucida (A127)	*D. glabella*
*P. phryganodes** (A128)	*Potentilla anserina* var. *groenlandica**
Eleocharis uniglumis	*P. multifida* (A124)
Carex glareosa (A33)	*Lathyrus japonicus* var. *pellitus** (A81)
C. mackenziei (A34)	*Hippuris vulgaris f. maritima*
*C. paleacea** (A37)	*Primula laurentiana*
C. recta	*Androsace septentrionalis* (A7)
C. salina	*Gentianopsis detonsa* (A70)
*C. subspathacea** (A39)	*G. nesophila* (A70)
*Salix brachycarpa**	*Lomatogonium rotatum* (A88)
Polygonum fowleri (A123)	*Mertensia maritima* (A97)
Atriplex glabriuscula (A28)	*Limosella aquatica* (A85)
A. patula	*Plantago maritima** (A117)
Montia lamprosperma	*Chrysanthemum arcticum* (A45)
Minuartia rubella (A99)	*Erigeron lonchophyllus* (A65)
Stellaria humifusa	*Senecio congestus**
Honkenya peploides (A75)	

Note: Numbers in parentheses refer to the map number in Appendix A.
*Species dominating coastal vegetation associations.

About half of the widespread coastal species are strictly halophytic, such as *Zostera marina* (Fig. A148), *Puccinellia phryganodes* (Fig. A128), *Eleocharis uniglumis*, *Carex glareosa* (Fig. A33), *Carex mackenziei* (Fig. A34), *Carex subspathacea* (Fig. A39), *Stellaria humifusa*, *Spergularia marina* (Fig. A142, only two Lowland sites), *Honkenya peploides* (Fig. A75), *Potentilla anserina* var. *groenlandica*, *Lathyrus japonicus* var. *pellitus* (Fig. A81), *Hippuris vulgaris* f. *maritima*, *Mertensia maritima* (Fig. A97), *Rumex salicifolius, Polygonum prolificum, Atriplex patula*, and *Plantago maritima.*

The remainder are a combination of estuarine wetland species, facultative halophytes, northern maritime species (also rarely found in back-beach freshwater wetlands southward), and species of exposed beach ridges. They include *Zannichellia palustris* (Fig. A147), *Dupontia fisheri* (Fig. A58), *Elymus mollis* (Fig. A62), *Ranunculus cymbalaria, Ranunculus pedatifidus, Ranunculus pallasii, Draba aurea, Draba glabella* (Fig. A54), *Potentilla multifida, Androsace septentrionalis* (Fig. A7), *Gentianopsis detonsa* (Fig. A70), *Lomatogonium rotatum* (Fig. A88), *Limosella aquatica* (Fig. A85), *Erigeron lonchophyllus* (Fig. A65), *Senecio congestus*, and *Chrysanthemum arcticum* (Fig. A45).

This group of widespread coastal species includes, in general, the characteristic coastal species of comparable habitats at Chesterfield on western Hudson Bay, Coral Harbour on Southampton Island, and Wolstenholme on the Quebec peninsula (Polunin 1948).

Southwest James Bay coastal species

There are 21 species in the Lowland restricted to the southwest James Bay coast (Table 11). These include a number of salt marsh species, freshwater estuarine species at their northern range limits, and coastal beach ridge species. Among these and another 19 James Bay species, there is a clear pattern of long-distance range disjunction (Table 12).

Some of these 30 species are disjunct directly to southern James Bay: *Aster brachyactis* (Fig. A21), *Glaux maritima* (Fig. A72), *Ruppia maritima* (Fig. A132), *Scirpus maritimus* (Fig. A138), *Spergularia canadensis* (Fig. A141), *Lycopus asper* (Fig. A93), *Suaeda calceoliformis* (Fig. A143), *Malaxis monophyllos* (Fig. A94), *Poa eminens* (Fig. A121), and *Juncus ensifolius* (Fig. A78). The others are also known from rare populations on southern Hudson Bay, but are also significantly disjunct from their nearest populations, including *Gentianopsis macounii* (Fig. A69) and *Conioselinum chinense* (Fig. A48) (Table 12). Overall, only *Ranunculus cymbalaria* is disjunct less than 500 km, and the others vary in disjunction up to 2100 km (*Juncus ensifolius*).

Twenty-one of the species occur in intertidal or supertidal meadow-marsh along the coast; 10 species occur on coastal beach ridges; one species is marine (*Zostera marina*); one occurs on river gravels within 10 km of the coast; and the other in willow–alder thickets adjacent to the shoreline.

Seven of these species have their closest populations around the Gulf of St. Lawrence (unidirectional disjunctions), such as *Carex paleacea* (Fig. A37), *Carex recta*, *Carex salina*, *Salicornia europaea* (Fig. A133), *Atriplex glabriuscula* (Fig. A28), *Gentianopsis nesophila* (Fig. A70), and *Ligusticum scothicum* (Fig. A84). Eleven of them are disjunct from both the Gulf of St. Lawrence and from the west (bidirectional disjunctions); for example, *Poa eminens* (Fig. A121) and *Eleocharis kamtschatica* (Fig. A59). The remaining 12 species are disjunct from native populations in the near west only (3 species), the

Table 11. Species restricted to the southwestern James Bay coast, R(JB).

Botrychium matricariaefolium	*Ceratophyllum demersum* (A43)
Ruppia maritima (A32)	*Ligusticum scothicum* (A84)
Poa eminens (A121)	*Angelica atropurpurea*
*Scirpus maritimus** (A138)	*Glaux maritima* (A72)
Juncus ensifolius (A78)	*Cynoglossum boreale*
Malaxis monophyllos (A94)	*Lycopus asper* (A93)
Polygonum ramosissimum	*Antennaria parvifolia*
Rumex orbiculatus	*Aster brachyactis* (A21)
R. salicifolius var. *subarcticus*	*Bidens cernua*
Suaeda calceoliformis (A143)	*B. hyperborea*
Spergularia canadensis (A141)	

Note: Numbers in parentheses refer to the map number in Appendix A.
*Species dominating coastal vegetation associations.

Table 12. Disjunct species of southwest James Bay (JB) and southern Hudson Bay (sHB): habitats and nearest populations.

Taxa	Location	Bay	Intertidal beach ridges	Supertidal beach ridges	River gravels	Willow-alder	"Weedy" in range	Distance/direction to closest native population (km)	Distance/direction to closest "weedy" population (km)	Direction to main body of distribution	Vegetative propagation
Zostera marina (A148)	JB,sHB	X	X					800SE, 1200NW		E,W	X
Aster brachyactis (A21)	JB		X				X	1000W?	800SW	W,S	
Carex mackenziei (A34)	JB,sHB		X					800SE, 2000W		E,W	X
Carex paleacea (A37)	JB,sHB		X					800SE		E	X
Carex recta	JB,sHB		X					800SE		E	X
Carex salina	JB,sHB		X					800SE		E	X
Eleocharis kamtschatica (A59)	JB,sHB		X					700NE, 2800W		E,W	X
Glaux maritima (A72)	JB		X					800SE, 1000W		E,W	X
Ranunculus cymbalaria	JB,sHB		X					450SW to 800SE		E,W,S	X
Ruppia maritima (A132)	JB		X					800SE		E,W,S	X
Salicornia europaea (A133)	JB,sHB		X				X	800SE	1000+S		E
Scirpus maritimus (A138)	JB		X					800SE, 1200W		E,W	X
Scirpus rufus	JB,sHB		X					800SE, 600W		E,W	X
Spergularia canadensis (A141)	JB		X					800SE, 2800W		E,W	
Spergularia marina (A142)	JB,sHB		X				X	850SE, 1100W	800S	E,W,S	
Lycopus asper (A93)	JB		X				X	1000W	700SE	WX	
Suaeda calceoliformis (A143)	JB		X					800SE, 1100W		E,W	
Zannichellia palustris (A147)	JB,sHB		X					750SE		E,W,S	X
Atriplex glabriuscula (A28)	JB,sHB		X	X				850E		E	
Gentianopsis nesophila (A70)	JB,sHB		X	X				1000E		E	
Ligusticum scothicum (A84)	JB,sHB		X	X				800SE		E	
Polygonum fowleri (A123)	JB,sHB		X	X				650SE, 2200W		E,W	
Artemisia tilesii (A19)	JB,sHB			X				900+NW		NW	
Chenopodium glaucum var. *salinum* (A44)	JB,sHB		X	X			X	500+W?	300S	W	
Erigeron lonchophyllus (A65)	JB,sHB			X			X	700+W, SW	700SW	E,W	
Gentianopsis macounii (A68)	JB			X				600SW, 1150E, W		E,W	
Malaxis monophyllos (A94)	JB*			X				550S		E,W,S	
Poa eminens (A121)	JB			X				800SE, 2800W		E,W	
Juncus ensifolius (A78)	JB*				X			2100+W		W	
Conioselinum chinense (A48)	JB,sHB*					X		700E		E,S	

Note: Numbers in parentheses refer to the map number in Appendix A.

*Not coastal species.

east–southwest (5 species), the far west only (*Juncus ensifolius*, Fig. A78), the northwest (*Artemisia tilesii*, Fig. A19), and the southeast (*Conioselinum chinense*, Fig. A48). There is no correlation between the direction of closest native population and species' habitat on James Bay (Table 12).

A variety of hypotheses have been proposed to explain this pattern of disjunction; all of which accept that plant movement did not necessarily originate from the currently closest populations. These hypotheses include: (1) migration of species, particularly non-obligative halophytes, via the shores of intervening proglacial lakes and marine incursions; (2) circum-dispersal around the Quebec–Labrador peninsula during the late hypsithermal; and (3) ongoing accumulative dispersal events, particularly those resulting from the migration of waterfowl (Svenson 1927; Potter 1932; Porsild 1932; Boivin 1952; Cody 1953; Dutilly et al. 1958; Schofield 1958; Rousseau 1974; Riley and McKay 1980).

The floristic patterns of the vast majority of other Lowland species support a conservative view of their immigration into the Lowland. Past climates, available environments, and individual migration abilities were responsible for the migration of most of the flora, with subsequent environmental changes consolidating or disrupting the ranges of those species. However, in the case of species disjunct to southwest James Bay, there are no intervening stations for the species in question in areas south, east, or west of the Lowland.

There is also no paleoenvironmental or macrofossil evidence of temperate or subarctic halophytic conditions in those intervening areas. For these reasons, the majority of these disjunctions likely occur as the result of long-distance dispersal mechanisms, such as grazing and migrating waterfowl. The coincidence of these disjunctions with major North American waterfowl and shorebird migration routes, particularly to the Gulf of St. Lawrence, is striking, but the evidence remains circumstantial. The newly emergent coast is also likely to have unusual receptivity to colonization events, given its open, low-competition habitats. Some of the disjuncts may be very recent immigrants, such as *Spergularia marina* (Fig. A142), which may be an unsuccessful introduction already extirpated at its point of introduction on James Bay (Riley and McKay 1980).

Some of the other southwest James Bay species (Table 11) are estuarine or marsh species at their northern range limits, such as *Ceratophyllum demersum* (Fig. A43), *Angelica atropurpurea, Typha angustifolia, Bidens cernua*, and *Bidens hyperborea*. A few species are restricted in the Lowland to this area, but will probably be found southward in the Lowland's Moose River basin, such as *Cynoglossum boreale, Botrychium matricariaefolium*, and *Polygonum ramoissimum*. Other coastal species occur as minor disjunctions within the Lowland itself. The rare southward occurrence of species typical of the Hudson Bay coast, such as *Saxifraga aizoides*, *Minuartia rubella*, *Ranunculus pallasii* (Fig. A130), and *Montia lamprosperma*, is another minor theme in the phytogeography of the Lowland.

Arctic species of the Hudson Bay Lowland

Maritime arctic species

A maritime tundra zone occurs as a band 20–40 km wide along Hudson Bay, with tundra locally extending farther towards the interior, such as at Cape Henrietta Maria and in

Manitoba. The occurrence of arctic ecosystems so far south is the direct result of the cold waters of Hudson Bay, in particular the north-to-south summer circulation of the ice cover in the bay, with the latest ice (usually late August) remaining in the Fort Severn – Cape Henrietta Maria area. Prevailing onshore, northwest winds cross the bay and its ice, and lower the spring and summer temperatures and cause fogs that decrease the effective solar radiation in the area (Lardner 1968). Floristically, this zone has been previously defined by some of its coincident plant distribution patterns (Riley 1979*b*).

Climatic definition of this zone is problematic because of the lack of meteorological data. However, the climate of the treeless tundra can be generally characterized as having, for example, a growing season of less than 65 days. This maritime tundra zone also occurred on at least the coldest shores of the postglacial Tyrrell Sea, where the analogous community, based on pollen stratigraphy from the 8200 BP Sutton Ridges island, was a sparse tundra dominated by *Dryas*, willows, grasses and sedges, with coastal species like *Plantago maritima*. This continued until 6500 BP, after which the uplift of the landscape drained the shoreline down to its modern position (McAndrews et al. 1982).

Of the 55 species confined to the maritime tundra zone within the Lowland, a few are primarily coastal species elsewhere in their ranges (Table 13), such as *Carex maritima* (Fig. A35) and *Cochlearia officinalis*. The others are largely arctic or high-subarctic species of arctic tundra or of dry, sandy or gravelly shores or uplands. As a group of species, they are more prevalent nearer the coast, especially coastward of the erratic maritime tree line. Beyond 40 km from the coast, there regularly occur large permafrost peat plateaus dominated by *Ledum decumbens* and *Rhododendron lapponicum*, permafrost fen plateaus dominated by mosses like *Scorpidium turgescens*, and extensive thermokarst lake plains.

These north-coast species share similarities with those restricted to the northwest (Manitoba) coast (23 taxa, Table 14) and to Cape Henrietta Maria (8 taxa, Table 15). Overall, 45% of all these species occur in the high arctic, and 84% are species of the low arctic elsewhere in their ranges (Table 3). Almost two-thirds of these species occur in Greenland, over 80% occur on the Labrador – northeast Newfoundland coast, and over 90% occur in the Alaska–Yukon area. Over 80% of these species are circumpolar, while about 9% only occur in North America.

Lowland populations of these species did not originate from southern or southwestern ice-margin tundra refugia; although there may have been some later westward ice-front movement of tundra species from the east coast or the Gulf of St. Lawrence into the vicinity of proglacial Lake Ojibway, and subsequently, to the James Bay area (Richard 1977). No data show that ice-margin tundra communities existed between the upper Great Lakes and James Bay following the Algonquin Stadial (Saarnisto 1974, 1975), and there is no pattern of arctic species occurring as disjunct populations between the Great Lakes – St. Lawrence River and the Hudson Bay Lowland.

Many of these arctic species are distributed erratically southward in the western cordillera, especially the Rockies, including *Kobresia myosuroides* (Fig. A79), *Luzula multiflora* (Fig. A91), *Ranunculus hyperboreus* (Fig. A129), *Juncus arcticus*, *Draba nivalis* (Fig. A56), *Saxifraga oppositifolia, Gentianella propinqua* (Fig. A71), *Salix alaxensis* (Fig. A134), *Pedicularis labradorica* (Fig. A106), *Pedicularis flammea* (Fig. A103),

Table 13. Species restricted to the maritime tundra coast in the Hudson Bay Lowland, R(NC).

Picea glauca var. *porsildii**	*Draba alpina*
*Alopecurus alpinus** (A5)	*D. nemorosa*
*Arctagrostis latifolia** (A12, A13)	*D. nivalis* (A56)
*Arctophila fulva** (A14)	*Saxifraga caespitosa*
*Calamagrostis deschampsioides** (A30)	*S. cernua*
Festuca saximontana	*S. oppositifolia*
*Hierochloe pauciflora**	*S. tricuspidata* (A137)
Puccinellia langeana	*Parnassia kotzebuei*
*P. vaginata**	*Potentilla crantzii*
Trisetum spicatum var. *spicatum*	*P. pulchella** (A126)
Eriophorum callitrix	*Dryas integrifolia*
Kobresia myosuroides (A79)	*Lathyrus japonicus* (typ.)
*Carex atrofusca**	*Linum lewisii* var. *lepagei* (A86)
*C. marina**	*Kalmia microphylla*
C. maritima (A35)	*Arctostaphylos alpina* var. *alpina*
C. misandra	*Armeria maritima*
C. nardina	*Gentianella propinqua* (A71)
*C. williamsii**	*Pedicularis flammea* (A102, A103)
Juncus arcticus	*P. labradorica* (A106)
Luzula multiflora (A91)	*P. lapponica** (A107)
Salix calcicola	*P. sudetica** (A109)
S. reticulata	*Rhinanthus borealis*
Koenigia islandica	*Utricularia ochroleuca*
*Stellaria edwardsii**	*Artemisia borealis*
Anemone richardsonii	*Aster alpinus* (A20)
Ranunculus hyperboreus (A129)	*Erigeron humilis*
Cochlearia officinalis	*Tripleurospermum phaeocephalum** (A144)
Barbarea orthoceros	

Note: Numbers in parentheses refer to the map number in Appendix A.
*Arctic species not occurring southward in the western cordillera or south of Labrador in the east.

and *Aster alpinus* (Fig. A20). However, even in these cases, their Lowland populations probably originated in the northwest or northeast.

Thirty percent of these arctic species occur only in the far north, without any southward Cordilleran radiant populations or any populations south of Labrador in the east. Their absence from these areas largely precludes the possibility that they survived the late-Wisconsinan south of the ice sheet. Among these are *Hierochloe pauciflora, Arabis arenicola, Draba cinerea* (Fig. A53), *Alopecurus alpinus* (Fig. A5), *Arctagrostis latifolia* (Figs. A12 and A13), *Calamagrostis deschampsioides* (Fig. A30), *Carex atrofusca, Arctophila fulva* (Fig. A14), *Potentilla pulchella* (Fig. A126), *Pedicularis lapponica* (Fig. A107), *Pedicularis sudetica* (Fig. A109), *Tripleurospermum phaeocephalum* (Fig. A144), *Gentianella tenella*, and *Carex marina*.

Table 14. Species restricted to the northwest coast in the Hudson Bay Lowland, R(NWC).

Potamogeton obtusifolius	*Ranunculus nivalis**
P. strictifolius	*Eutrema edwardii**
Poa flexuosa	*Lesquerella arctica*
*Carex adelostoma**	*Hutchinsia procumbens* (A76)
*C. rotundata**	*Draba lactea*
C. rupestris	*Saxifraga rivularis*
*C. ursina**	*Oxytropis arctica**
Juncus biglumis	*O. campestris* var. *varians*
Salix alaxensis (A134)	*Loiseleuria procumbens* (A87)
S. athabascensis	*Veronica peregrina*
Chenopodium hybridum	*Campanula uniflora*
Suaeda maritima	

Note: Numbers in parentheses refer to the map number in Appendix A.
* Arctic species not occurring southward in the western cordillera or south of Labrador in the east.

Table 15. Species restricted in the Lowland to the Cape Henrietta Maria area, R(NE).

*Potamogeton subsibiricus**	*Draba cinerea** (A53)
Vahlodea atropurpurea	*Draba norvegica* (A57)
Festuca baffinensis	*Phyllodoce caerulea* (A111)
Carex capillaris var. *krausei*	*Gentianella tenella**

Note: Numbers in parentheses refer to the map number in Appendix A.
* Arctic species not occurring southward in the western cordillera or south of Labrador in the east.

Tree-line white spruce (*Picea glauca*) is usually considered to have recolonized the central boreal forest from south of the late-Wisconsinan ice sheet, but it is possible that white spruce also survived the Wisconsinan northwest of the ice sheet (Ritchie 1984). The coastward tree species of some sections of the Hudson Bay coast is the smooth-barked Yukon var. *porsildii* (Raup 1947), suggesting that a northwest genotype may have also migrated across the subarctic to Hudson Bay (Wilkinson et al. 1971).

This group of north-coast species includes one of central Canada's only endemics, *Linum lewisii* var. *lepagei* (Fig. A86). This white, large-flowered, self-pollinated, prostrate flax is over 750 km from the typical western populations of *Linum lewisii*, and has been described by Mosquin (1971*b*) as one of the few or only Canadian postglacial endemics.

The north coast of Akimiski Island, in James Bay, from the western tip of the island to Cape Duncan in the east, has affinities with the James Bay coast north of Lake River and with the Hudson Bay maritime tundra zone in general (Riley 1981). Species such as *Ranunculus hyperboreus, Dryas integrifolia, Linum lewisii* var. *lepagei*, and *Pedicularis sudetica* only occur along the north shore of Akimiski. The south shore supports numerous species characteristic of the southwest James Bay shore, such as *Glaux maritima, Spergularia canadensis, Ligusticum scothicum*, and *Suaeda calceoliformis*.

Table 16. Species restricted to the northern Hudson Bay Lowland, R(N).

Lycopodium annotinum var. *pungens**	*S. uralensis**
L. selago vars. *patens* and *appressum*	*Stellaria laeta**
Sparganium hyperboreum	*Draba incana* (A55)
*Agropyron violaceum** (A3)	*Arabis alpina**
*Calamagrostis purpurascens**	*A. arenicola* (typ.)* and var. *pubescens**
*Poa arctica** (A120)	*Saxifraga aizoides*
Eriophorum scheuchzeri	*S. hirculus*
Kobresia simpliuscula (A80)	*Chrysosplenium tetrandum**
*Carex bicolor**	*Potentilla nivea** (A125)
*C. bigelowii** (A31)	*P. palustris* var. *parvifolia**
*C. capitata**	*Oxytropis deflexa* vars. *foliosa and sericea*
C. glacialis (A32)	*O. viscida**
C. microglochin (A36)	*O. splendens** (A101)
*C. norvegica**	*Hedysarum mackenzii* (A73)
*C. rossii**	*Epilobium latifolium**
*Juncus castaneus** (A77)	*Pyrola rotundifolia**
Coeloglossum viride (typ.)*	*Ledum decumbens** (A82, A83)
*Salix arbusculoides** (A135)	*Rhododendron lapponicum* (A131)
S. arctica	*Andromeda polifolia*
*S. arctophila**	*Arctostaphylos alpina* var. *rubra*
S. ballii	*Primula egaliksensis**
S. vestita	*Castilleja raupii** (A40)
Betula glandulosa	*Bartsia alpina** (A29)
*Cerastium alpinum** (A42)	*Pinguicula villosa** (A112)
*Minuartia stricta**	*Agoseris glauca* (A1, A2)
Arenaria humifusa	*Arnica attenuata* (A16)
*Silene involucrata** (A139)	*Taraxacum lacerum**

Note: Numbers in parentheses refer to the map number in Appendix A.

*Species restricted to the extreme north, excluding species with minor disjunctions southward.

Non-maritime arctic species

An additional 54 Lowland species (Table 16) are restricted to the area north of a line running from north of the Ekwan River westward to the lower Fawn River and the Gillam area of Manitoba (dashed line, Fig. 13). Another eight species occur only on the Precambrian uplands of the Sutton Ridges (Table 17). About 70% of all of these are even more restricted northward, to north of a line running across the Lowland farther north of the Ekwan River west to York Factory and beyond. These species also have strong arctic affinities, with over 43% occurring in the high arctic and almost 90% characteristic of the low arctic elsewhere in their ranges.

More than a quarter of the species restricted to the Maritime Tundra zone (Fig. 13) do not occur south of the subarctic, while less than 10% of these more general northern species

Table 17. Species restricted to Precambrian exposures of the Sutton Ridges area, R(LS).

Lycopodium sitchense	*Sparganium fluctuans*
Dryopteris fragrans	*Hierochloe alpina* (A74)
Gymnocarpium jessoense	*Minuartia groenlandica* (A98)
Woodsia glabella	*Saxifraga paniculata* (A136)

Note: Numbers in parentheses refer to the map number in Appendix A.

(Table 16) are similarly confined. This suggests that these general northern species are probably more able to succeed in subarctic woodland and discontinuous permafrost habitats, persisting southward in suitable sites as the Tyrrell Sea shore shifted northward. Phytosociological data suggest that some of these taxa, such as *Ledum decumbens* (Fig. 4; Figs. A82 and A83), *Empetrum nigrum, Rubus chamaemorus* and *Vaccinium uliginosum*, "attain highest frequencies in the communities of the forest-tundra ecotone" (Larsen 1971).

More than 80% of these species also occur in the Alaska–Yukon area, and almost 90% of them occur in the northeast along the Labrador – northeast Newfoundland coasts or in the Gulf of St. Lawrence area. They could have redispersed into the Lowland from the northwest, the north, or the northeast. The taxa restricted to the Sutton Ridge uplands (Table 17) have distinctly eastern affinities. For example, *Minuartia groenlandica* (Fig. A98), *Hierochloe alpina* (Fig. A74), and *Saxifraga paniculata* (Fig. A136) are restricted to northeastern continental distributions.

Almost half of these northern species also currently occur southwest of the former late-Wisconsinan ice sheet as radiants in the western cordillera, where they are generally confined to mesic, colder-than-normal sites on east–west oriented mountains (Weber 1961; Packer and Vitt 1974). These include *Poa arctica* (Fig. A120), *Kobresia simpliuscula* (Fig. A80), *Carex bigelowii* (Fig. A31), *Carex glacialis* (Fig. A32), *Carex microglochin* (Fig. A36), *Juncus castaneus* (Fig. A77), *Silene involucrata* (Fig. A139), *Silene uralensis, Potentilla nivea* (Fig. A125), *Agoseris glauca* (Figs. A1 and A2), and *Hedysarum mackenzii* (Fig. A73). Hultén stated: "When the ice retreated, the arctic species [in this southern latitude] were able, as fossils show, to follow to a certain extent, but the increasingly warmer and drier conditions, combined with the rapidly spreading forests, soon exterminated them" (1961).

There is little doubt that the southern Hudson Bay coast lies in the "real" arctic (Polunin 1951; Ahti 1964; Rousseau 1974; Maycock 1974). Over 84% of the 170 species considered as arctic indicators in Young's general phytogeographic treatment of the arctic flora occur in the Hudson Bay Lowland (1971), confirmation that coastal Hudson Bay is typically arctic, or more precisely, maritime arctic.

In the upper Great Lakes region, Given and Soper (1981) documented the arctic-alpine species on Lake Superior. They were generally growing in rocky, non-forested habitats exposed to harsh shore, cliff, or canyon microclimates with low summer temperatures and frequent fog. As discussed earlier, these species probably survived in that area since ice-margin tundra developed there before 10 000 BP (Saarnisto 1974). All but five of these 50 species occur in the Hudson Bay Lowland (Table 18), even though the histories of their movements to these areas were almost certainly completely different.

Table 18. Hudson Bay Lowland arctic-alpine species.

(*a*) Species disjunct to the upper Great Lakes region. Predominantly found in open heath or herbfield communities on rock, talus, cliffs, sand, or gravel (Given and Soper 1981; Oldham 2002).

Lycopodium annotinum var. *pungens*	*Saxifraga paniculata* (A136)
L. sitchense	*S. cernua*
Elymus mollis (A62)	*Dryas integrifolia*
Poa alpina (A119)	*Potentilla crantzii*
Trisetum spicatum var. *pilosiglume*	*P. multifida* (A124)
Carex glacialis (A32)	*P. nivea* (A125)
C. saxatilis (A38)	*P. pensylvanica*
C. scirpoidea	*Astragalus alpinus* (A23)
C. supina	*Lathyrus japonicus* var. *pellitus* (A81)
Eriophorum chamissonis var. *albidum*	*Empetrum nigrum*
Tofieldia pusilla	*Pyrola grandiflora*
Listera borealis	*Euphrasia arctica*
Polygonum viviparum	*Arnica attenuata* (A16)
Arenaria humifusa	*A. lonchophylla*
Cerastium alpinum (A42)	*Artemisia canadensis*
Anemone parviflora	*Solidago multiradiata*
Draba aurea	*Tanacetum huronense* (A46)
D. incana (A55)	*Taraxacum ceratophorum*
D. norvegica (A57)	

(*b*) Species distributed in a more or less widespread manner, in suitable habitat, between the Lowland and upper Great Lakes region.

Selaginella selaginoides	*Vaccinium uliginosum*
Lycopodium selago	*Castilleja septentrionalis*
Scirpus cespitosus	*Pinguicula vulgaris*

(*c*) Other species sometimes considered arctic-alpine indicators in the Great Lakes region, but also more or less widespread in the northern prairie provinces at temperate latitudes, in suitable habitats.

Poa glauca	*Saxifraga tricuspidata* (A137)
Cypripedium passerinum (A50, A51)	*Hedysarum alpinum*
Salix myrtillifolia	*Vaccinium vitis-idaea*
Anemone multifida	*Senecio congestus*
Ranunculus lapponicus	

Note: Numbers in parentheses refer to the map number in Appendix A.

Overall, the Hudson Bay Lowland supports a very diverse arctic flora. This diversity is especially notable because parts the northern coast of the Lowland, for example, Cape Henrietta Maria, are isolated from other arctic environments — northward by more than 150 km of open ocean, eastward by James Bay, and westward by the narrowness of the maritime tundra zone. The Lowland is also located farther from extraglacial refugia than any other part of the arctic and should be expected, as a result, to be species depauperate by comparison. However, even at the isolated, recently emerged Cape Henrietta Maria, an additional eight arctic species occur (Table 15).

The Lowlands is perhaps a particularly receptive terrain for the immigration of new species, through long-distance dispersal, because of the climate and the prevalence of permafrost and peatland. Also the emergent coastal ecosystems have never allowed closed forest to dominate the landscape. This situation is analogous to early postglacial environments farther away from the centres of the Laurentide Ice Sheet.

Other themes in the flora of the Hudson Bay Lowland

The Precambrian Sutton Ridges support eight species disjunct from outside of the Hudson Bay Lowland (Table 17); for example, disjunct from east of James Bay are *Minuartia groenlandica* (Fig. A98) and *Hierochloe alpina* var. *orthantha* (Fig. A74). Comparable exposures of discontinuous Precambrian bedrock occur in the Churchill area (Bostock 1968), but these eastern species do not occur there. Several other Precambrian disjuncts occur both on the Sutton Ridges and in the Churchill area, such as *Agrostis mertensii* (Fig. A4) and *Luzula confusa* (Figs. A89, A90).

Other areas will probably be found that, like the Sutton Ridges, exert particular edaphic or microclimatic influences on the local flora. For example, on north-facing, spray-zone cliff sites along the Attawapiskat River, a number of small arctic niches support species that are slightly disjunct from more typically northern habitats, such as *Arctostaphylos alpina* var. *rubra*, *Agoseris glauca* (Fig. A1), *Arnica attenuata* (Fig. A16), *Taraxacum ceratophorum*, *Tofieldia pusilla*, *Salix brachycarpa*, and *Carex glacialis* (Fig. A32). *Cypripedium parviflorum* var. *planipetalum* (Fig. A49) also occurs here, at one of its only known locations west of the Gulf of St. Lawrence.

The species restricted to the northwest (Manitoba) Lowland in general (Table 19) and those restricted in the northwestern interior (Table 20) also deserve note. They include rare

Table 19. Species restricted to the northwest Hudson Bay Lowland, R(NW).

Bromus pumpellianus	*Astragalus adsurgens*
Luzula wahlenbergii (A92)	*Collomia linearis*
Polygonum lapathifolium (typ.)	*Phacelia franklinii*
Arabis divaricarpa	

Note: Numbers in parentheses refer to the map number in Appendix A.

Table 20. Species restricted to the northwestern interior of the Hudson Bay Lowland, R(NWI).

Koeleria macrantha	*Arabis holboellii*
Carex backii	*Physostegia virginiana*
Juncus vaseyi	*Potentilla arguta*
Betula neoalaskana	*Achillea sibirica*
Cerastium nutans	*Solidago decumbens*

western and arctic taxa such *as Luzula wahlenbergii* (Fig. A92), *Phacelia franklinii, Astragalus adsurgens, Koeleria macrantha*, and *Carex backii*.

Species introduced into the flora of the Hudson Bay Lowland

The coasts of Hudson and James bays were among the first areas explored by Europeans in central North America. The first documented visits were by Hudson (in 1610), Button (in 1612, searching for Hudson), Munck (in 1619), and Fox and James (in 1631). It was not until 1668 that further investigations of the area resumed. At that time, Fort Charles was established by Gillam at the mouth of the Rupert River.

Following the enfranchisement of the Hudson Bay Company in 1670, trading posts in the Lowland were established: York Factory in 1670, Moose Factory in 1671, Fort Albany in 1683, Fort Severn in 1685, and Churchill in 1717. The English fur traders were content with tide-water trading operations until forced by French competition to establish interior posts, such as Cumberland House (1773), and other posts such as Charlton Island, the transhipment depot for the shallow-water posts of southern James Bay. From the coast of Hudson Bay, the commercial occupation of central Canada extended to the Rocky Mountains, with animals and plants being central to the trade.

Later trading settlements established in Ontario included those at Attawapiskat (after 1903); the southern James Bay railhead at Moosonee; Keschawan; Winisk (site of a military installation in the 1950s and resettled as Peawanuck at the confluence of the Winisk and Shamattawa Rivers); Ogoki; Martin Falls; Sutton Lake (a year-round outfitting camp); Mammamattawa (abandoned by early 1950s); and Lake River (also abandoned). In Manitoba, Churchill, Shamattawa, and Fox Lake are the major settlements. Akimiski Island is not settled. The First Nations are predominantly Cree, although the Chipewyan also have communities in Manitoba.

An important event in the development of this handful of communities was the completion of the transcontinental railroad to Winnipeg in 1885. This completely displaced the economic role of York Factory as a shipment depot for points west. Until that time, York Factory transhipped most of the westbound goods and personnel from Europe, via York boats, to the Red River settlements and to the network of trading posts in the west and northwest. By 1920, the last York boat disappeared from York Factory.

Churchill was linked to the south by rail in 1927. The railhead's grain elevators were completed in 1930, and the shipment of grain along this route resulted in a rich adventive flora in the Lowland, as many as 75 species (Beckett 1959). Few if any of the alien species arriving since that time originated directly from European sources, but rather, arrived by rail from the south. It has been observed that abandoned settlements, such as across the river from modern Churchill at Fort Prince of Wales and elsewhere in the Lowland at Lake River and Mammamattawa, have decreasing habitats for weedy species and are being overtaken by the surrounding native flora.

In 1935, the Hudson Bay Company boats stopped service to southern James Bay because of the new rail link to Moosonee, from which barges could service coastal settlements. Introduced species have invaded naturally disturbed habitats along the coast near the mouth of the Moose River: *Bromus inermis*, *Festuca pratensis*, *Phleum pratensis*, *Poa pratensis* (probably native in some Lowland sites), *Ranunculus acris*, *Trifolium pratense*,

Trifolium hybridum, *Trifolium repens*, *Capsella bursa-pastoris*, *Matricaria matricarioides*, *Galeopsis tetrahit*, and *Plantago major* (also probably native elsewhere in the Lowland). *Taraxacum officinale* and *Vicia cracca* are more widespread, now occurring along the nearby coast (Riley and McKay 1980).

The development of the many radar and radio sites of the Mid-Canada Line during the Cold War of the 1950s increased the amount of disturbed habitat in the Lowland interior. However, these sites were supported by air or by winter road, and the number of introduced species persisting since their abandonment is negligible. The grading of beach deposits to build airstrips, such as at Winisk and Site 415, has produced habitats that have allowed low-density native beach ridge associations to establish. The winter roads between communities do not support any non-native species that have been noted to date.

The cultivation of land and the grazing of livestock were always of marginal practicality at the posts at which they were attempted (Moose Factory, Albany, and others). The attempts, where they happened, were usually the result of self-sufficiency directives from profit-minded Hudson Bay Company officials without experience in the region. The Hudson Bay Company sent out seeds for Hudson Bay post gardens as early as 1674 (Moodie 1978), and in the first century livestock husbandry was attempted at Churchill, York Factory, and Fort Severn.

The variety of vegetables grown (or attempted) in that period was impressive: cabbage (coleworts and collards), turnips, potatoes, buckwheat, carrots, onions, lettuce, parsley, radish, cress, purslane, peas, celery, beans, sallets, spinach, mustard, horse radish, cucumbers, sorrel, savoy, dandelions, nettles, flax, hemp, and a variety of barley from the Orkney Islands. The efforts to be successful were often elaborate but only the first two of these vegetables appear to have been hardy and to have produced predictable crops (Moodie 1978).

The absence of modern large-scale development in the Lowland, the stable and low overall population (est. 1 person/50 km^2, concentrated in less than a dozen towns and settlements), and the present dependence on air and barge transportation have resulted in a very minor non-native flora. Of the total number of introduced species (98, Tables 21, 22),

Table 21. Non-native species introduced only into the Manitoba Hudson Bay Lowland.

Avena fatua	*Lepidium bourgeauanum**
Lolium persicum	*L. ramosissimum**
Setaria viridis	*Sisymbrium loesellii*
Urtica urens	*Erucastrum gallicum*
Rumex crispus	*Conringia orientalis*
*Chenopodium pratericola**	*Descurainia sophioides**
*Monolepis nuttalliana**	*Arabis glabra*
Axyris amaranthoides	*Thermopsis rhombifolia**
*Chenopodium rubrum**	*Gentiana affinis**
Salsola kali	*Solanum triflorum*
Amaranthus retroflexus	*Artemisia ludoviciana**
Silene pratensis	*Iva xanthifolia*
S. noctiflora	*Helianthus annuus**
Saponaria vaccaria	

* Indicates native North American species.

Table 22. Non-native species introduced widely into the Hudson Bay Lowland.

Agropyron repens	*Sisymbrium altissimum*
Agrostis gigantea	*Sedum telephium*
A. stolonifera	*Medicago lupulina*
Avena sativa	*M. sativa*
*Bromus inermis**	*Melilotus alba*
B. japonicus	*M. officinalis*
Dactylis glomerata	*Trifolium hybridum*
Festuca elatior	*T. pratense*
F. pratensis	*T. repens*
F. trachyphylla	*Vicia cracca*
Phleum pratense	*Lathyrus pratensis*
Poa annua	*Carum carvi*
P. compressa	*Pastinaca sativa*
P. trivialis	*Lappula squarrosa*
Triticum ×aestivum	*Myosotis laxa*
Salix pentandra	*Galeopsis tetrahit* var. *bifida*
Polygonum convolvulus	*Linaria vulgaris*
P. persicaria	*Chaenorrhinum minus*
Chenopodium album	*Plantago lanceolata*
Arenaria serpyllifolia	*Galium verum*
Stellaria graminea	*Conyza canadensis*
S. media	*Achillea millefolium* (typ.)
Cerastium vulgatum	*A. ptarmica*
Silene cserei	*Arctium minus*
S. vulgaris	*Artemisia absinthium*
Ranunculus acris	*Chrysanthemum leucanthemum*
*Lepidium densiflorum**	*Cirsium arvense*
Thlaspi arvense	*Conyza canadensis**
Brassica juncea	*Crepis tectorum*
B. kaber	*Hieracium aurantiacum*
B. rapa	*Matricaria matricarioides*
Barbarea vulgaris	*Senecio vulgaris*
Capsella bursa-pastoris	*Sonchus arvensis*
Camelina microcarpa	*S. uliginosus*
*Descurainia richardsonii**	*Taraxacum officinale*
D. sophia	

* Indicates native North American species.

only 31 are commonly or occasionally found in most settlements. The remainder are rare, and many of them have been seasonal waifs or once-cultivated taxa unable to survive, such as *Avena sativa, Avena fatua, Triticum aestivum*, and *Helianthus annuus*. About 15% of the introduced species are of North American origin, and 10 of these 14 North American taxa are western species occurring almost exclusively in the Churchill area, such as *Lepidium bourgeauanum, Lepidium ramosissum*, and *Descurainia sophioides*. Twenty-eight percent of the introduced taxa are known only from Churchill, and most of these are attributable to the railhead and the shipment of grain through the Churchill seaport.

Rare species of the Hudson Bay Lowland

Almost 23% of the native species of the flora of the Hudson Bay Lowland are known currently from three or less sites (191; Table 23). For comparison, in the northern Clay Belt to the south, about 18% of species were considered scarce or rare, based on an undefined field assessment (Baldwin 1958). Over 55% of the rare species in the Lowland are restricted to the south. Fifty of the taxa are eastern North American in their distribution; 22 of the rare species are western. This listing of rare species is offered as a guide to further work in the Lowland, and for planners and surveyors as documentation of regionally significant vascular plants in the Lowland.

Table 23. Rare native species of the Hudson Bay Lowland. Taxa currently known from three or less locations.

*Lycopodium sitchense**	*P. subsibiricus**	*Puccinellia langeana*
Isoetes echin ospora	*P. strictifolius**	*P. vaginata*
*Equisetum hyemale**	*Ruppia maritima* (A132)	*Vahlodea atropurpurea**
*Botrychium matricariaefolium**	*Elodea nuttallii*	*Carex adelostoma**
*B. multifidum**	X*Agrohordeum macounii*	*C. arcta**
*Cystopteris bulbifera**	*Andropogon gerardii* (A6)	*C. arctata**
Dryopteris cristata	*Bromus kalmii*	*C. atherodes*
*D. expansa**	*Calamagrostis*	*C. capillaris* var. *krausei**
D. fragrans	*deschampsioides* (A30)	*C. crawfordii*
*D. intermedia**	*C. purpurascens*	*C. intumescens**
*Gymnocarpium jessoense**	*Festuca baffinensis**	*C. leptonervia**
*Onoclea sensibilis**	*Hierochloe alpina* (A74)	*C. loliacea*
*Pteridium aquilinum**	*Koeleria macrantha*	*C. misandra**
Woodsia glabella	*Panicum capillare*	*C. nardina*
Picea glauca var. *porsildii*	*P. lanuginosum* var. *implicatum*	*C. ormostachya**
*Pinus resinosa** (A115)	*P. lindheimeri*	*C. peckii*
*P. strobus** (A116)	*P. virgatum**	*C. pedunculata*
*Taxus canadensis**	*Phleum alpinum*	*C. richardsonii**
*Sparganium fluctuans**	*Phragmites australis** (A110)	*C. rossii**
Potamogeton epihydrus	*Poa eminens* (A121)	*C. rotundata**
*P. oakesianus**	*P. flexuosa**	*C. rupestris*
*P. obtusifolius**	*P. saltuensis**	*C. sartwellii*

Table 23. (*concluded*).

*C. scoparia**	*Polygonum lapathifolium* (typ.)	*A. tenellus* (A27)
C. umbellata	*Rumex orbiculatus*	*Lathyrus japonicus* (typ.)
*C. ursina**	*R. salicifolius* var. *subarcticus**	*Oxytropis arctica**
*Cladium mariscoides**	*Chenopodium giganto-*	*O. campestris* var. *varians**
*Rhynchospora fusca**	*spermum**	*O. deflexa* var. *sericea*
*Acorus calamus**	*Suaeda maritima**	*Oxalis acetosella**
*Xyris montana** (A146)	*Minuartia groenlandica* (A98)	*Rhus radicans*
Eriocaulon aquaticum	*Spergularia marina* (A142)	*Triadenum fraseri*
Juncus articulatus	*Nymphaea odorata**	*Viola pubescens*
*J. biglumis**	*N. leibergii**	*V. selkirkii*
*J. brevicaudatus**	*Nuphar microphyllum**	*Epilobium strictum*
J. ensifolius (A78)	*N. Xrubrodiscum**	*Oenothera biennis*
J. pelocarpus	*Ceratophyllum demersum** (A43)	*Myriophyllum alterniflorum*
J. pylaei	*Anemone richardsonii*	*Aralia hispida**
J. subtilis	*Caltha natans**	*Osmorhiza claytonia*
*J. tenuis**	*Clematis occidentalis*	*O. depauperata*
*Luzula acuminata**	*Ranunculus flammula* var.	*O. longistylis**
L. confusa (A89, A90)	*ovalis*	*Pyrola elliptica**
*L. wahlenbergii** (A92)	*R. nivalis**	*Chimaphila umbellata*
Prosartes trachycarpa	*Thalictrum sparsiflorum*	*Kalmia microphylla*
*Streptopus amplexifolius**	*Corydalis aurea*	*Loiseleuria precumbens** (A87)
*Coeloglossum viride**	*Arabis arenicola* (typ.)	*Phyllodoce coerulea* (A111)
Corallorhiza striata	*A. holboellii**	*Gentianella tenella**
Cypripedium parviflorum var.	*Armoracia lacustris**	*Collomia linearis*
planipetalum (A49)	*Barbarea orthoceros**	*Phacelia franklinii**
C. reginae (A52)	*Cochlearia officinalis*	*Cynoglossum boreale**
Listera auriculata	*Draba cinerea* (A53)	*Monarda fistulosa*
Malaxis monophyllos (A94)	*D. lactea*	*Lycopus asper* (A93)
*M. paludosa** (A95)	*D. nemorosa*	*Melampyrum lineare*
Platanthera orbiculata	*D. nivalis* (A56)	*Veronica peregrina**
P. psycodes	*D. norvegica** (A57)	*Utricularia Xochroleuca**
Populus X*jackii**	*Eutrema edwardii**	*Campanula uniflora**
*Salix alaxensis** (A134)	*Hutchinsia procumbens** (A76)	*Antennaria parvifolia*
S. arbusculoides (A135)	*Lesquerella arctica**	*A. rosea* (A9)
Salix athabascensis	*Saxifraga cernua*	*Aster alpinus** (A20)
*S. ballii**	*S. paniculata* (A136)	*A. lateriflorus*
S. discolor	*S. rivularis**	*Bidens hyperborea*
S. petiolaris	*Sorbus americana*	*Erigeron humilis*
*Betula neoalaskana**	*Prunus pumila*	*Lactuca biennis*
Corylus cornuta	*Potentilla crantzii*	*Megalodonta beckii**
Ulmus americana	*Aronia prunifolia* (A18)	*Solidago decumbens**
Asarum canadense	*Astragalus adsurgens**	*S. nemoralis*
Koenigia islandica	*A. agrestis* (A22)	*S. rugosa*

Note: Numbers in parentheses refer to the map number in Appendix A.
*Single locations.

CONCLUSION

The vascular plant flora of the Hudson Bay Lowland includes at least 816 native species (857 ecologically and geographically distinct taxa). This is 110 more species than previously suggested as a maximum by Hustich (1957), which he considered at that time to represent "a very rich flora considering the climatic position of the region." The Lowland flora also includes 98 non-native, adventive species.

These floristic data were assembled over many years; during this time over 350 sites were visited and over 7000 vascular plant collections were made by the author. Another 3000 collections and many more site records were forwarded to the author for review and identification. These collected data were complemented by detailed field notes on the distribution of Lowland species, a complete literature review (Sims et al. 1979), and herbarium searches to verify previous reports and earlier collections.

Eleven areas were established for the collection and cataloguing of data to ensure even data collection during field surveys, to define local floras within the Lowland, and to provide a data base sufficiently dense to assess floristic change across the Lowland. These data collection areas were based on the results of earlier coarse-scale mapping of the landscape, climate, and vegetation of northern Quebec, Ontario, and Manitoba. They were selected to reflect broad biogeoclimatic trends, to allow reasonably equitable floristic comparison between areas, and to be accurate for field use (Fig. 3).

The analysis of the latitudinal and longitudinal distributions of each individual Lowland species throughout its North American range allowed for the quantitative comparison of the geographic affinities of the floras of each data collection area, and of selected groups of species with similar distribution patterns within the Lowland (Tables 2 and 3). This illustrated the dominance of transcontinental and subarctic species in the Lowland, with subarctic species comprising a clear majority in each data collection area. Transcontinental elements were concentrated along the coastal zones and northward in the interior, paralleling the distribution of arctic species in the Lowland flora. A strong temperate influence was identified in the Lowland interior south of the Albany River, reflecting the strong eastern character of the flora in that area.

A calculation of the degree of change in geographic affinities indicated that the boundaries separating the most dissimilar floras were those separating the data collection areas along Hudson Bay and Akimiski Island from the adjacent interior zones, and those separating the area south of the Attawapiskat River watershed. This analysis was extended to the floras on either side of the boundaries that marked the distributional limits of geographically distinct species groups within the Lowland. This allowed the mapping of coarse-scale zones of maximum floristic dissimilarity (Fig. 13; Riley 1980). The most distinct floristic boundaries parallel the coasts of Hudson Bay and James Bay, separated at Akimiski Island, and separate the southern interior from the northern interior. These boundaries define the *Maritime Tundra* coast (low arctic; a 20–40+ km wide area along the Hudson Bay coast); the narrower *Southwest James Bay* coast (low subarctic); and the southern *Boreal Peatland* plain (high temperate). A distinctive, but more gradual, floristic change occurs south-to-north in the central interior. The northern portion of the interior is termed the *Peat Plateau and Woodland* plain (high subarctic), and the central interior is summarized as the *Peatland and Woodland* plain (low subarctic). These zones are generally similar to those proposed by Coombs (1952) and Hare (1954), but are more precise and have a legacy of stratified floristic data against which to test such zones.

The Hudson Bay Lowland is the ecological and geological region embedded at the centre of the world's largest continental ice sheet, the Laurentide. As a result, the origins of its flora are the most distant from postglacial refugia of species, and the most recent.

The origins of the flora of the Hudson Bay Lowland are, on this basis, interpreted from the present distribution of species beyond the Lowland; the continent's geological and ecological setting; and paleoecological studies of the postglacial revegetation of North America. A relatively homogeneous boreal ecosystem existed beyond the late-Wisconsinan glacier across a broad longitudinal zone (Ritchie 1987). The more southern, temperate vegetation types were limited in their glacial refugia; their migrational responses were individualistic and their coincident ranges were not homogeneous (Delcourt and Delcourt 1987). Equally as complex, but less well documented, is the development of the distinctive arctic and western floras of central and northern Canada. In measurable terms, the floras and vegetation of Beringia, the high arctic, the southern Cordillera, the southwestern interior, and the northeastern North American seaboard, all contributed to the development of the flora of the Hudson Bay Lowland.

Two general patterns of plant migration are clear in this region: the conservative (and relict) species of gradual migration (and subsequent range disruption); and the saltatory or vanguard species of long-distance dispersal. Both patterns are constrained by the same postglacial chronology of events, by the reproductive and migrational abilities of individual species, and by the myriad of random and process-related interactions that take place between the vegetation and its supporting environment.

The species that are presently widespread within the Lowland (35% of native taxa) have comparatively broad latitudinal distributions outside the Lowland as well as within. They are largely transcontinental and dominate both the Lowland and the transcontinental boreal and subarctic regions. Many of these species currently occupy areas that were extraglacial during the late-Wisconsinan (or are known there from macrofossil evidence), and most of them have probably persisted in the same or nearby areas since that time, even if in very different proportions. The climatic parameters that influenced them after deglaciation generally shifted towards the areas left by the decaying ice sheet, and most species were unlikely to migrate against those gradients. These assumptions may not apply to every individual species, but their application to large groups of species establishes the confidence with which to interpret the extraglacial origins of species (Hultén 1937).

About 80% of these widespread taxa may have survived the late-Wisconsin in multiple, mostly southern, extraglacial areas, from which multiple or frontal redispersal into deglaciated areas occurred. Some Lowland species have derived from the refugia closest to the Lowland, such as those south of the Great Lakes region, but others may have migrated longer distances because of more effective vectors of dispersal or because their particular routes did not encounter problematic migrational hurdles, such as the Precambrian Canadian Shield.

There is a strong eastern North American element in the Lowland flora (15% of Lowland species), and over three-quarters of the eastern species could have survived the late-

Wisconsinan in widespread refugia from south of the Upper Great Lakes east to the Gulf of St. Lawrence and the Atlantic coast. There is a strong correlation in the Lowland between eastern species and restricted southern Lowland species, reflecting the most frequently traveled migration routes into the Lowland. Particular species-diverse areas south of the Lowland suggest possible routes for this frontal migration of species from the Great Lakes and St. Lawrence, through the northern Clay Belt, into the southern Lowland. The disjunct occurrence of several eastern species into the southern interior Lowland from south of the Precambrian Shield suggests the possibility of more recent long-distance dispersal, as does the disjunction of 30 coastal species into southern James Bay and Hudson Bay across distances of 550 to 2100 km. The regular recurrence of disjunct species within such a youthful regional flora suggests that such disjunctions have been typical of the early development of floras in other glaciated regions.

Western species comprise 8% of all Lowland taxa. This subarctic group has a high proportion of species that are successful in all or part of their ranges as adventives. Forty percent of the western species have current distributions both northwest and southwest of the late-Wisconsinan ice sheet, and three-quarters of these have disjunct distributions eastwards in a manner suggestive of a conservative or relict migration history. Other western species apparently redispersed only from the northwest or only the southwest, and most of these have continuous ranges eastwards, suggestive of less disrupted postglacial migration histories and, in some cases, more recent migration eastwards.

The Lowland supports a distinctive coastal flora. Forty-seven coastal species are widespread and many of them are the dominant species of coastal ecosystems. Almost all of them are transcontinental and circumpolar, and migrated into the Lowland from one or more of the following environments: northern archipelago refugia; northwestern refugia; and closer Atlantic coast populations. Over 100 other species are also restricted to coastal habitats, but are not widespread. Some occur only along the Hudson Bay coast; others only at Cape Henrietta Maria; another group occurs only along the northwesternmost coast; and another group, including a disproportionate number of disjunct species, occurs only along the southwestern James Bay coast.

The northern species (coastal and non-coastal) that are restricted to within 20 to 40 km of the Hudson Bay coast (south to Lake River on northern James Bay) comprise about 10% of Lowland taxa. A further 7% of Lowland taxa are restricted to the northern Lowland, but also persist farther south into the interior. These taxa also have the same tendencies (strictly coastal vs. occurring southward): they are predominantly transcontinental and circumpolar, and where they occur outside the Lowland, they are also arctic and high-subarctic taxa. Northern refugia, both east and west, were most important to their redispersal into the Lowland, although a suggestive minority are known southward in eastern and western North America as disjunct populations and as macrofossils.

The dominant Lowland flora and the immigration of nearly 70% of Lowland species are the result of the gradual migration of boreal, subarctic species into the Lowland from the east, south, southwest, and northwest. Ongoing migration and the slow consolidation of the modern vegetation would have disrupted what were once more widespread distributions of western, subarctic species, causing the modern disjunct occurrences of these species. However, the occurrence of numerous other geographically restricted species, about 50 of them with disjunctions over 500 km, is most probably the result of recent long-distance dispersal. The Hudson Bay Lowland is the region of North America most remote from the vegetated margins of late-Wisconsinan ice sheets, and its patterns of postglacial revegetation provide a very youthful analogue for the probable development of the floras of other, older glaciated regions of North America.

ACKNOWLEDGEMENTS

The Ontario Ministry of Natural Resources' former Provincial Remote Sensing Office and Forest Research Branch; the Ontario Geological Survey; the Canadian Forest Service, Canadian Wildlife Service and former Lands Directorate of Environment Canada; and the Royal Ontario Museum provided most of the field support for this study. I would particularly like to thank S. Pala, A.N. Boissonneau, R.A. Sims, J.K. Jeglum, D.W. Cowell, R.K. Ross, R.I. Morrison, and W.A. Glooschenko in this regard.

P.W. Ball of the University of Toronto assisted with the identification of *Carex* specimens. G.W. Argus of the National Museum of Natural Sciences in Ottawa verified many *Salix* collections. The curators of numerous collections assisted greatly by facilitating access and use of collections and unpublished notes, most notably the National Museum of Natural Sciences in Ottawa (CAN), the Biosystematics Research Institute of Agriculture Canada (DAO), the University of Manitoba (WIN), the Royal Ontario Museum (TRT), and the Herbier Louis-Marie at Université Laval (QFA). P.F. Maycock and J.C. Ritchie generously allowed access to their herbaria. C.E. Punter and B. Ford (WIN, Winnipeg), and M.J. Oldham (Peterborough) were especially helpful in sharing their recent field studies in the northern Lowlands.

I also want to express my sincere thanks for the suggestions and constructive criticism offered by G.W. Argus, P.M. Catling, T. Dickinson, M.D. Engstrom, J.H. McAndrews, R. Staniforth, and B.G. Warner, and for the editorial assistance of Carol McKinley, NRC Research Press. This study would never have concluded except for the patient support and interest of my family.

Finally, the early encouragement by Ptre Ernest Lepage, J.C. Ritchie, and W.K.W. Baldwin, whose efforts laid the foundation for botanical studies in the Lowland, meant a great deal to me.

REFERENCES

Abraham, K.F., and Jeffries, R.L. 1997. High goose populations: causes, impacts and implications. *In* Arctic ecosystems in peril. *Edited by* B.D.J. Batt. Arctic Goose Habitat Working Group, Joint Venture Publication. United States Fish and Wildlife Service, Washington, and Canadian Wildlife Service, Ottawa, pp. 7–72.

Ahti, T. 1964. The macrolichens and their zonal distribution in boreal and arctic Ontario, Canada. Annua Botanica Fennica, **1**: 1–35.

Andrews, J.T. 1968. The pattern and interpretation of restrained, postglacial and residual rebound in the area of Hudson Bay. Geological Survey of Canada, Paper 69-53, pp. 49–62.

Andrews, J.T. 1970. Present and postglacial rates of uplift for glaciated northern and eastern North America from postglacial uplift curves. Canadian Journal of Earth Sciences, **7**: 703–715.

Andrews, J.T. 1987. The late Wisconsinan glaciation and deglaciation of the Laurentide Ice Sheet. *In* North America and adjacent oceans during the last glaciation. *Edited by* W.F. Ruddiman and H.E. Wright, Jr. Geology of North America. Vol. K-3. Geological Survey of America, Boulder, CO, pp. 13–37.

Argus, G.W. 1965. The taxonomy of the *Salix glauca* complex in North America. Contributions from the Gray Herbarium of Harvard University, No. 196, pp. 1–142.

Argus, G.W. 1986. Studies of *Salix lucida* and *Salix reticulata* complexes in North America. Canadian Journal of Botany, **64**: 541–551.

Argus, G.W., and White, D.J. 1978. The rare vascular plants of Alberta. Syllogeus, Vol. 17, pp. 1–37.

Argus, G.W., Pryer, K.M., White, D.J., and Keddy, C.J. (*Editors*). 1982–1987. Atlas of the rare vascular plants of Ontario. 4 parts. National Museum of Natural Sciences, Ottawa, ON.

Arseneault, D., and Payette, S. 1992. A postfire shift from lichen-spruce to lichen-tundra vegetation at tree line. Ecology, **73**: 1067–1081.

Baldwin, H.I. 1979. The distribution of *Pinus banksiana* Lamb in New England and New York. Rhodora, **81**: 549–565.

Baldwin, W.K.W. 1948. List of vascular plants from east coast of Hudson Bay. National Museum of Natural Sciences, Vascular Plant Herbarium, Ottawa, ON. [Manuscript]

Baldwin, W.K.W. 1953. Plants from two small islands in James Bay. National Museum of Canada, Bulletin 128, pp. 154–167.

Baldwin, W.K.W. 1958. Plants of the Clay Belt of northern Ontario and Quebec. National Museum of Canada, Bulletin 156, pp. 1–324.

Baldwin, W.K.W., Lepage, E., Terasmae, J., Maclean, D.W., and Bassett, I.J. 1959. Botanical excursion to the boreal forest region in northern Quebec and Ontario. National Museum of Canada, Ottawa, ON. 119 pp.

Barneby, R.C. 1952. Revision of the North American species of *Oxytropis* DC. Proceedings of the California Academy of Sciences, **27**: 177–309.

Barneby, R.C. 1964. Atlas of North American *Astragalus* Parts 1, 2. Memoirs of the New York Botanical Garden, Vol. 13, pp. 1–1188.

Barnett, P.J. 1992. Quaternary geology of Ontario. *In* Geology of Ontario. *Edited by* P.C. Thurston, H.R. Williams, R.H. Sutcliffe, and G.M. Scott. Ontario Geological Survey, Special Vol. 4, Part 2. Ontario Ministry of Northern Development and Mines, Toronto, ON, pp. 1011–1088.

Bassett, I.J. 1973. The plantains of Canada. Canada, Department of Agriculture, Research Branch Monograph No. 7, pp. 1–46.

Bassett, I.J., and Crompton, C.W. 1978. The genus *Suaeda* (Chenopodiaceae) in Canada. Canadian Journal of Botany, **56**: 581–591.

Bates, D.N., and Simkin, D. 1969. Vegetation patterns of the Hudson Bay Lowland.

Ontario Ministry of Natural Resources, Research Branch, Map 3269.

Beckett, E. 1959. Adventitious plants at Churchill, Manitoba. Canadian Field-Naturalist, **73**: 169–173.

Bergeron, Y., Bouchard, A., and Massicotte, G.N. 1978. Additions à la flore de l'Abitibi, Québec. Naturaliste canadien, **105**: 479–484.

Bertie, A.A. 1975. Paleobotany of Wisconsin interstadials, eastern Great Lakes region, North America. Quaternary Research, **5**: 591–619.

Bjorck, S. 1985. Deglaciation chronology and revegetation in northwestern Ontario. Canadian Journal of Earth Sciences, **22**: 850–871.

Blaney, C.S., and Kotanen, P.M. 2001. The vascular flora of Akimiski Island, Nunavut Territory, Canada. Canadian Field-Naturalist, **115**: 88–98.

Böcher, T.W., Holmen, K., and Jacobsen, K. 1968. The flora of Greenland. *Translated by* I. Frederiksen, T.T. Elkington, and M.C. Lewis. P. Haase, Copehagen, Denmark. 312 pp.

Boissonneau, A.N. 1966. Glacial history of northeastern Ontario. I. The Cochrane–Hearst area. Canadian Journal of Earth Sciences, **3**: 559–578.

Boivin, B. 1952. The distribution of *Arnica wilsonii* Rydb., and its significance. Rhodora, **54**: 200–205.

Boivin, B. 1966–1967. Énumération des plantes du Canada. Université Laval, Québec, QC. Provancheria 6, Mémoires de l'Herbier Louis-Marie. [Extrait du Naturaliste canadien, **93–94**.]

Boivin, B. 1967–1981. Flora of the prairie provinces. Vols. I–V. Provancheria 2, 3, 4, 5, 12, Mémoires de l'Herbier Louis-Marie. Université Laval, Ste-Foy, QC.

Boivin, B. 1968–1969. Flora of the prairie provinces. Part II. Provancheria 3. Université Laval, Québec, QC. 293 pp.

Boivin, B. 1972. Flora of the prairie provinces. Part III. Provancheria 4. Université Laval, Québec, QC. 224 pp.

Boivin, B. 1979. Flora of the prairie provinces. Part IV. Provancheria 5, Université Laval, Québec, QC. 189 pp.

Bostock, H.H. 1968. Precambrian sedimentary rocks of the Hudson Bay Lowlands. *In* Earth Science Symposium on Hudson Bay. *Edited by* P.J. Hood. Geological Survey of Canada, GSC Paper 68-53, pp. 206–214.

Boulton, G.S. 1979. A model of Weichselian glacier variation in the North Atlantic region. Boreas, **8**: 373–395.

Bowden, W.M. 1964. Cytotaxonomy of the species and interspecific hybrids of the genus *Elymus* in Canada and neighbouring areas. Canadian Journal of Botany, **42**: 547–601.

Brokx, P.A.J. 1965. The Hudson Bay Lowlands as caribou habitat. M.Sc. thesis, University of Guelph, Guelph, ON. 244 pp.

Brown, R.J.E. 1967. Permafrost in Canada. Geological Survey of Canada Map 1246A.

Brown, R.J.E. 1973. Permafrost distribution and relation to environmental factors in the Hudson Bay Lowland. *In* Proceedings of the Symposium on the Physical Environment of the Hudson Bay Lowland, March 1973, University of Guelph, Guelph, ON, pp. 35–68.

Bryson, R.A. 1966. Air masses, streamlines and the boreal forest. Geographical Bulletin, **8**: 228–269.

Burger, D. 1993. Revised site regions of Ontario: concepts, methodology and utility. Ontario Forest Research Institute, Sault Ste. Marie, ON. 25 pp.

Canada (Meteorological Branch and National Research Council (Division of Building Research)). 1953. Climatological atlas of Canada. *Prepared by* M.K. Thomas. N.R.C. No. 3151. 253 pp.

Canada (Department of Transport, Meteorological Branch). 1970. Canada Department of Transport, Meteorological Branch, Climatic Maps, Series 8, Sheet 1 (T56-3667/8-1).

Canada (Surveys and Mapping Branch). 1974. The National Atlas of Canada. 4th ed. Macmillan Co., Ottawa, ON, in association with the Department of Energy, Mines and Resources and Information Canada.

Canada (Environment Canada). 1981. Environment Canada, Canadian climate normals 1951–80. Department of Energy, Mines and Resources, Ottawa, ON.

Canada (Environment Canada). 1989. Ecoclimatic regions of Canada. Canadian Committee on Ecological Land Classification, Ecological Land Classification Series No. 23. 118 pp.

Chapman, L.J., and Thomas, M.K. 1968. The climate of northern Ontario. Canada, Department of Transport, Meteorological Branch, Climatological Studies, No. 6, pp. 1–58.

Cody, W.J. 1953. *Phyllodoce coerula* in North America. Canadian Field-Naturalist, **67**: 131–134.

Cody, W.J. 1982. A comparison of the northern limits of distribution of some vascular plant species in southern Ontario. Canadian Field-Naturalist, **109**: 63–90.

Coombs, D.B. 1952. The Hudson Bay Lowland, a geographical study. M.A. thesis, McGill University, Montréal, QC. 227 pp.

Cowell, D.W., and Riley, J.L. 1979. Nature reserve proposals for the Hudson Bay Lowland. *A*. Kinoje Lakes area. *B*. Lower Attawapiskat River. Manuscript prepared for the Ontario Ministry of Natural Resources, Parks Planning Branch, Toronto, ON, 106 pp.

Cowell, D.W., Sims, R.A., and Wickwatre, G.M. 1982. Frozen beach ridge soils in the Hudson Bay lowland, Ontario. Canadian Journal of Soil Science, **62**(2): 421–425.

Cowell, F.N. 1968. List of determinations of the material submitted by Mr. F.N. Cowell, collected in Polar Bear Provincial Park, near Cape Henrietta Maria, 1968. (*Identified by* I.M. Brodo and W.K.W. Baldwin, National Museum of Canada, Ottawa, ON.) [Manuscript]

Cowell, F.N. 1969. List of determinations of the material submitted by Mr. F.M. Cowell, collected in Polar Bear Provincial Park, 1969. [Manuscript]

Craig, B.G. 1968. Late-glacial and post-glacial history of the Hudson Bay region. Geological Survey of Canada, Paper 68-53, pp. 66–77.

Craig, B.G., and Mcdonald, B.C. 1968. Quaternary geology, Operation Winisk, Hudson Bay Lowland. Geological Survey of Canada, Paper 68-1A, pp. 161–162.

Crins, W.J. 2000 (updated 2002). Ecozones, ecoregions and ecodistricts of Ontario. Ontario Ministry of Natural Resources, Ontario Parks, Peterborough, ON. Prepared for Ecological Land Classification Working Group. [Manuscript]

Critchfield, W.B., and Little, E.L. 1966. Geographic distribution of the pines of the world. United States Department of Agriculture, Forest Service, Miscellaneous Publication 991, pp. 1–97.

Dadswell, M. J. 1974. Distribution, ecology and postglacial dispersal of certain crustaceans and fishes in eastern North America. National Museum of Natural Sciences, Ottawa, Publications in Zoology, No. 11, pp. 1–110.

Dahl, E. 1946. On different types of unglaciated areas during the ice ages and their significance to phytogeography. New Phytologist, 45: 225–242.

Dahl, E. 1955. Biogeographic and geological indications of unglaciated areas in Scandinavia during the glacial ages. Geological Society of America Bulletin, **66**: 1499–1519.

Dalla Torre, K.W., von. 1958. Register zu de Dalla Torre et Harms, Genera Siphonogarum. H.R. Engelmann (J. Cramer), Weinheim, Germany. 568 pp. [Original edition 1908.]

Davis, M.B. 1975. Phytogeography and palynology of northeastern United States. *In* The Quaternary of the United States. *Edited by* H.E. Wright Jr. and D.G. Frey. Princeton University Press, Princeton, NJ, pp. 377–401.

Davis, M.B. 1976. Pleistocene biogeography of temperate deciduous forests. Geoscience and Man, **13**: 13–26.

Davis, M.B. 1983. Quaternary history of deciduous forests in eastern North America and Europe. Annals of the Missouri Botanical Garden, **70**: 550–563.

Davis, R.B., Bradstreet, T.E., Stuckenrath, R., Jr., and Borns, H.W., Jr. 1975. Vegetation and associated environments during the past 14,000 years near Moulton Pond, Maine. Quaternary Research, **5**: 436–465.

Dawson, J.E. 1973. A systematic and ecological study of the Ontario taxa of *Parnassia*. M.Sc. thesis, University of Toronto, Toronto, ON. 162 pp.

Dawson, J.E. 1979. A biosystematic study of *Rumex* section *Rumex* in Canada and the United States. Ph.D. thesis, Carleton University, Ottawa, ON. 366 pp.

Delcourt, H.R., and Delcourt, P.A. 1991. Quaternary ecology. A paleoecological perspective. Kluwer Academic Publishers, Chapman and Hall, New York, NY. 242 pp.

Delcourt, P.A., and Delcourt, H.R. 1987. Long-term forest dynamics of the temperate zone. Ecological Studies 63. Springer-Verlag, New York, NY.

Dore, W.G. 1967. Canadian milk-vetch, *Astragalus canadensis*. Trail & Landscape, **1**: 84–87.

Dore, W.G., and McNeill, J. 1980. Grasses of Ontario. Agriculture Canada, Research Branch Monograph No. 26, pp. 1–566.

Downie, S.R. 1988. Morphological, cytological and flavonoid variability of the *Arnica angustifolia* aggregate (Asteraceae). Canadian Journal of Botany, **66**: 24–39.

Dredge, L.A. 1992. Field guide to the Churchill region Manitoba: glaciations, sea level changes, permafrost landforms, and archeology of the Churchill and Gillam areas. Geological Survey of Canada, Ottawa, ON, Miscellaneous Report 53.

Dredge, L.A., and Cowan, W.R. 1989. Quaternary geology of the southwestern Canadian Shield. *In* Quaternary geology of Canada and Greenland. *Edited by* R.J. Fulton. Geological Survey of Canada, Geology of Canada, No. 1, pp. 214–249.

Dredge, L.A., and Nixon, F.M. 1992. Glacial and environmental geology of northeastern Manitoba. Geological Survey of Canada, Ottawa, ON, Memoir 432.

Dreimanis, A. 1977. Correlation of Wisconsin glacial events between the eastern Great Lakes and the St. Lawrence Lowlands. Géographie physique et Quaternaire, **31**: 37–51.

Drury, W.H. 1956. Bog flats and physiographic processes in the upper Koskokwim River region, Alaska. Contributions from the Gray Herbarium of Harvard University, No. 178, pp. 1–130.

Drury, W.H. 1969. Plant persistence in the Gulf of St. Lawrence. *In* Essays in plant geography and ecology. *Edited by* K.N.H. Greenidge. Nova Scotia Museum, Halifax, NS, pp. 105–148.

Dutilly, A., and Lepage, E. 1948. Coup d'ouil sur la flore subarctique du Québec. Catholic University of America, Contribution of the Arctic Institute 1F. 170 pp.

Dutilly, A., and Lepage, E. 1952. Exploration sommaire de la rivière Harricana. Catholic University of America, Contribution of the Arctic Institute 3F. 39 pp.

Dutilly, A., and Lepage, E. 1963. Contribution à la flore du versant sud de la baie James, Québec–Ontario. Catholic University of America, Contribution of the Arctic Institute 12F. 198 pp.

Dutilly, A., Lepage, E., and Duman, M. 1954. Contribution à la flore du versant occidental de la baie James, Ontario. Catholic University of America, Contribution of the Arctic Institute 5F. 144 pp.

Dutilly, A., Lepage, E., and Duman, M. 1958. Contribution à la flore des îles (T.N.O.) et du versant orientale de la baie James. Catholic University of America, Contribution of the Arctic Institute 9F. 199 pp. [DLD 1958.]

Dutilly, A., Lepage, E., and Duman, M. 1959. A collection of plants from Winisk. Naturaliste canadien, **86**: 214–218.

Dyke, A.S., and Prest, V.K. 1987*a*. Late Wisconsinan and Holocene retreat of the Laurentide Ice Sheet. Geological Survey of Canada, Ottawa, ON, Map 1702A.

Dyke, A.S., and Prest, V.K. 1987*b*. Paleogeography of northern North America 18 000 – 5000 years ago. Geological Survey of Canada, Ottawa, ON, Map 1703A.

ESWG (Ecological Stratification Working Group). 1995. A national ecological framework for Canada. Agri-Food, Ottawa–Hull. 125 pp.

Fagerstrom, L. 1948. Vascular plants collected by I. Hustich in Ontario and Quebec, Canada. Memoranda Societatis Pro Fauna et Flora Fennica, **24**: 194–212.

Fernald, M.L. 1925. Persistence of plants in unglaciated areas of boreal America. Memoirs of the American Academy of Arts and Science, **15**: 237–342.

Fernald, M.L. 1935. Critical plants of the upper Great Lakes region of Ontario and Michigan. Rhodora, **37**: 197–222.

Fernald, M.L. 1950. Gray's manual of botany. 8th ed. American Book, New York, NY. 1632 pp.

Ford, B.A., and Ball, P.W. 1992. The taxonomy of the circumpolar short-beaked taxa of *Carex* sect. *Vesicariae* (Cyperaceae). Systematic Botany, **17**: 620–639.

Frederiksen, S. 1982. *Festuca brachyphylla*, *F. saximontana* and related species in North America. Nordic Journal of Botany, **2**: 525–536.

Fulton, R.J. (*Editor*). 1989. Quaternary geology of Canada and Greenland. Geological Survey of Canada, Geology of Canada, No. 1. 839 pp.

Gagnon, A.S., and Gough, W.A. 2002. Hydroclimatic trends in the Hudson Bay region, Canada. Canadian Water Resources Journal, **27**: 245–262.

Gajewski, K., Payette, S., and Ritchie, J.C. 1993. Holocene vegetation history at the boreal-forest – shrub-tundra transition in northwestern Quebec. Journal of Ecology, **81**: 433–443.

Gardner, G. 1937. Liste annotée des espèces de ptéridophytes, de phanérogames et de l'algues récoltées sur la côte du Labrador, à la baie d'Hudson et dans le Manitoba nord, en 1930 et 1935. Bulletin de la Société Botanique de France, **84**: 19–47.

Gardner, G. 1946. Liste des plantes recoltées sur la côte du Labrador, dans le détroit d'Hudson, à la baie James et dans de Manitoba nord. Bulletin de la Société Botanique de France, **93:** 162–200.

Gaudreau, L. 1972. Extensions d'aire et additions à la flore de la zone d'argile, au Québec. Naturaliste canadien, **99:** 509–514.

Gillett, J.M. 1960. The flora of the vicinity of the Mereweather Crater, northern Labrador. Canadian Field-Naturalist, **74:** 8–27.

Gillett, J.M. 1963. The gentians of Canada, Alaska and Greenland. Canada, Department of Agriculture, Research Branch Publication No. 1180, pp. 1–99.

Gillett, J.M. 1971. The native Rhododendrons of Canada and Alaska. Greenhouse, Garden, Grass, **10**(3): 35–45.

Given, D.R., and Soper, J.H. 1981. The arctic-alpine element of the vascular flora at Lake Superior. National Museum of Natural Sciences, Ottawa, ON, Publications in Botany No. 10. 70 pp.

Gleason, H.A. 1922. The vegetational history of the Middle West. Association of American Geographers Annals, **12**: 39–85.

Gleason, H.A., and Cronquist, A. 1991. Manual of the vascular plants of northeastern United States and adjacent Canada. 2nd ed. New York Botanical Garden, New York, NY. 910 pp.

Grant, D.R. 1977. Glacial style and ice limits, the Quaternary stratigraphic record and changes of land and ocean levels in the Atlantic provinces, Canada. Géographie physique et Quaternaire, **31**: 247–260.

Gussow, W.C. 1933. Contribution to the knowledge of the flora of northern Manitoba and the North-western Territories, Dominion of Canada. Canadian Field-Naturalist, **47**: 116–119.

Halliday, W.E.D. 1937. A forest classification for Canada. Canada Department of Mines and Resources, Forest Service Bulletin 89.

Hanson, H.C., and Smith, R.H. 1950. Canada geese of the Mississippi Flyway, with special reference to an Illinois flock. Illinois Natural History Survey Bulletin, No. 25, pp. 67–210.

Hare, F.K. 1950. Climate and zonal divisions of the boreal forest formation in eastern Canada. Geographical Review, **40**: 615–635.

Hare, F.K. 1954. The boreal conifer zone. Geographical Studies, **1**: 4–18.

Hare, F.K., and Ritchie, J.C. 1972. The boreal bioclimates. Geographical Review, **62**: 333–365.

Haynes, R.R. 1974. A revision of North American *Potamogeton* subsection *Pusilli* (Potamogetonaceae). Rhodora, **76**: 564–649.

Henderson, N.C. 1962. A taxonomic revision of the genus *Lycopus* (Labiatae). American Midland Naturalist, **68**: 95–138.

Hills, G.A. 1959. A ready reference to the description of the land of Ontario and its productivity. Ontario Department of Lands and Forests, Research Division, Toronto, ON. 141 pp.

Hills, G.A. 1962. Soil–vegetation relationships in the boreal Clay Belts of eastern Canada. *In* Report on botanical excursion to the boreal forest region of northern Quebec and Ontario. *Edited by* W.K.W. Baldwin et al. National Museum of Canada, Ottawa, ON. pp. 39–53.

Hills, G.A. 1966. The ecological basis for planning. Ontario Department of Lands and Forests, Research Branch, Research Report 46. 210 pp.

Hills, G.A. 1976. Mapping the site regions and landscape districts of Ontario, 1944–1975. Ontario Forest Research Institute, Sault Ste. Marie, ON. 48 pp. [Manuscript]

Hosie, R.C. 1969. Native trees of Canada. 7th ed. Canadian Forestry Service, Ottawa, ON. 380 pp.

Hughes, T. 1987. Ice dynamics and deglaciation when ice sheets collapsed. *In* North America and adjacent oceans during the last glaciation. *Edited by* W.F. Ruddiman and H.E. Wright, Jr. Geology of North America. Vol. K-3. Geological Survey of America, Boulder, CO, pp. 183–220.

Hultén, E. 1937. Outline of the history of arctic and boreal biota during the Quaternary period. Bokförlags aktiebolaget Thule, Stockholm, Sweden. 168 pp. [Reprinted 1972, Verlag von J. Cramer, Lehre.]

Hultén, E. 1958. The amphi-Atlantic plants and their phytogeographical connections. Almqvist and Wiksell, Stockholm, Sweden. 340 pp.

Hultén, E. 1961. Ranges of circumpolar arctic-montane plants. *In* 9th International Botanical Congress: Recent Advances in Botany, Montréal, 1959. Vol. 1. University of Toronto Press, Toronto, ON, pp. 907-911.

Hultén, E. 1962. The circumpolar plants. Vol. 1. Almqvist and Wiksell, Stockholm, Sweden. 275 pp.

Hultén, E. 1968. Flora of Alaska and neighbouring territories: a manual of the vascular plants. Stanford University Press, Stanford, CA. 1008 pp.

Hultén, E. 1971. The circumpolar plants. Vol. 2. Almqvist and Wiksell, Stockholm, Sweden. 463 pp.

Hultén, E. 1973. Supplement to Flora of Alaska and neighbouring territories. Botaniska Notiser **126**: 459–512.

Hunter, G.T. 1970. Post-glacial uplift at Fort Albany, James Bay. Canadian Journal of Earth Sciences, **7**: 547–548.

Hustich, I. 1949. Phytogeographical regions of Labrador. Arctic, **2**: 36–43.

Hustich, I. 1955. Forest botanical notes from the Moose River area, Canada. Acta Geographica, **13**: 1–50.

Hustich, I. 1957. On the phytogeography of the subarctic Hudson Bay lowland. Acta Geographica, **16**: 1–6.

Iltis, H. 1965. The genus *Gentianopsis* (Gentianaceae): transfers and phytogeographic comments. SIDA Contributions to Botany, **2**: 129–153.

Ives, J.D. 1963. Field problems in determining the maximum extent of Pleistocene glaciation along the eastern Canadian seabord a geographer's point of view. *In* North Atlantic biota and their history. *Edited by* A. Löve and E. Löve. Pergamon Press, Oxford, pp. 337–354.

Ives, J.D. 1978. The maximum extent of the Laurentide ice sheet along the east coast of North America during the last glaciation. Arctic, **312**: 24–53.

Ives, J.D., Andrews, J.T., and Barry, R.G. 1975. Growth and decay of the Laurentide ice sheet and comparisons with Fenno-Scandinavia. Die Naturwissenschaften, **62**: 118–125.

Jeglum, J.K. 1971. Personal collection notes of fieldwork at Cape Henrietta Maria, Winisk and Black Duck River, specimens at Great Lakes Forest Research Centre, Sault Ste. Marie, ON. [Manuscript]

Johnson, E.J. 1949. Pollen analysis of peat underlying a treeless heath area in the forest–tundra transition near Churchill Manitoba. M.A. thesis, McMaster University, Hamilton, ON. 28 pp.

Johnson, K.L. 1987. Wildflowers of Churchill and the Hudson Bay Region. Manitoba Museum of Man and Nature, Winnipeg, MB. 400 pp.

Johnson, M.D., Armstrong, D.K., Sanford, B.J., Telford, P.G., and Rutka, M.A. 1992. Paleozoic and Mesozoic geology of Ontario. *In* Geology of Ontario. *Edited by* P.C. Thurston, H.R. Williams, R.H. Sutcliffe, and G.M. Stott. Ontario Geological Survey, Special Vol. 4, pp. 907–1008.

Kalela, A. 1962. Notes on the forest and peatland vegetation in the Canadian Clay Belt region and adjacent areas. Communicationes Instituti Forestal Fenniae, **55.33**: 1–14.

Kallunki, J.A. 1976. Population studies in *Goodyera* (Orchidacese). Brittonia, **28**: 53–75.

Kapp, R.O. 1977. Late Pleistocene and post-glacial plant communities. *In* Geobotany. *Edited by* R.C. Romans. Plenum Press, New York, NY, pp. 1–26.

Karrow, P.F. 1989. Quaternary geology of the Great Lakes subregion. *In* Quaternary geology of Canada and Greenland. *Edited by* R.J. Fulton. Geological Survey of Canada, Geology of Canada, No. 1, pp. 326–350.

Kershaw, K.A., and Rouse, W.R. 1971. Studies on lichen-dominated systems I. The water relations of *Cladonia alpestris* in spruce-lichen woodland in northern Ontario. Canadian Journal of Botany, **49**: 1389–1399.

Kershaw, K.A., and Rouse, W.R. 1973. Studies on lichen-dominated systems. V. A primary survey of a raised-beach system in northwestern Ontario. Canadian Journal of Botany, **51**: 1285–1307.

Kor, P.S.G., Shaw, J., and Sharpe, D.R. 1991. Erosion of bedrock by subglacial meltwater, Georgian Bay, Ontario: a regional overview. Canadian Journal of Earth Sciences, **28**: 623–642.

Kotanen, P.M., and Jeffries, R.L. 1997. Long-term destruction of wetland vegetation by Lesser Snow Geese. Ecoscience, **4**: 1895–1898.

Kral, R. 1966. *Xyris* (Xyridaceae) of the continental United States and Canada. Sida Contributions to Botany, **2**: 177–260.

Lakela, O. 1965. A flora of northeastern Minnesota. University of Minnesota Press, Minneapolis, MN. 541 pp.

Lardner, M.P. 1968. The ice. *In* Sciences, history and Hudson Bay. *Edited by* C.S. Beals. Energy, Mines and Resources, Ottawa, ON. 2 vols. 1057 pp.

Larsen, J.A. 1971. Vegetational relationships with air mass frequencies: boreal forest and tundra. Arctic, **24**: 178–194.

Lepage, E. 1966. Apercu floristique de secteur nord-est de l'Ontario. Naturaliste canadien, **93**: 207–246.

Lepage, E., Baldwin, W.K.W., and Bassett, I.J. 1962. Floristic notes—spermatophyta. *In* Report on botanical excursion to the boreal forest region of northern Quebec and Ontario. *Edited by* W.K.W. Baldwin et al. National Museum of Canada, Ottawa, ON. pp. 88–107.

Lindroth, C.H. 1963. The fauna history of Newfoundland illustrated by caribid beetles. Opuscula Entomologica Supplementum, **23**: 1–96.

Liu, K.-B. 1990. Holocene paleoecology of the boreal forest and Great Lakes forest in northern Ontario. Ecological Monographs, **60**: 179–212.

Luer, C.A. 1975. The native orchids of the United States and Canada, excluding Florida. New York Botanical Garden, New York, NY. 360 pp.

Macoun, J. 1903. List of plants collected by W.J. Wilson along the southwestern shore of James Bay and along the Kapiskau River. Summary Report of the Geological Survey Department of Canada for the Calendar Year 1902, Sessional Paper 26, pp. 239–241. [Also published: Geological Survey of Canada. 1906. GSC Annual Report (New Series) Vol. 15 (1902–1903): 239A–241A.]

Macoun, J. 1906. List of plants collected by W.J. Wilson along the shores of James Bay and in the valley of the Kapiskau River. Annual Report of the Geological Survey of Canada, **15**: 241–243a.

Maher, R.V., Argus, G.W., Harms, V.L., and Hudson, J.H. 1979. The rare vascular plants of Saskatchewan. Syllogeus, Vol. 20, pp. 1–55.

Mandrak, N.E., and Crossman, E.J. 1992. Postglacial dispersal of freshwater fishes into Ontario. Canadian Journal of Zoology, **70**: 2247–2259.

Marie-Victorin, Fr. 1938. Phytogeographical problems of eastern Canada. American Midland Naturalist, **19**: 489–558.

Marie-Victorin, Fr. 1964. Flore laurentienne. 2nd ed. *Revised by* E. Rouleau. Les Presses de l'Université Montréal, Montréal, QC. 925 pp.

Martin, P.S. 1958. Taiga–tundra and the full-glacial period in Chester County, Pennsylvania. American Journal of Science, **256**: 470–502.

Martin, P.S. 1959. How many logs make a forest? Ohio Journal of Science, **59**: 221–222.

Maxwell, B. 1992. Arctic climate: potential for change under global warming. *In* Arctic ecosystems in a changing climate. *Edited by* F.S. Chapin, R.L. Jeffries, J.F. Reynolds, G.R. Shaver, and J. Svoboda. Academic Press, Toronto, ON, pp. 11–34.

Maycock, P.F. 1968. The flora and vegetation of the southern Manigounuk Islands, southeast Hudson Bay, and a consideration of phytogeographical relationships in the region. Naturaliste canadien, **95**: 423–468.

Maycock, P.F. 1974. Preliminary draft of a brief description of the subarctic tundra region of Ontario. University of Toronto, Mississauga, ON. 25 pp. [Manuscript]

Maycock, P.F. 1979. A preliminary survey of the vegetation of Ontario with special reference to the establishment of nature reserves. Ontario Ministry of Natural Resources, Parks and Natural Heritage Policy Branch, Site Region Matrices, Toronto, ON.

McAndrews, J.H. 1993. 6ka mean July temperature in eastern Canada from Barlein and Webb's (1985) pollen transfer functions: comments and illustrations. *In* Proxy climate data and models of the six thousand years before present time interval. *Edited by* A. Telka. Geological Survey of Canada, Canadian Global Change Program, Incidental Report Series No. IR93-3, pp. 22–25.

McAndrews, J.H., Riley, J.L., and Davis, A.M. 1982. Vegetation history of the Hudson Bay Lowland: a postglacial pollen diagram from the Sutton Ridge. Naturaliste canadien, **109**: 597–608.

McDonald, B.C. 1968. Glacial and interglacial stratigraphy, Hudson Bay Lowland. Geological Survey of Canada, Paper 68-53, pp. 78–99.

McNeill, J. 1976. Nomenclature of four perennial species of *Bromus* in eastern North America, with a proposal for the listing of *B. purgans* L. as a rejected name under Article 69. Taxon, **25**: 611–616.

Miller, N.G., and Benninghoff, W.S. 1969. Plant fossils from a Cary-Port Huron interstade deposit and their paleoecological interpretation. Geological Society of America, Special Paper 123, pp. 225–248.

Mills, G.F., Veldhuis, H., Forester, D.B., and Schmidt, R. 1976. A guide to biophysical land classification in Manitoba. Department of Resources and Transportation Services, and Manitoba Soil Surveys, Winnipeg, MB, Northern Resource Information Programme, Biophysical Land Classification Mapsheet 54C (Hayes River, Manitoba; 1 : 125 000). 25 pp.

Mobberley, D.G. 1956. Taxonomy and distribution of the genus *Spartina*. Iowa State College Journal of Science, **30**: 471–574.

Moir, D.R. 1958. A floristic survey of the Severn River drainage basin, northwestern Ontario. Ph.D. thesis, University of Minnesota, Minneapolis, MN. 261 pp.

Moodie, D.W. 1978. Gardening on Hudson Bay: the first century. The Beaver Magazine, **309**(1): 54–59.

Morisset, P. 1971. Endemism in the vascular plants of the Gulf of St. Lawrence region. Naturaliste canadien, **98**: 167–177.

Morton, J.K., and Venn, J.M. 1990. A checklist of the flora of Ontario vascular plants. University of Waterloo, Waterloo, ON, Biology Series No. 34. 218 pp.

Mosquin, T. 1966. Reproductive specialization as a factor in the evolution of the Canadian flora. *In* The evolution of Canada's flora. *Edited by* R.L. Taylor and R.A. Ludwig. University of Toronto Press, Toronto, ON, pp. 43–65.

Mosquin, T. 1971*a*. Biosystematic studies in the North American species of *Linum*, section *Adenolinum* (Linaceae). Canadian Journal of Botany, **49**: 1379–1388.

Mosquin, T. 1971*b*. Evolutionary aspects of endemism. Naturaliste canadien, **98**: 121–130.

Mulligan, G.A. 1970. Cytoaxonomic studies of *Draba glabella* and its close allies in Canada and Alaska. Canadian Journal of Botany, **48**: 1431–1437.

Mulligan, G.A. 1971. Cytotaxonomic studies of the closely allied *Draba cana, D. cinerea* and *D. groenlandica* in Canada and Alaska. Canadian Journal of Botany, **49**: 89–93.

Mulligan, G.A. 1974. Cytotaxonomic studies of *Draba nivalis* and its close allies in Canada and Alaska. Canadian Journal of Botany, **52**: 1793–1801.

Neal, M.W., and Kershaw, K.A. 1973. Studies on lichen-dominated systems. III. Phytosociology of a raised-beach system near Cape Henrietta Maria, northern Ontario. Canadian Journal of Botany, **51**: 1115–1125.

Norris, A.W., and Sanford, B.V. 1968. Paleozoic and Mesozoic geology of the Hudson Bay Lowlands. *In* Earth Science Symposium on Hudson Bay. *Edited by* P.J. Hood. Geological Survey of Canada, GSC Paper 68–53, pp. 169–205.

Norris, A.W., Sanford, B.V., and Bostock, H.H. 1967. Geology, Hudson Bay Lowlands. Geological Survey of Canada, Ottawa, ON, Map 17-1967.

NWWG (National Wetland Working Group). 1988. Wetlands of Canada. Canadian Wildlife Service, Environment Canada, Ecological Land Classification Series, No. 24. 452 pp.

Oldham, M.J. 2002. 2001 botanical highlights. Field Botanists of Ontario Newsletter, **14**(4): 1–9.

Packer, J.G., and Vitt, D.H. 1974. Mountain Park: a plant refugium in the Canadian Rocky Mountains. Canadian Journal of Botany, **52**: 1393–1409.

Pala, S., and Boissonneau, A. 1982. Wetland classification maps for the Hudson Bay Lowland. Naturaliste canadien, **109**: 653–659.

Polunin, N.V. 1948. Botany of the Canadian eastern arctic. Part III. Vegetation and ecology. National Museum of Canada, Bulletin 104, pp. 1–304.

Polunin, N. 1951. The real arctic: suggestions for its delimitation, subdivision and characterization. Journal of Ecology, **39**: 309–315.

Polunin, N. 1955. Long-distance dispersal in the north polar regions. Nature, **176**: 22–24.

Porsild, A.E. 1932. Notes on the occurrence of *Zostera* and *Zannichellia* in arctic North America. Rhodora, **34**: 90–94.

Porsild, A.E. 1955. The vascular plants of the western Canadian arctic archipelago. National Museum of Canada, Bulletin 135, pp. 1–226.

Porsild, A.E. 1957. Illustrated flora of the Canadian Arctic Archipelago. National Museum of Canada, Bulletin 146, pp. 1– 209.

Porsild, A.E. 1964. Illustrated flora of the Canadian arctic archipelago. 2nd ed. National Museum of Canada, Bulletin 146, pp. 1–218.

Porsild, A.E., and Cody, W.J. 1980. Vascular plants of continental Northwest Territories. National Museums of Canada, Ottawa, ON. 667 pp.

Potter, D. 1932. Botanical evidence of post-Pleistocene marine connection between Hudson Bay and the St. Lawrence basin. Rhodora, **34**: 69–89, 101–112.

Potter, D. 1934. Plants collected in the southern region of James Bay. Rhodora, **36**: 274–284.

Prest, V.K. 1970. Quaternary geology in Canada. *In* Geology and economic minerals of Canada. *Edited by* R.J.W. Douglas. 5th ed. Geological Survey of Canada, Economic Geology Report 1, pp. 676–764.

Price, J.S., Ewing, K., Woo, M., and Kershaw, K.A. 1988. Vegetation patterns in James Bay coastal marshes. II. Effects of hydrology on salinity and vegetation. Canadian Journal of Botany, **66**: 2586–2594.

Protz, R., Ross, G.J., and Martini, I.P. 1984. Rate of podzolic soil formation near Hudson Bay, Ontario. Canadian Journal of Soil Science, **64**: 31–49.

Railton, J.B., and Sparling, J.H. 1973. Preliminary studies on the ecology of palsa mounds in northern Ontario. Canadian Journal of Botany, **51**: 1037–1044.

Raup, H.M. 1943. The willows of the Hudson Bay region and the Labrador peninsula. Sargentia, **4**: 81–135.

Raup, H.M. 1947. The botany of southwestern Mackenzie. Sargentia, **6**: 1–275.

Raup, H.M. 1959. The willows of boreal western America. Contributions from the Gray Herbarium of Harvard University, No. 185, pp. 3–95.

Raveling, D.G., and Lumsden, H.G. 1977. Nesting ecology of Canada Geese in the Hudson Bay Lowlands of Ontario: evolution and population regulation. Ontario Ministry of Natural Resources, Fish and Wildlife Research Report 98, pp. 1–77.

Raymond, M. 1950. Esquisse phytogéographique de Québec. Mémoires du Jardin Botanique de Montréal No. 5.

Reeves, B.O.K. 1973. The nature and age of the contact between the Laurentide and Cordilleran ice sheets in the western interior of North America. Arctic and Alpine Research, **5**: 1–16.

Repentigny, L.-G. 1978. Extension d'aire: *Malaxis monophyllos* var. *brachypoda* (Gray) F. Morris. Naturaliste canadien, **105**: 117–118.

Reznicek, A.A. 1979. Personal communications, 1979, concerning observations at Bull's Bay, Missinaibi River, 1974.

Reznicek, A.A., and Ball, P.W. 1980. The taxonomy of *Carex* section Stellulatae in North America north of Mexico. University of Michigan Herbarium, Contribution 14, pp. 153–203.

Richard, P. 1977. Végétation tardiglaciaire au Québec méridonal et implications paléoclimatiques. Géographie physique et Quaternaire, **31**: 161–176.

Riley, J.L. 1978. The vegetation of the Pierre-Montreuil Park Reserve, 1974–75. Ontario

Ministry of Natural Resources, Provincial Parks Branch, Northern Region, Timmins, ON. 219 pp.

Riley, J.L. 1979*a*. The vegetation of Kap-Kig-Iwan Provincial Park, 1978. Ontario Ministry of Natural Resources, Provincial Parks Branch, Northern Region, Timmins, ON. 118 pp.

Riley, J.L. 1979*b*. Some new and interesting vascular plant records from northern Ontario. Canadian Field-Naturalist, **93**: 355–362.

Riley, J.L. 1980. The Flora and phytogeography of the Hudson Bay Lowland. M.Sc. thesis, Botany Department, University of Toronto, Toronto, ON. 436 pp.

Riley, J.L. 1981. The vascular flora of Akimiski Island, James Bay, N.W.T. Naturaliste canadien, **108**: 229–235.

Riley, J.L. 1982. Hudson Bay Lowland floristic inventory, wetlands catalogue and conservation strategy. Naturaliste canadien, **109**: 543–555.

Riley, J.L. 1989*a*. Southern Ontario bogs and fens off the Canadian Shield. *In* Proceedings, Wetlands: Inertia or Momentum, Toronto, ON, October 21–22, 1988. *Edited by* M.J. Bardecki and N. Patterson. Federation of Ontario Naturalists, Don Mills, ON, pp. 355–567.

Riley, J.L. 1989*b*. Peat and peatland resources of northeastern Ontario. Ontario Geological Survey, Miscellaneous Paper 144. 175 pp.

Riley, J.L., and Mckay, S.M. 1980. The vegetation and phytogeography of coastal southwestern James Bay. Royal Ontario Museum, Life Sciences Contributions No. 124, pp. 1–86.

Riley, J.L., and Moore, C. 1973. Preliminary vegetation survey of Shipsands Island Waterfowl Sanctuary 1972. Ontario Ministry of Natural Resources, Forest Research Branch, Toronto, ON, File Report 5-73. 44 pp.

Riley, J.L., and Walshe, S. 1985. New and interesting vascular plant records from northern Ontario. Canadian Field-Naturalist, **99**: 30–33.

Ritchie, J.C. 1956. The native plants of Churchill, Manitoba. Canadian Journal of Botany, **34**: 269–320.

Ritchie, J.C. 1960. The vegetation of northern Manitoba. V. Establishing the major zonation. Arctic, 13: 211–229.

Ritchie, J.C. 1962. A geobotanical survey of northern Manitoba. Arctic Institute of North America, Calgary, AB, Technical Paper 9, pp. 1–46.

Ritchie, J.C. 1966. Late-Pleistocene history of the Canadian flora. *In* The evolution of Canada's flora. *Edited by* R.L. Taylor and R.A. Ludwig. University of Toronto Press, Toronto, ON, pp. 66–80.

Ritchie, J.C. 1967. Holocene vegetation of the northwestern precincts of the glacial Lake Agassiz basin. *In* Proceedings of the 1966 Conference on Environmental Studies of the Glacial Lake Agassiz Region: Life, Land and Water. *Edited by* W.J. Mayer-Oakes. University of Manitoba Press, Winnipeg, MB, pp. 217–229.

Ritchie, J.C. 1976. The late-Quaternary vegetational history of the western interior of Canada. Canadian Journal of Botany, **54**: 1793–1818.

Ritchie, J.C. 1983. The paleoecology of the central and northern parts of the Lake Agassiz basins. *In* Glacial Lake Agassiz. Geological Association of Canada, Special Paper 26, pp. 157–170.

Ritchie, J.C. 1984. Vegetation of the far northwest of Canada. University of Toronto Press, Toronto, ON. 251 pp.

Ritchie, J.C. 1987. Postglacial vegetation of Canada. Cambridge University Press, Cambridge, England. 178 pp.

Ritchie, J.C. 1989. History of the boreal forest in Canada. *In* Quaternary geology of Canada and Greenland. *Edited by* R.J. Fulton. Geological Survey of Canada, Geology of Canada, No. 1, pp. 508–512.

Roulet, N., Jano, A., Kelly, C., Klinger, L., Moore, T., Protz, R., Ritter J., and Rouse, W. 1994. Role of the Hudson Bay lowland as a source of atmospheric methane. Journal of Geophysical Research, **99**: 1439–1454.

Rouse, W.R. 1991. Impacts of Hudson Bay on the terrestrial climate of the Hudson Bay Lowlands. Arctic and Alpine Research, **23**: 24–30.

Rousseau, C. 1974. Géographie floristique de Québec–Labrador: Distribution des principales espèces vasculaires. Travaux et Documents du Centre d'Études Nordiques No. 7, pp. 1–799.

Rousseau, J. 1968. The vegetation of the Quebec–Labrador Peninsula between 55 and 60 degrees North. Naturaliste canadien, **95**: 49–169.

Rowe, J.S. 1972. Forest Regions of Canada. Canada, Department of the Environment, Canadian Forestry Service Publication 1300. 172 pp.

Ruddiman, W.F. 1987. Synthesis; the ocean/ice sheet record. *In* North America and adjacent oceans during the last glaciation. *Edited by* W.F. Ruddiman and H.E. Wright, Jr. Geology of North America. Vol. K-3. Geological Survey of America, Boulder, CO, pp. 463–478.

Saarnisto, M. 1974. The deglaciation history of the Lake Superior region and its climatic implications. Quaternary Research, **4**: 316–339.

Saarnisto, M. 1975. Stratigraphical studies on the shoreline displacement of Lake Superior. Canadian Journal of Earth Sciences, **12**: 300–319.

Sanford, B.V., Norris, A.W., and Ball, R.T. 1968. Geology of the Hudson Bay Lowlands (Operation Winisk). Geological Survey of Canada, Paper 67-60, pp. 1–118.

Savile, D.B.O. 1956. Known dispersal rates and migratory potential as clues to the origin of the North American biota. American Midland Naturalist, **56**: 434–453.

Savile, D.B.O. 1972. Arctic adaptations in plants. Canada, Department of Agriculture, Research Branch Monograph No. 6, pp. 1–81.

Schofield, W.B. 1958. The salt marsh vegetation of Churchill, Manitoba, and its phytogeographic implications. National Museum of Canada, Bulletin 160, pp. 107–132.

Scoggan, H.J. 1957. Flora of Manitoba. National Museum of Canada, Bulletin 140, pp. 1–619.

Scoggan, H.J. 1969. The native flora of Churchill, Manitoba. Department of Northern Affairs and National Resources and National Museum of Canada, Ottawa, ON. 51 pp.

Scoggan, H.J. 1978–1979. The flora of Canada. Parts 1–4. National Museum of Natural Sciences, Ottawa, Publications in Botany No. 7, pp. 1–1711.

Scott, P.A., Hansell, R.I.C., and Erickson, W.R. 1993. Influences of wind and snow on northern tree-line environments at Churchill, Manitoba, Canada. Arctic, **46**: 316–323.

Seymour, F.C. 1960. Flora of Lincoln County, Wisconsin. Taunton, MA. 363 pp.

Shoemaker, E.M. 1991. On the formation of large subglacial lakes. Canadian Journal of Earth Sciences, **28**: 1975–1981.

Sims, R.A., Riley, J.L., and Jeglum, J.K. 1979. Vegetation, flora and vegetational ecology of the Hudson Bay Lowland: a literature review and annotated bibliography. Canada, Department of the Environment, Canadian Forestry Service, Report O-X-297. 177 pp.

Sims, R.A., Wickware, G.A., and Cowell, D.W. 1987*a*. A study of coastal vegetation at a site on Hudson Bay near Winisk, Ontario. Canadian Field-Naturalist, **101**: 335–345.

Sims, R.A., Wickware, G.A., and Cowell, D.W. 1987*b*. Wetlands of the southern Hudson Bay coast in Ontario. *In* Proceedings, Symposium 1987: Wetlands/Peatlands. *Edited by* C.D.A. Rubec and R.P. Overend. Environment Canada, Edmonton, AB, pp. 435–442.

Sjörs, H. 1959. Bogs and fens in the Hudson Bay Lowlands. Arctic, **12**: 3–19.

Sjörs, H. 1961. Forest and peatland at Hawley Lake, northern Ontario. National Museum of Canada, Bulletin 171, pp. 1–31.

Sjörs, H. 1963. Bogs and fens on the Attawapiskat River, northern Ontario. National Museum of Canada, Bulletin 186, pp. 45–133.

Small, E. 1968. The systematics of autopolyploidy in *Epilobium latifolium* (Onagraceae). Brittonia, **20**: 169–181.

Snyder, L.L., and Lapworth, E.D. 1953. A comparative study of two Canadian races of Red-wings. Canadian-Field Naturalist, **67**: 143–147.

Soper, J.H. 1962. Some genera of restricted range in the Carolinian flora of Canada. Transactions of the Royal Canadian Institute No. 34, pp. 2–56.

Soper, J.H., and Heimburger, M.L. 1982. Shrubs of Ontario. Royal Ontario Museum, Toronto, ON, Life Science Miscellaneous Publication. 495 pp.

Soper, J.H., and Maycock, P.F. 1963. A community of arctic-alpine plants on the east shore of Lake Superior. Canadian Journal of Botany, **41**: 183–198.

Stanfield, R., Riley, J., Mackey, B., Burdett, R.C., Arthur, M., and Watt, I. 1972. Biological studies of the Onakawana area. Ontario Ministry of the Environment, Task Force Onakawana, Working Paper 3. 60 pp.

Staniforth, R.J., and Scott, P.A. 1991. Dynamics of weed populations in a northern subarctic community. Canadian Journal of Botany, **69**: 814–821.

Stormer, P. 1933. Plants collected by F. Johansen in 1929, at Hudson Bay railway and Port Churchill in arctic Canada. Nytt Magasin for Naturvidenskapene, Bind **73**: 259–272.

Stuckey, R.L. 1969. The introduction and spread of *Lycopus asper* (western water horehound) in the western Lake Erie and Lake St. Clair region. Michigan Botanist, **8**: 111–120.

Svenson, H.K. 1927. Studies of interior distribution of maritime plants. I. Effects of post-Pleistocene marine submergence in eastern North America. Rhodora, **29**: 41–48, 57–72, 87–93, 105–114.

Tarnocai, C. 1974. Peat landforms and associated vegetation in Manitoba. *In* Proceedings of the Canadian Soil Survey, Organic Soil Mapping Workshop. *Edited by* J.H. Day. Soil Research Institute, Ottawa, ON, pp. 3–21.

Teller, J.T. 1987. Postglacial lakes and the southern margin of the Laurentide Ice Sheet. *In* North America and adjacent oceans during the last glaciation. *Edited by* W.F. Ruddiman and H.E. Wright, Jr. Geology of North America. Vol. K-3. Geological Survey of America, Boulder, CO, pp. 39–69.

Terasmae, J., and Anderson, T.W. 1970. Hypsithermal range extension of white pine (*Pinus strobus*) in Quebec. Canadian Journal of Earth Sciences, **7**: 406.

Terasmae, J., and Hughes, O.L. 1960. A palynological and geological study of Pleistocene deposits in the James Bay Lowlands, Ontario. Geological Survey of Canada, Bulletin 62, pp. 1–15.

Thien, L.B., and Marcks, B.G. 1972. The floral biology of *Arethusa bulbosa*, *Calopogon tuberosus* and *Pogonia ophioglossoides* (Orchidaceae). Canadian Journal of Botany, **50**: 2319–2325.

Verdoorn, F. (*Editor*). 1938. Manual of pteridology. Martinus Nijhoff, The Hague. 640 pp.

Vesper, S.J., and Stuckey, R.L. 1977. The return of aquatic vascular plants into the Great Lakes region after late-Wisconsin glaciation. *In* Geobotany. *Edited by* R.C. Romans. Plenum Press, New York, NY, pp. 283–299.

Vincent, J.-S. 1973. A palynological study for the Little Clay Belt, northwestern Quebec. Naturaliste canadien, **100**: 59–70.

Vincent, J.-S., and Hardy, L. 1979. The evolution of glacial Lakes Barlow and Ojibway, Quebec and Ontario. Geological Survey of Canada, Bulletin 316, pp. 1–18.

Voss, E.G. 1972. Michigan flora. Part 1: Gymnosperms and monocots. Cranbrook Institute of Science, Bulletin 55, pp. 1–488.

Voss, E.G. 1985. Michigan flora. Part 2: Dicots (to Cornacase). Cranbrook Institute of Science, Bulletin 59, pp. 1–724.

Voss, E.G. 1996. Michigan flora. Part 3: Dicots (Pyrolaceae to Compositae). Cranbrook Institute of Science, Bulletin 61, pp. 1–622.

Walker, B.H., and Coupland, R.T. 1968. An analysis of vegetation-environment relationships in Saskatoon sloughs. Canadian Journal of Botany, **46**: 509–522.

Watts, W.A. 1967. Late-glacial plant macrofossils from Minnesota. *In* Quaternary paleoecology. *Edited by* E.J. Cushing and H.E. Wright, Jr. Yale University Press, New Haven, CT, pp. 89–97.

Webber, P.J., and Richardson, J.W. 1970. Postglacial uplift and age at Cape Henrietta Maria, southeastern Hudson Bay, Canada. Canadian Journal of Earth Sciences, **7**: 313–325.

Weber, W.A. 1961. Some features of the distribution of arctic relicts at their austral limits. *In* 9th International Botanical Congress: Recent Advances in Botany, Vol. 1, Montréal, 1959. University of Toronto Press, Toronto, ON, pp. 912–914.

Westgate, J.A., Fritz, P., Kalas, L., Greene, R., and Matthews, J.V., Jr. 1974. Geochronology and palaeoecology of Mid-Wisconsin sediments in westcentral Alberta, Canada. International Geological Congress, Ottawa, ON. [Abstract]

White, D.J., and Johnson, K.L. 1980. The rare vascular plants of Manitoba. Syllogeus, Vol. 27, pp. 1–52.

Whitehead, D.R. 1972. Approaches to disjunct populations: the contribution of palynology.

Annals of the Missouri Botanical Garden, **59**: 125–137.

Wiken, E.B. 1986. Terrestrial ecozones of Canada. Environment Canada, Hull, QC, Ecological Land Classification Series No. 19. 26 pp. + map.

Wilkinson, R.C., Hanover, J.W., Wright, J.W., and Flake, R.H. 1971. Genetic variation in monoterpene composition of white spruce. Forest Science, **17**: 83–90.

Wright, H.E. 1964. Aspects of the early post-glacial forest succession in the Great Lakes region. Ecology, **45**: 439–448.

Wynne-Edwards, V.C. 1937. Isolated arctic-alpine floras in eastern North America: a discussion of their glacial and recent history. Transactions of the Royal Society of Canada, 3rd Ser., Vol. 31, Sect. 5, pp. 33–58.

Wynne-Edwards, V.C. 1939. Some factors in the isolation of rare alpine plants. Transactions of the Royal Society of Canada, 3rd Ser., Vol. 33, Sect. 5: 33–42.

Yeatman, C.W. 1967. Biogeography of jack pine. Canadian Journal of Botany, **45**: 2201–2211.

Young, S.B. 1971. The vascular flora of Saint Lawrence Island, with special reference to floristic zonation in the arctic regions. Contributions from the Gray Herbarium of Harvard University, No. 201, pp. 11–115.

Zenkert, C. 1934. The flora of the Niagara Frontier region. Buffalo Society of Natural Sciences, Bulletin 16, pp. 1–328. [Supplement, 1975, Bulletin 16]

Zoltai, S.C., Pollett, F.C., Jeglum, J.K., and Adams, G.D. 1974. Developing a wetland classification for Canada. Proceedings of the 4th North American Forest Soil Conference, Québec City, August 1973. 17 pp. [Manuscript]

PLATES

Plates 1–6

1. Patterned bogs and fens dominate the southern and central reaches of the Hudson Bay Lowland. More than 90% of the Lowland is covered by a sea of saturated peat, such as here in the Albany river area.

2. Some of the southern Lowland landscapes, such as the saturated peatlands here in the Albany River area, are dominated by open water ponds only minorly divided and terraced by bog and fen vegetation.

3. Swarms of storm beaches have been thrown up along Hudson Bay and James Bay and are covered in tundra heath, lichen spruce woodland, and dune communities. Here, along the shores of western James Bay are a mixture of intertidal marshes and supertidal meadow-marshes and thickets, interrupted by widely spaced storm beaches.

4. Precambrian cuestas outcrop above the lowlands at the Sutton Ridges, which extend from the Winisk to the Ekwan rivers about 40–60 km from the coast. These dramatic cliffs, joints, and crests support a number of arctic species only known in the Lowland from the Sutton Ridges.

5. Cape Henrietta Maria, at the junction of Hudson and James bays, is rising out of the ocean faster than any other area in North America, rebounding from the released weight of the last continental glacier. Exposed to a 1000-km fetch of ocean from the northwest, it is a scene of chaotic storm beaches, with stunted arctic tundra in sheltered depressions.

6. Like most rivers draining off the Canadian Shield into Hudson and James bays, the Attawapiskat River has eroded deep down through the surface clays and tills, down to the limestone bedrock. The Attawapiskat has revealed limestone reefs and bioherms, with dry cliffs and rims, and with karst sinkholes draining interior peatlands into the river.

1

2

3

4

5

6

Plates 7–12

7. A broad zone of continuous permafrost occurs back from Hudson Bay, expressed in beach ridges and river levees, and in interior peatlands as raised palsas and peat plateaus. These raised areas support lichen woodlands and dwarf-shrub tundra that periodically burn and regenerate rapidly without significant permafrost decay.

8. The southern Lowlands in the Moose River basin are distinguished by broad, stunted tamarack fens, with richer black spruce swamps along the streams.

9. Vast wetlands occur along the Ontario–Manitoba border as much as 240 km from Hudson Bay. Pictured here are remarkable fens whose slopes are so regular and slight that vast "strings" have formed across the landscape, perpendicular to water flow. This natural terracing is the result of freeze–thaw cycles over the millennia.

10. On raised peat plateaus up to 120 km from Hudson Bay near the Winisk River, there are large thermokarst lake systems moving gradually downwind due to wave-erosion of the shoreline permafrost.

11. White Elm levees persist along the Kenogami River in the southernmost lowland, disjunct far to the north of other elms and safe (to date) from the Dutch elm disease. Other southern species occur in the southernmost lowlands, many of them separated from their nearest populations by the full width of the Canadian Shield.

12. Netscapes of peatland ponds and ridges cover large parts of the interior lowlands. The raised ridges often hold ice longer in the summer, and northward, discontinuous permafrost persists in them year-round.

7

8

9

10

11

12

Plates 13–22

13. The orchid *Cypripedium parviflorum* var. *planipetalum* is known from northern Newfoundland and easternmost Quebec on subarctic limestone, and is disjunct to the Lowlands on the wooded limestone rims of the Attawapiskat River (Fig. A49).

14. The False Dandelion, *Agoseris glauca*, is widespread throughout western North America but hops the Canadian Shield completely to reoccur throughout the northern Lowlands (Figs. A1, A2).

15. The northwestern Mackenzie Water-hemlock, *Cicuta mackenzieana*, is widespread in Alaskan marshes and extends eastward across the northern prairies and parklands to reach its easternmost limit on James Bay (Fig. A47).

16. The succulent Seabeach Sandwort, *Honkenya peploides*, is a common Lowland maritime species, occurring throughout the Atlantic and Arctic shores of northeastern North America (Fig. A75).

17. The inconspicuous Green Adder's-mouth orchid, *Malaxis unifolia*, extends northward through Ontario into the rich open peatlands of the Clay Belt and the southernmost Lowlands (Fig. A96).

18. Velvet Bells, *Bartsia alpina*, is a subarctic species of open tundra and heath, with an amphi-Atlantic distribution from northern Europe and Asia westward as far as the Hudson Bay Lowland (Fig. A29).

19. The Greenland Sandwort, *Minuartia groenlandica*, is disjunct from northern Quebec populations across James Bay to a few of the highest open crests of the Precambrian Sutton Ridge cuestas in the Lowland (Fig. A98).

20. A coastal species of north Atlantic shores, Scotch Lovage or *Ligusticum scothicum*, is disjunct to James Bay from nearest occurrences in the Gulf of St. Lawrence, typical of many coastal species of southern James Bay (Fig. A84).

21. Known from a few coastal estuaries in eastern Canada and the northeastern U.S., the Coastal Beggar-tick, *Bidens hyperborea*, was first described from southern James Bay.

22. A widespread species of Beringia and southward in the cordillera, *Hedysarum mackenzii* crosses the northern parklands and subarctic east to the Lowlands, and beyond to disjunct occurrences on Anticosti and western Newfoundland (Fig. A73).

13

14

15

16

17

18

19

20

21

22

Plates 23–32

23. The only known eastern North American occurrence of the Alpine Aster, *Aster alpinus*, is on an ephemeral river delta bar along the Hudson Bay coast (Fig. A20).

24. The Franklin Lady-slipper orchid, *Cypripedium passerinum,* is another species of the northwestern interior, from Alaska throughout the prairies, avoiding the Canadian Shield to reoccur on dry shores, slopes, and forests in the Lowland and disjunct to Lake Superior and the Gulf of St. Lawrence (Fig. A51).

25. The Lowland's only endemic plant, the Lepage Wild Flax, *Linum lewisii* var. *lepagei*, is a prostrate, white-flowered, coastal-beach variant of the western perennial flax (Fig. A86).

26. *Ledum decumbens*, the arctic's true Labrador-tea, is a species very much defining the extent of the Maritime Tundra zone in the Ontario Lowlands, and extends farther inland on permafrost in northern Manitoba and Ontario (Fig. A83).

27. The widespread Prickly Saxifrage, *Saxifraga tricuspidata*, of the North American arctic extends south to the Lowland's Hudson Bay coast, and also occurs farther south as an arctic-alpine disjunct, such as to Lake Superior (Fig. A137).

28. The transcontinental northern Small Butterwort, *Pinguicula villosa*, occurs on wet mossy seeps in the northernmost Lowland (Fig. A112).

29. The arctic River Beauty, *Epilobium latifolium*, occurs on river bars and shores close to Hudson Bay throughout the Lowland.

30. The Polar Daisy, *Chrysanthemum arcticum* var. *polaris*, occurs along Hudson Bay and James Bay shores, and then north on the shores of the western Arctic Ocean and Alaska (Fig. A45).

31. Widespread in the cordilleran interior and northern prairies, the America Milk Vetch, *Astragalus americanus*, persists eastward on dry uplands of the northwestern Lowlands, and is from there disjunct eastward to the Gaspé (Fig. A24).

32. The distinctive Elephant-heads, *Pedicularis groenlandica*, is widespread in the far west and occurs only in the Lowlands of Manitoba and Ontario, then eastward through northern Quebec and Labrador to Greenland (Fig. A105).

23

24

25

26

27

28

29

30

31

32

APPENDIX A

DISTRIBUTION MAPS

The following maps combine occurrence maps (dot maps) with range-limit maps to provide precise distributional data for the Hudson Bay Lowland and related areas, while still indicating the broader distribution of the species. Literature sources for range limit lines are indicated on each map, along with the herbarium sources of occurrence locations and the area for which specimens were mapped. Standard acronyms are used for herbarium collections. The maps are presented in alphabetical order by scientific name. They are numbered sequentially and are referred to in the text as Fig. A1, Fig. A2, etc.

Fig. A1. Distribution of *Agoseris glauca* in eastern Canada.

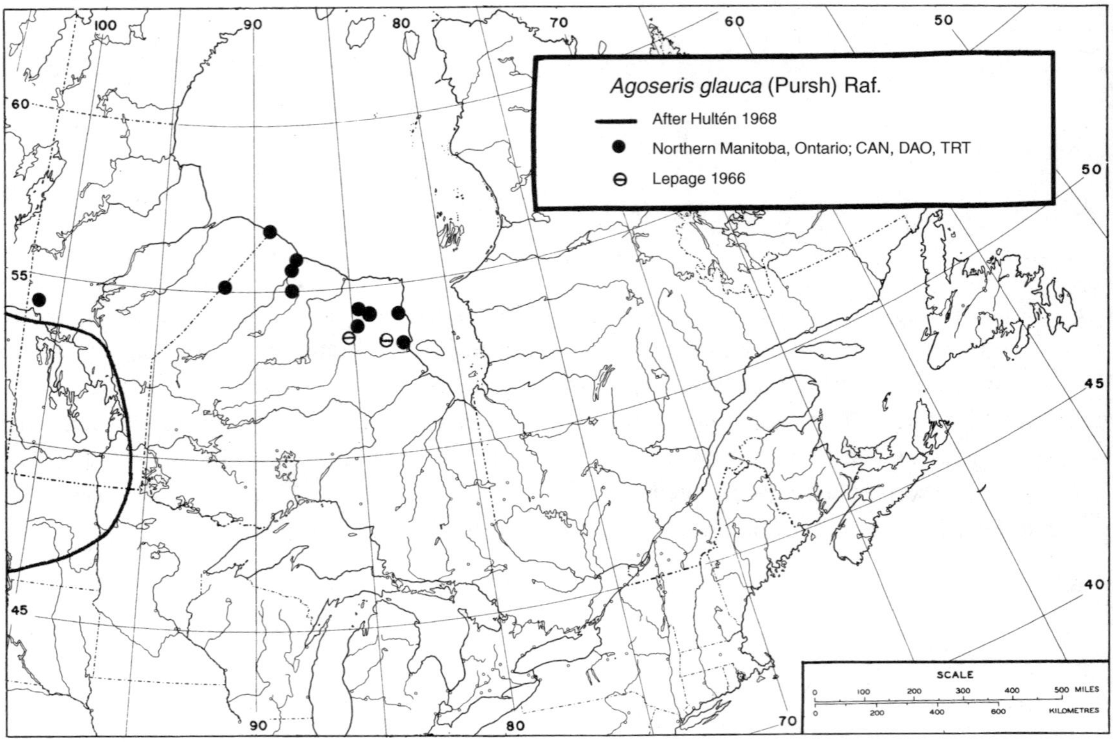

Fig. A2. Distribution of *Agoseris glauca* in Canada.

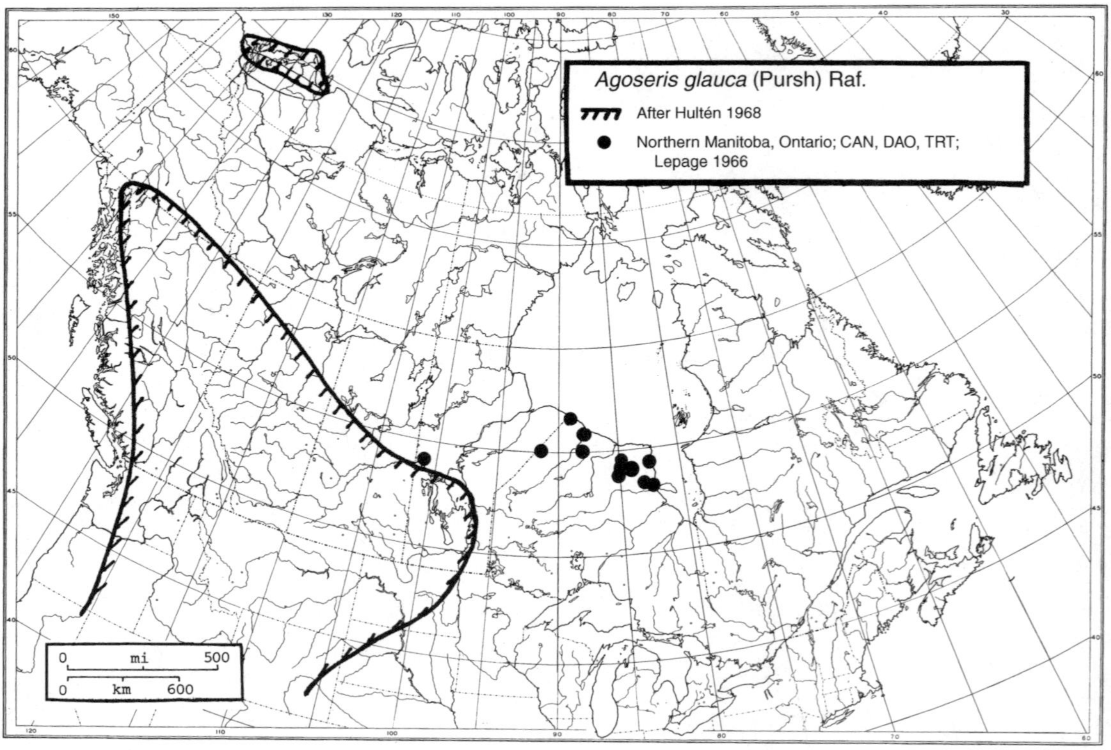

Fig. A3. Distribution of *Agropyron violaceu*m in Canada.

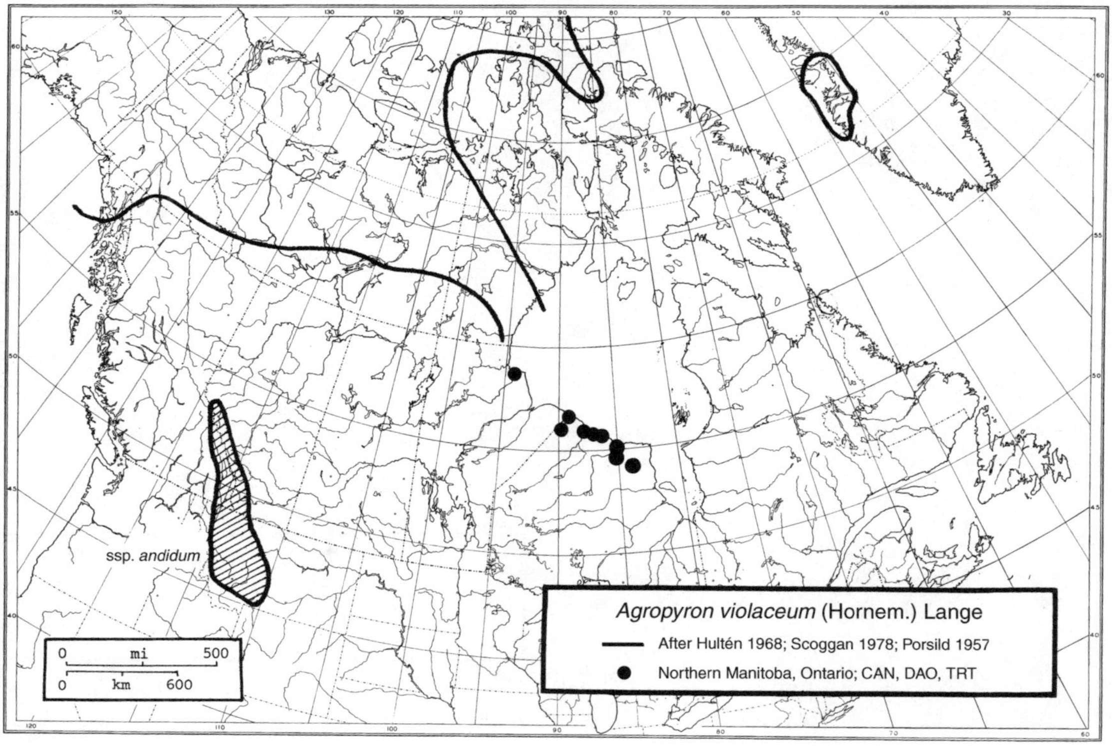

Fig. A4. Distribution of *Agrostis mertensii* in Canada.

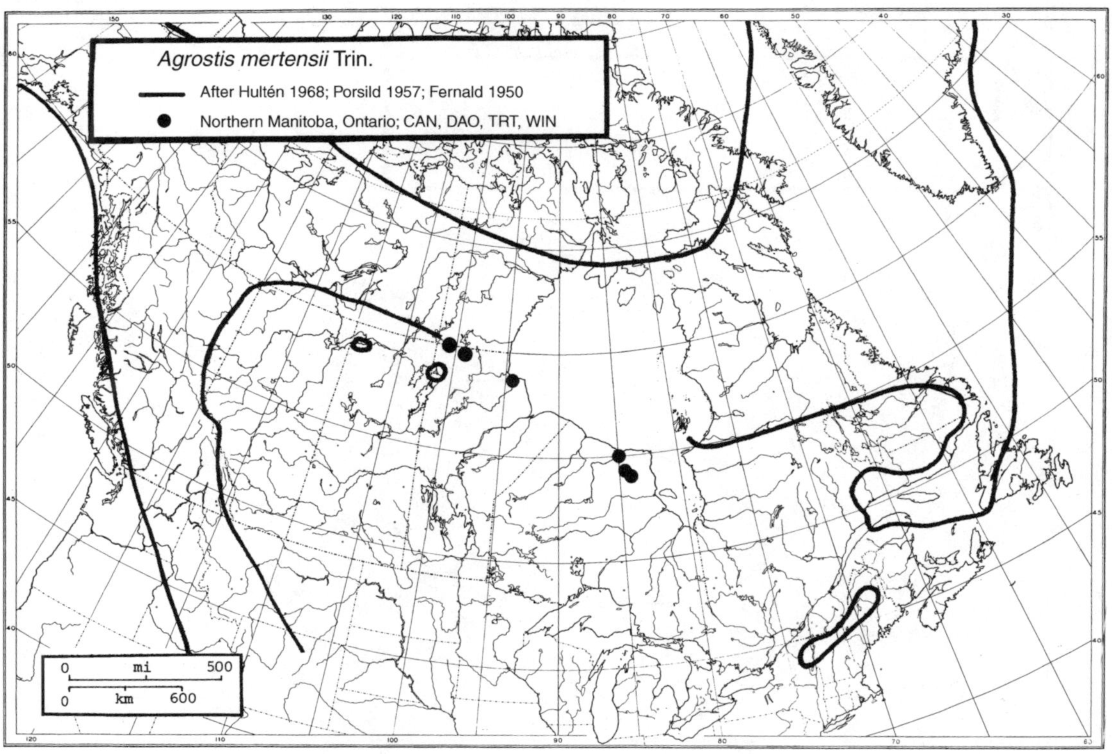

Fig. A5. Distribution of *Alopecurus alpinus* in eastern Canada.

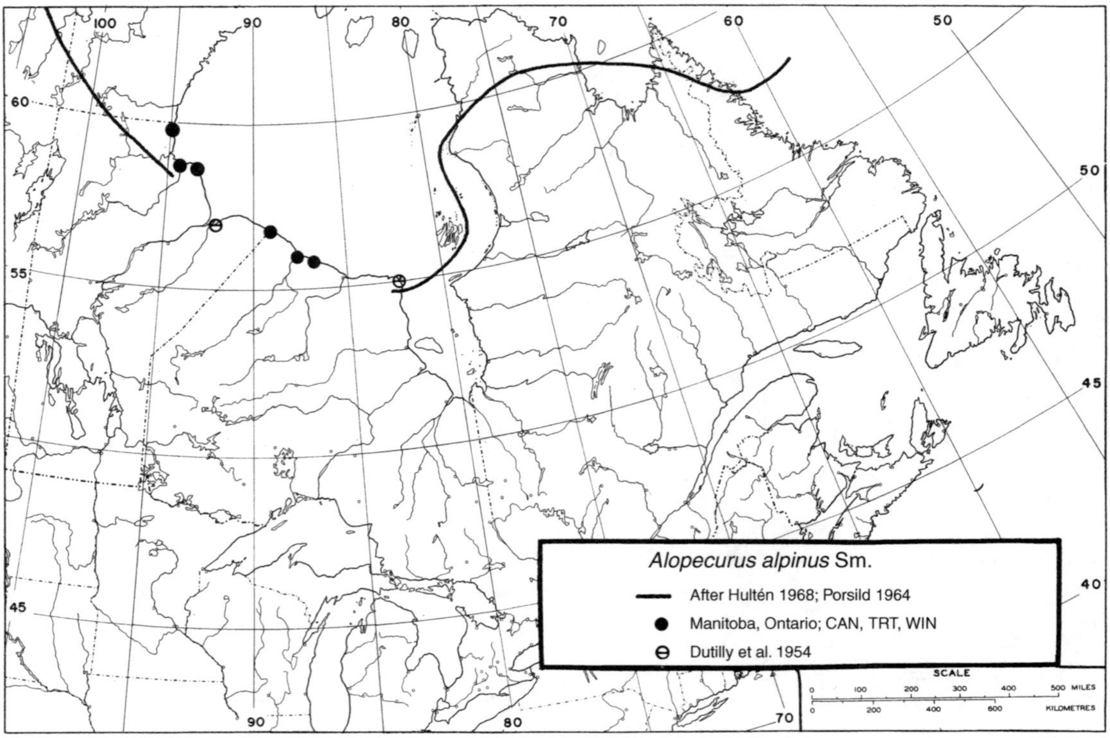

Fig. A6. Distribution of *Andropogon gerardii* in eastern Canada.

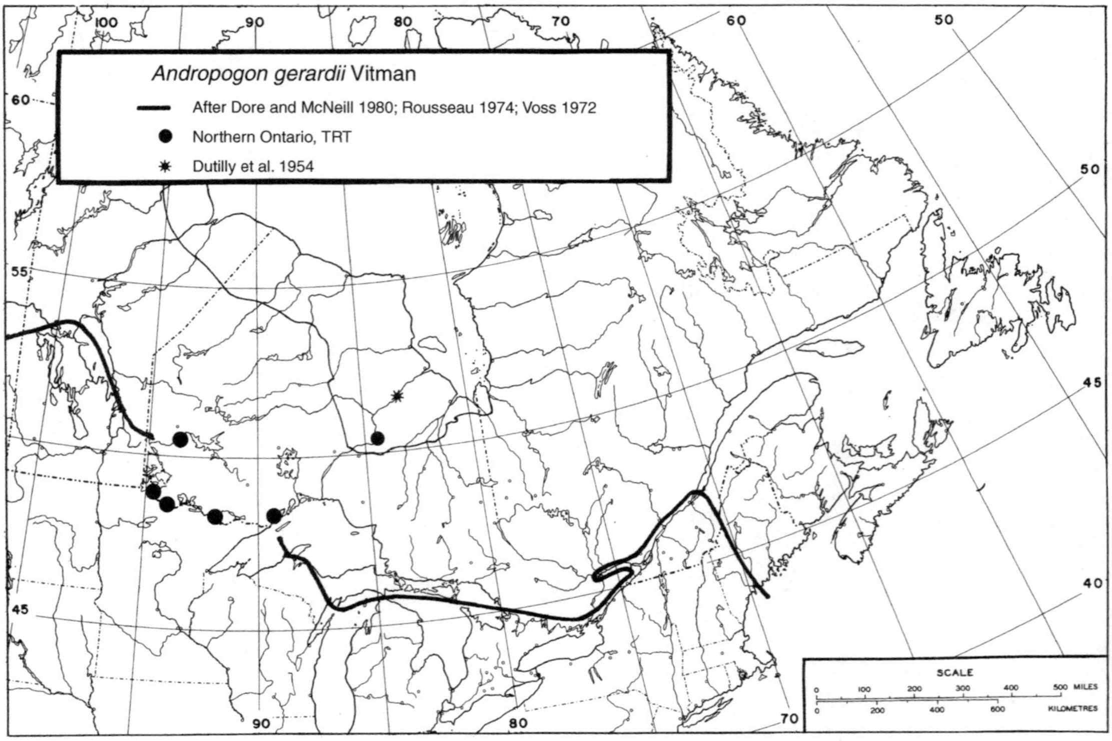

Fig. A7. Distribution of *Androsace septentrionalis* in eastern Canada.

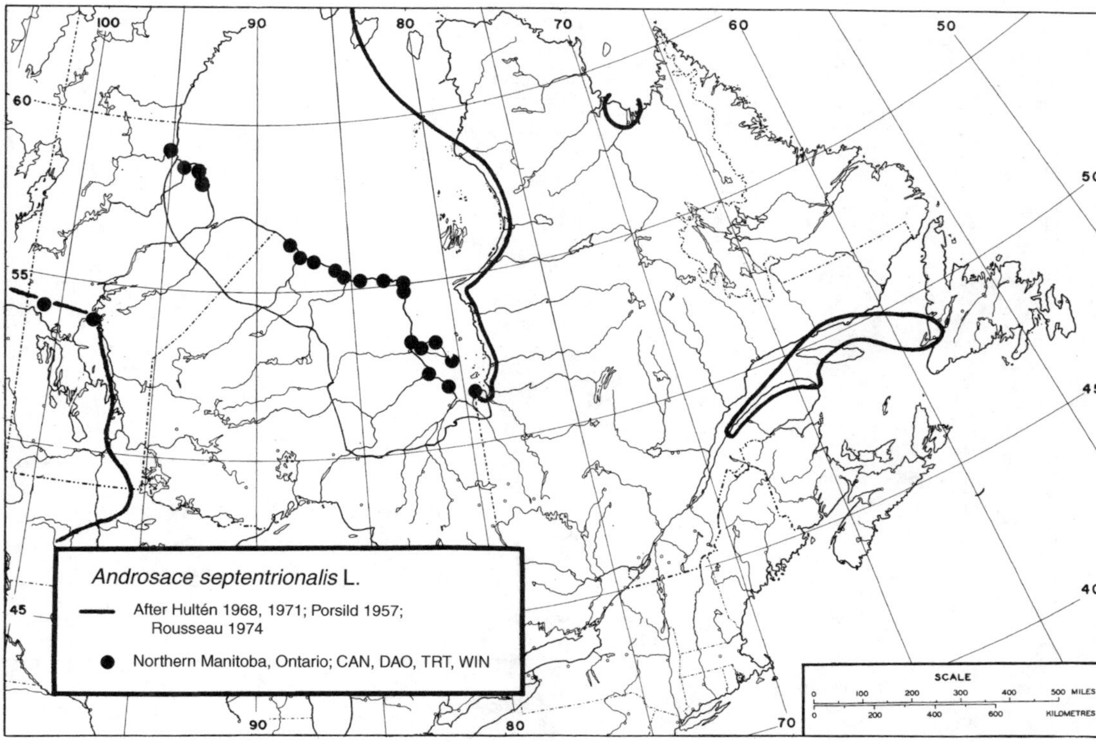

Fig. A8. Distribution of *Antennaria pulcherrima* in eastern Canada.

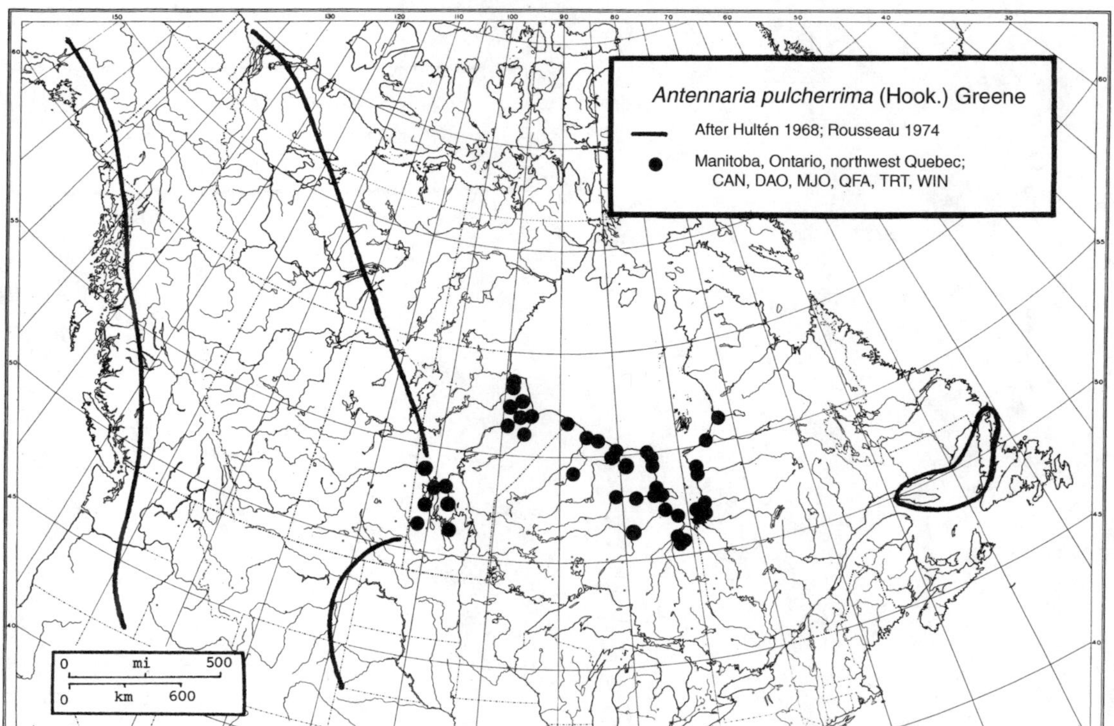

Fig. A9. Distribution of *Antennaria rosea* in eastern Canada.

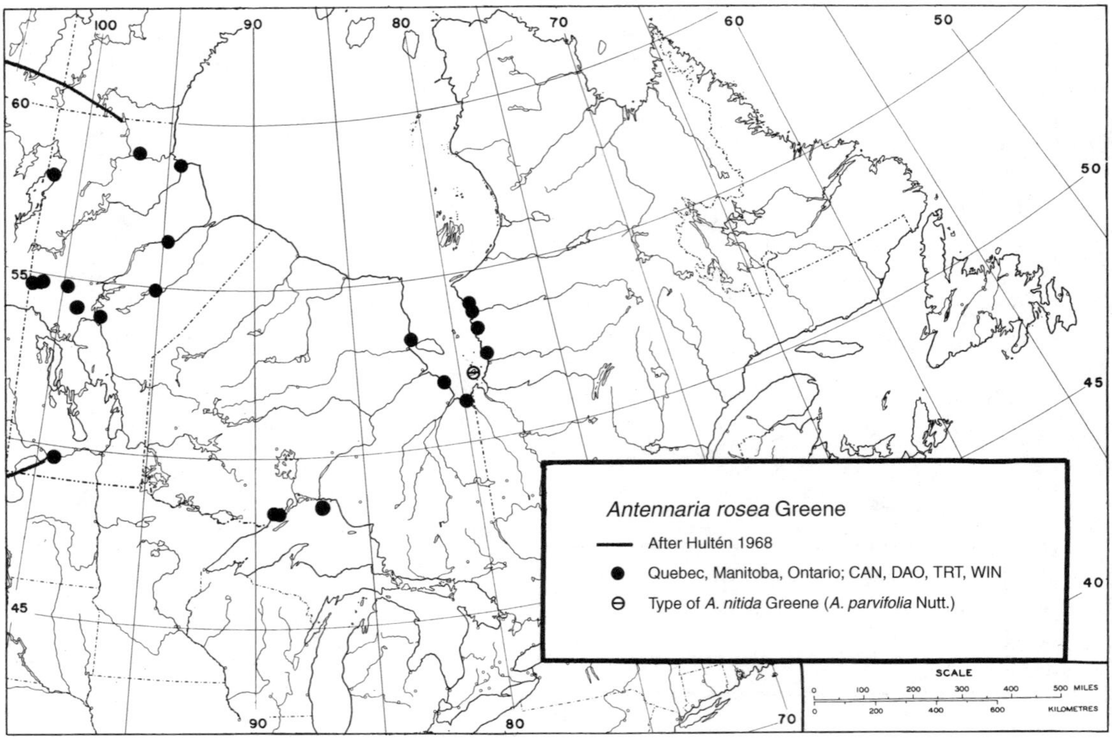

Fig. A10. Distribution of *Aquilegia brevistyla* in eastern Canada.

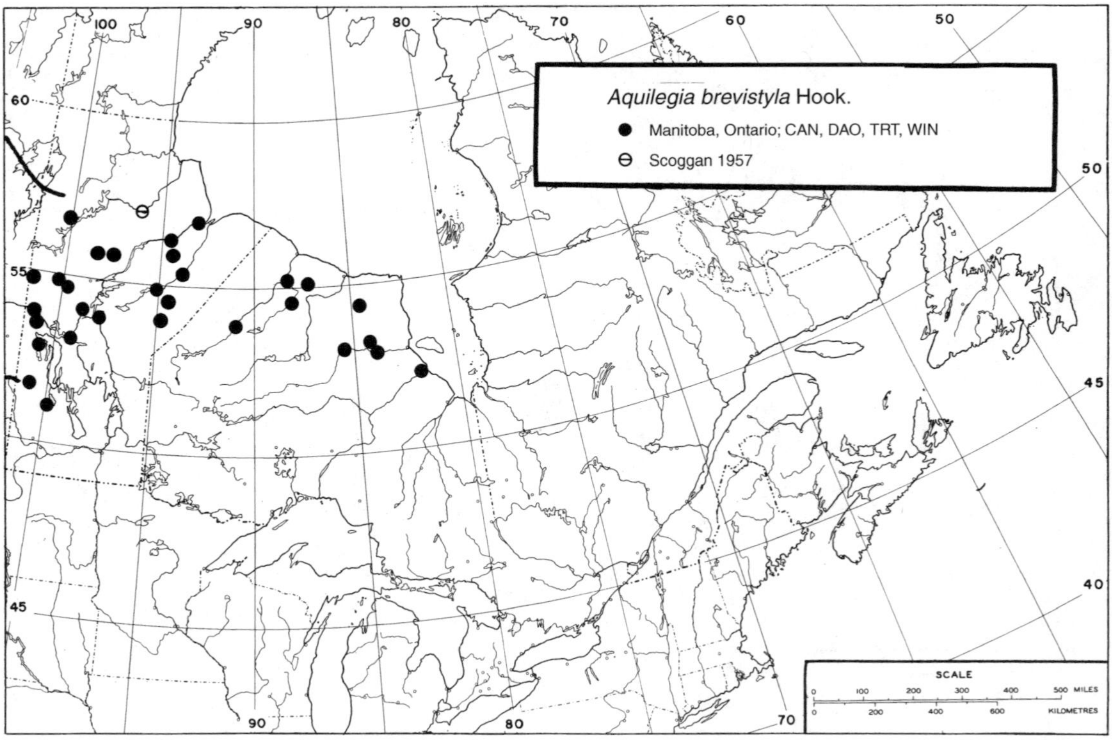

Fig. A11. Distribution of *Aquilegia brevistyla* in Canada.

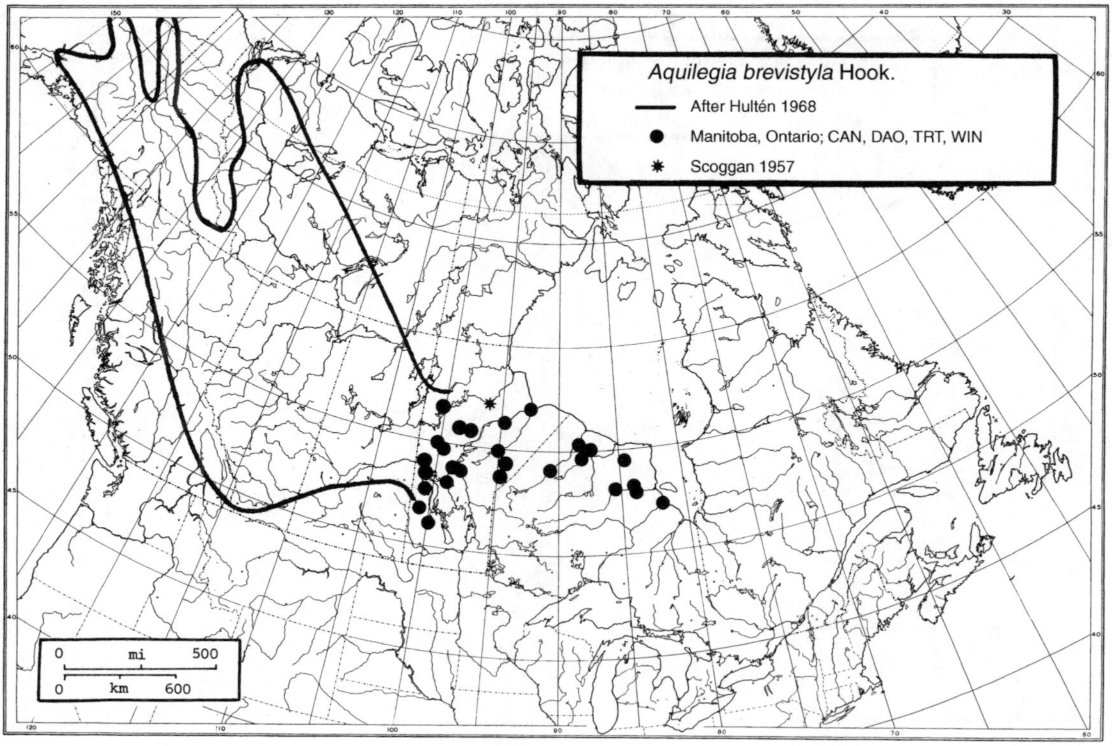

Fig. A12. Distribution of *Arctagrostis latifolia* in eastern Canada.

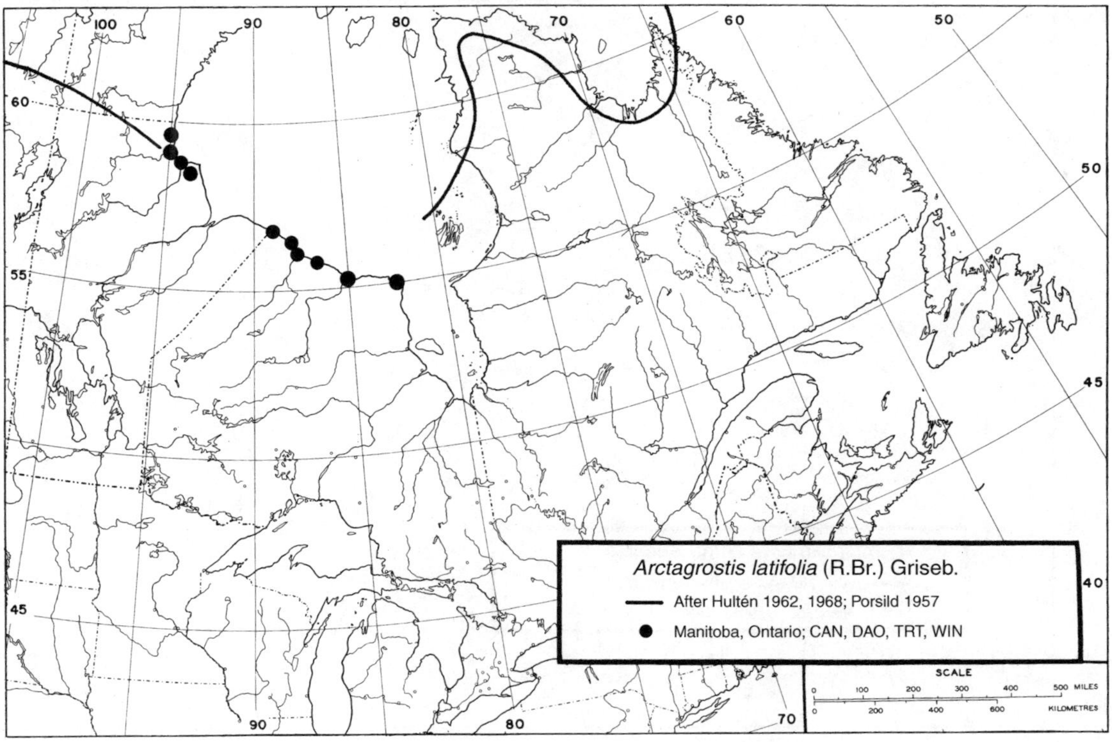

Fig. A13. Distribution of *Arctagrostis latifolia* in Canada.

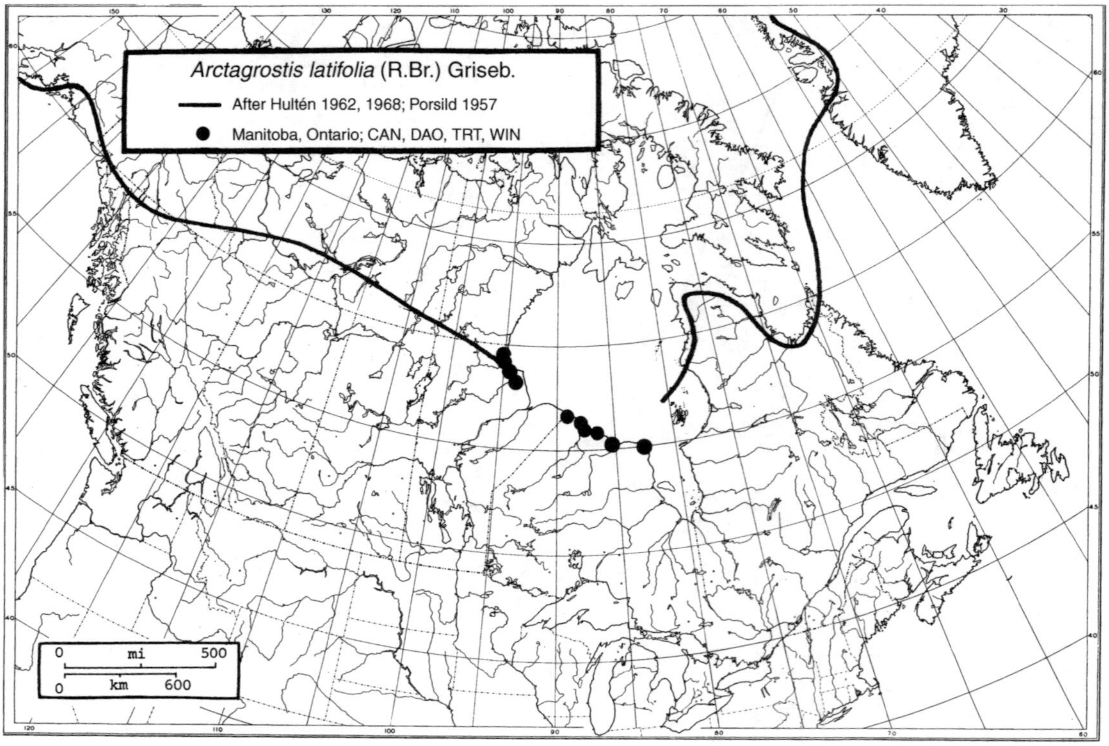

Fig. A14. Distribution of *Arctophila fulva* in eastern Canada.

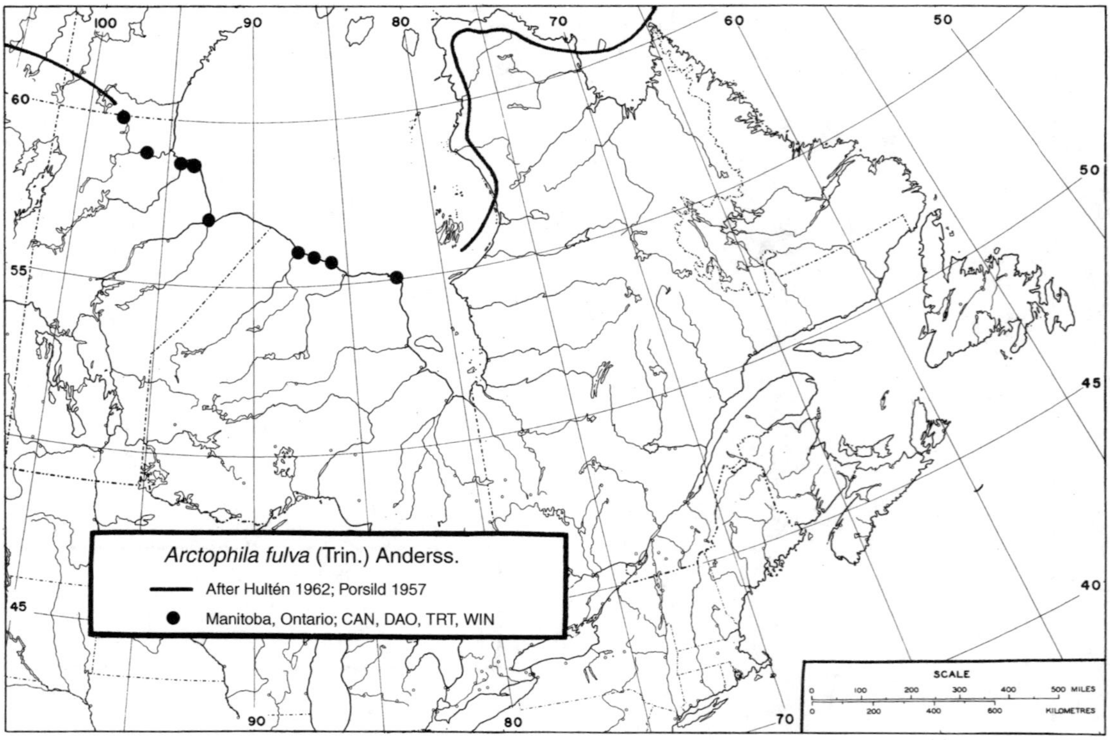

Fig. A15. Distribution of *Arethusa bulbosa* in Canada.

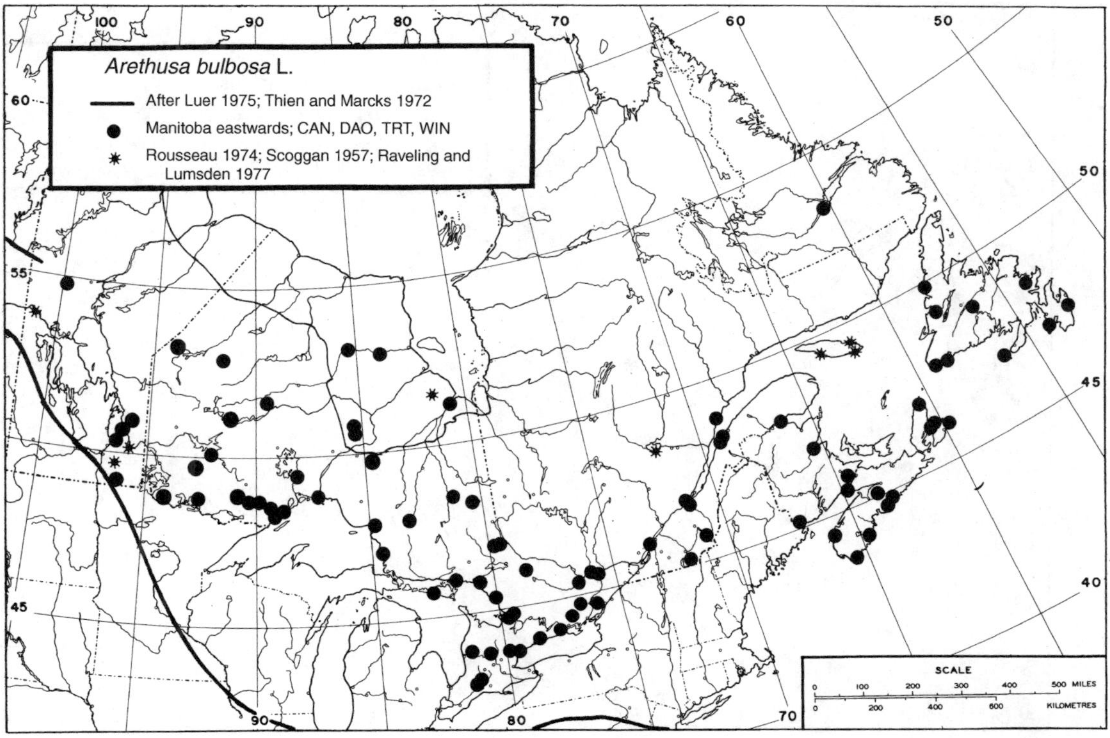

Fig. A16. Distribution of *Arnica attenuata* in eastern Canada.

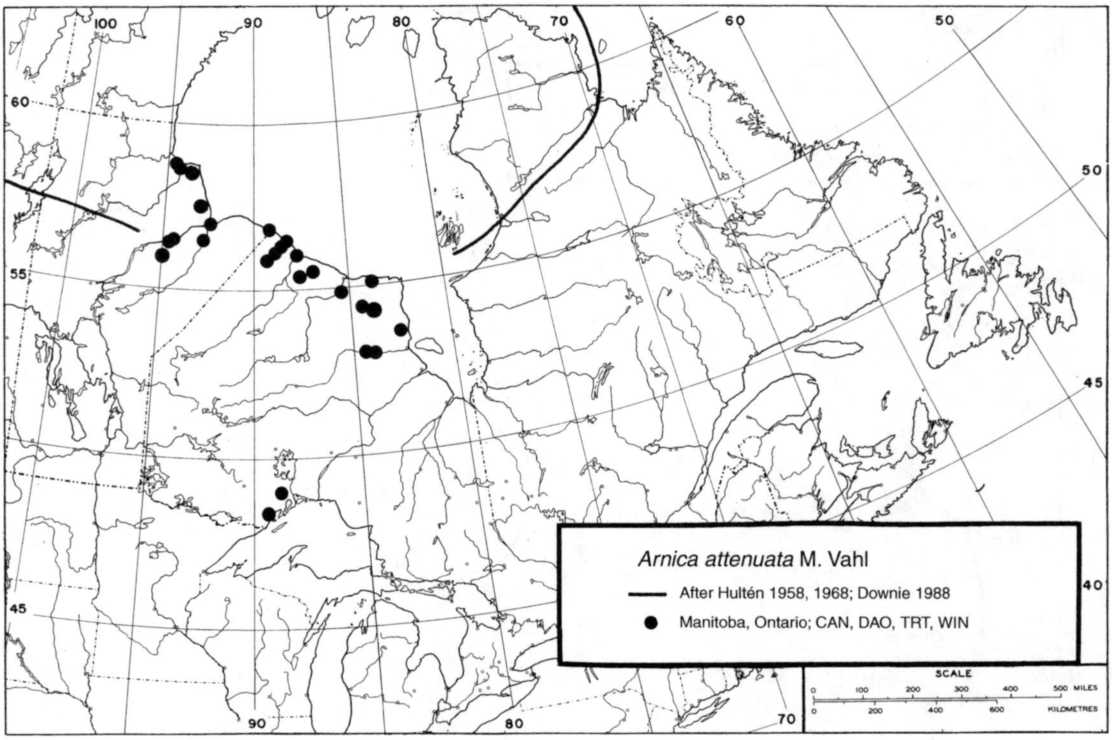

Fig. A17. Distribution of *Arnica chamissonis* in eastern Canada.

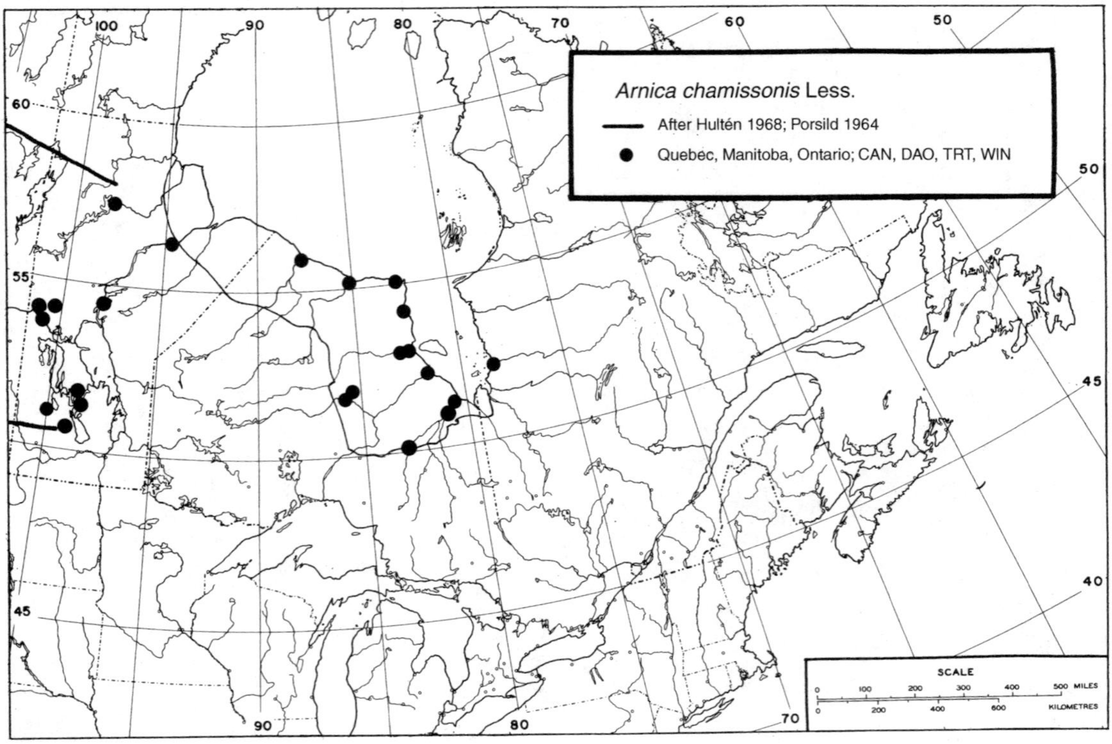

Fig. A18. Distribution of *Aronia prunifolia* in eastern Canada.

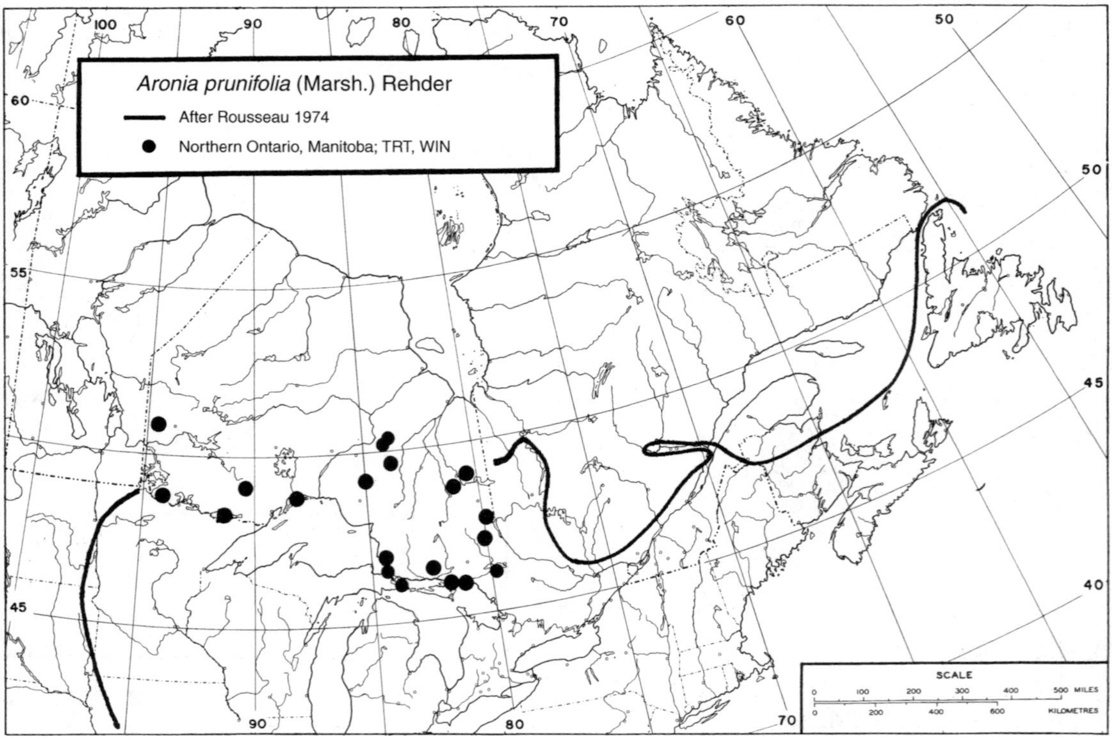

Fig. A19. Distribution of *Artemisia tilesii* in Canada.

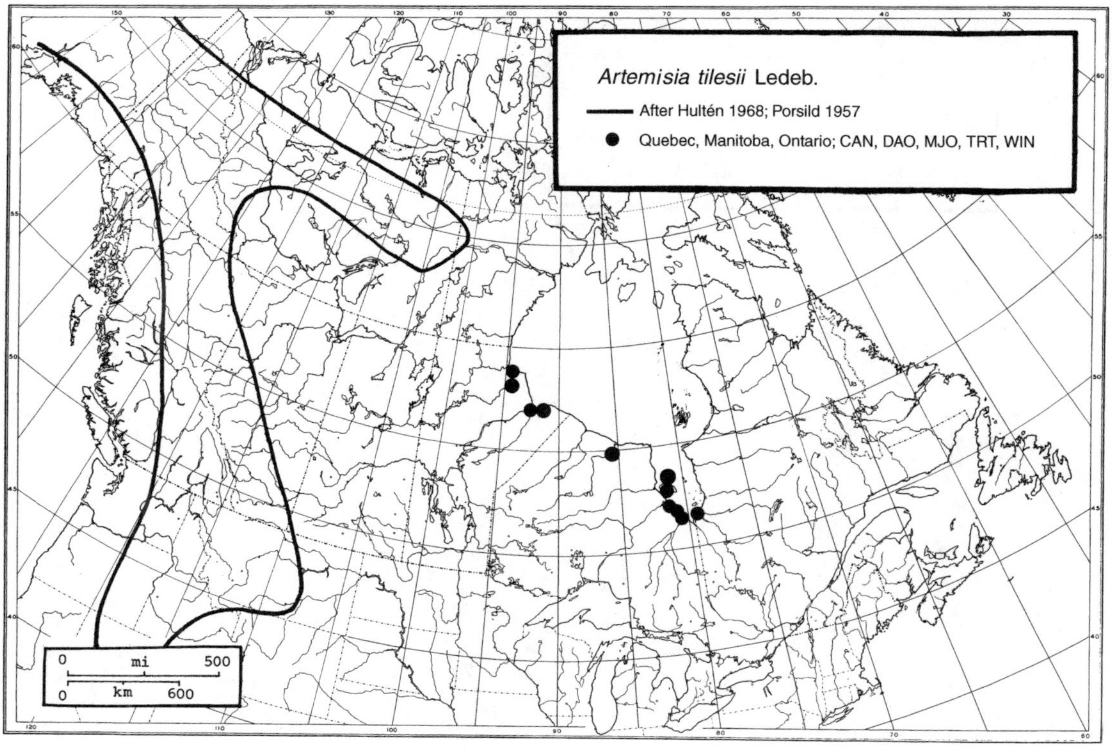

Fig. A20. Distribution of *Aster alpinus* in Canada.

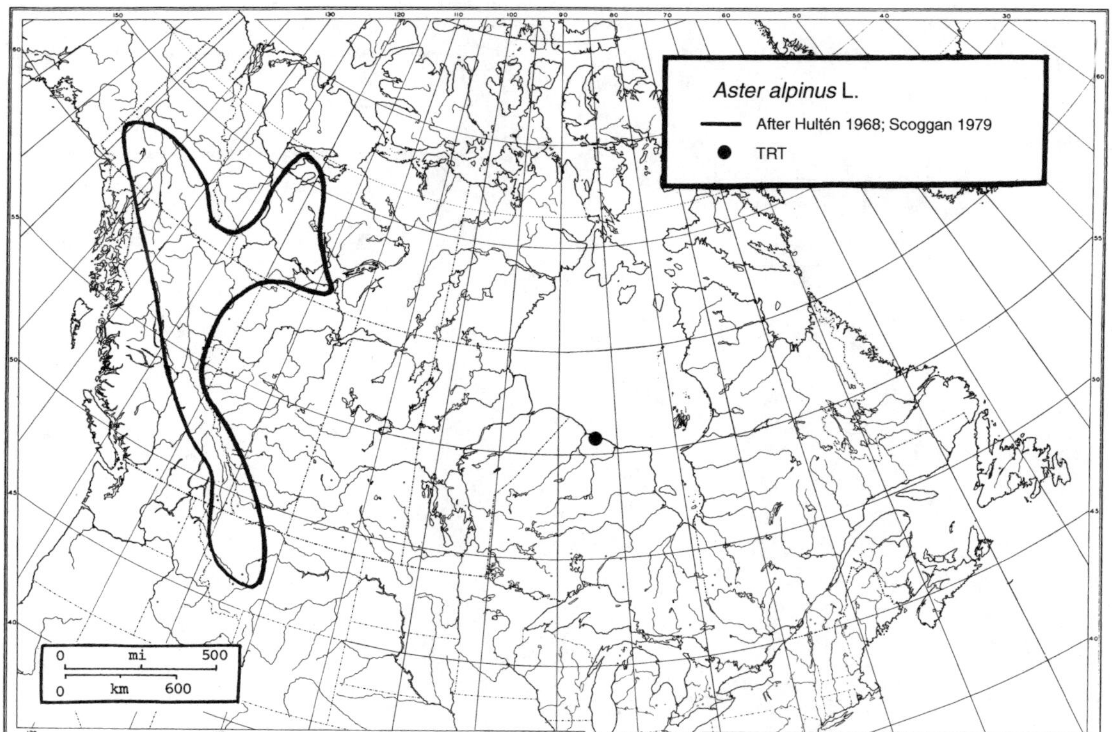

Fig. A21. Distribution of *Aster brachyactis* in eastern Canada.

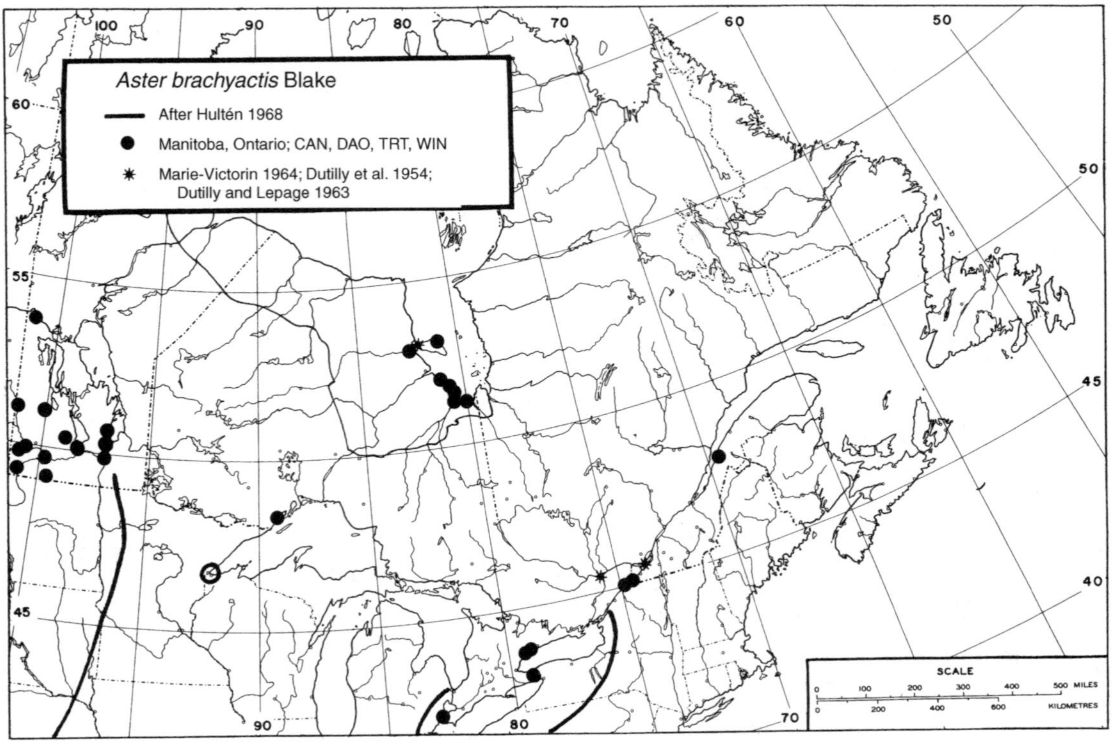

Fig. A22. Distribution of *Astragalus agrestis* in Canada.

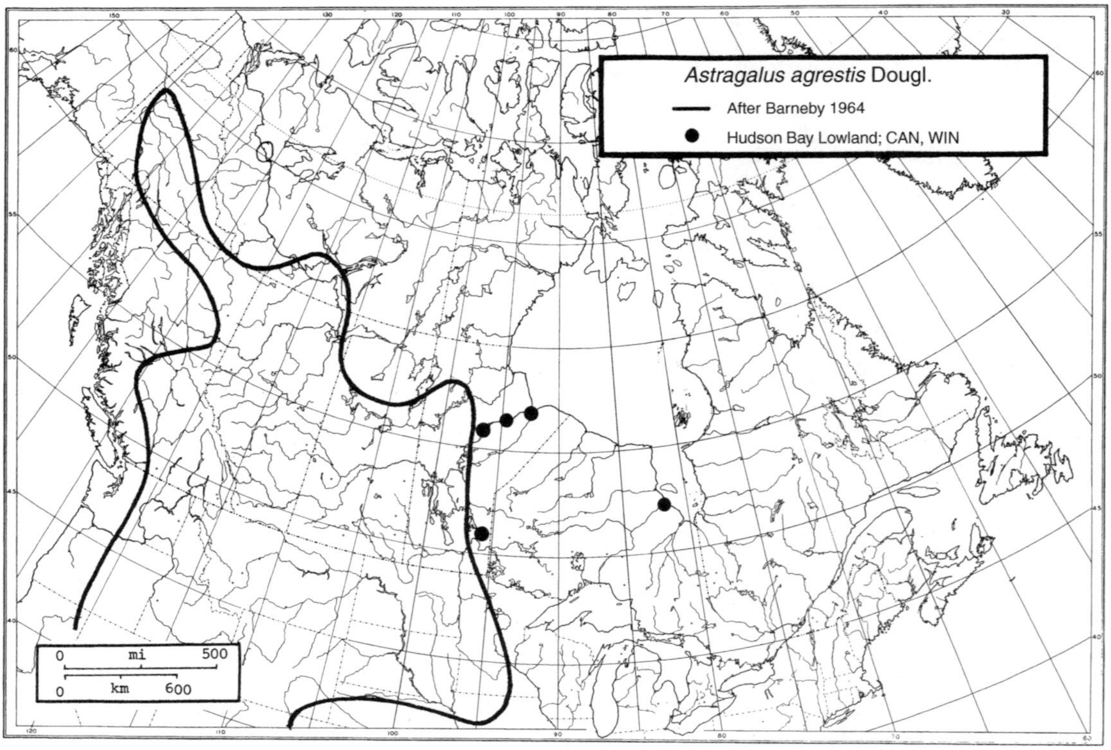

Fig. A23. Distribution of *Astragalus alpinus* in Canada.

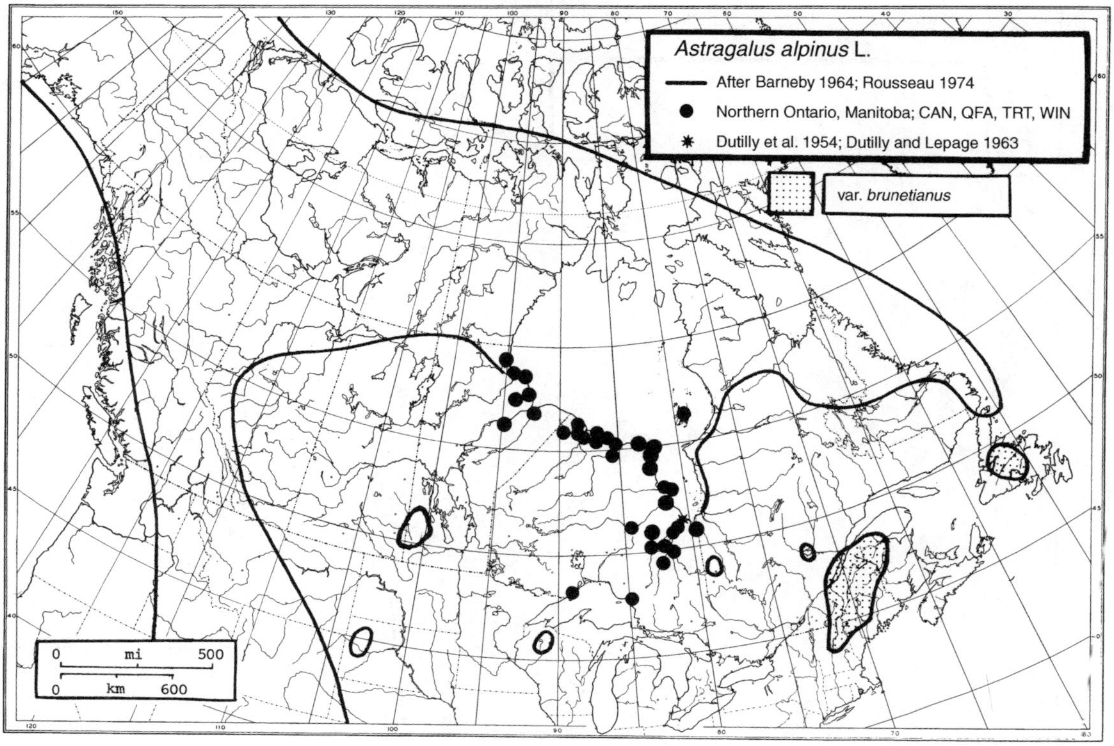

Fig. A24. Distribution of *Astragalus americanus* in Canada.

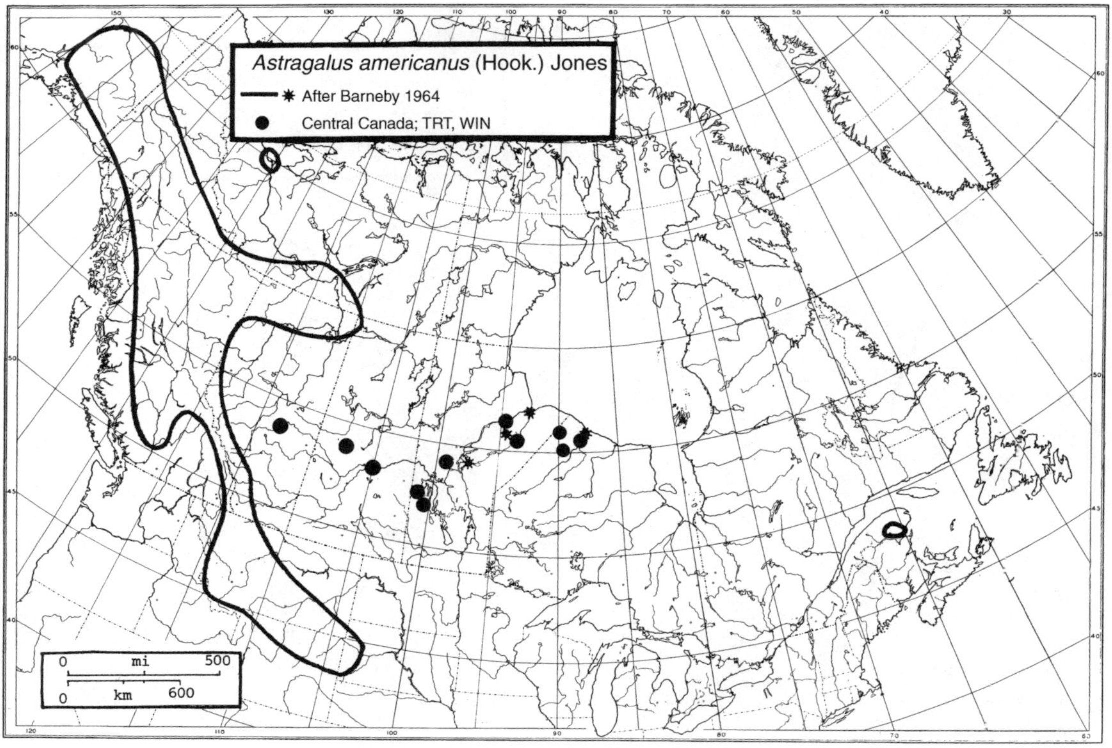

Fig. A25. Distribution of *Astragalus canadensis* in eastern Canada.

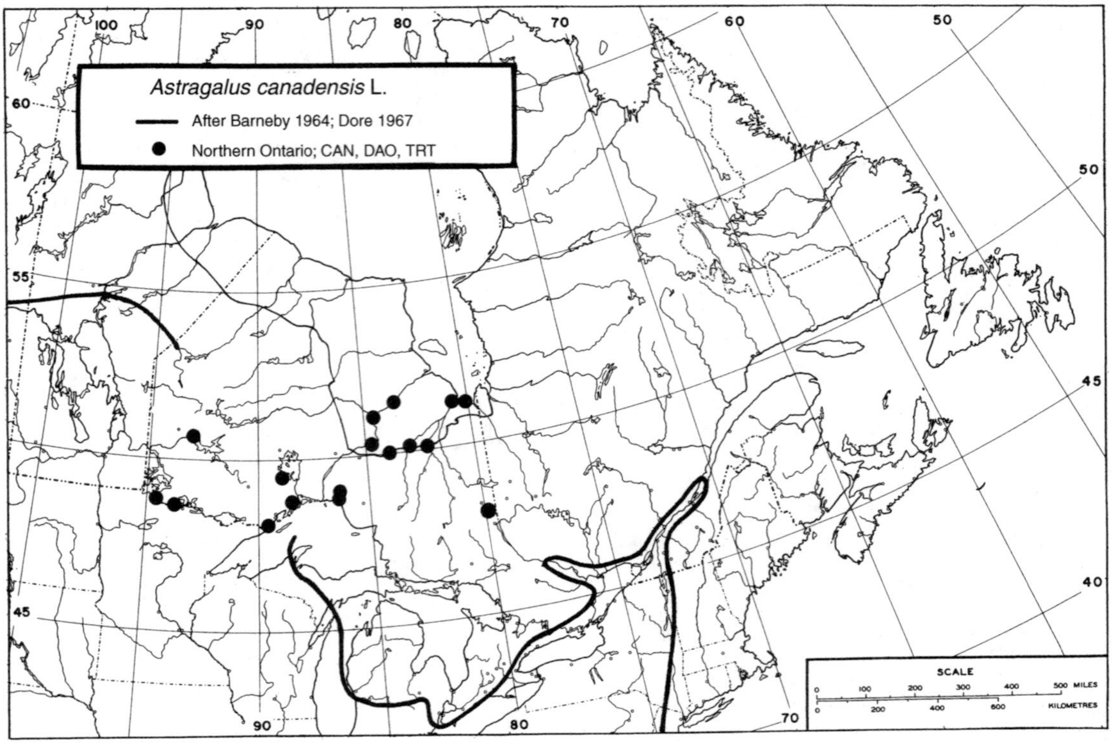

Fig. A26. Distribution of *Astragalus eucosmus* in Canada.

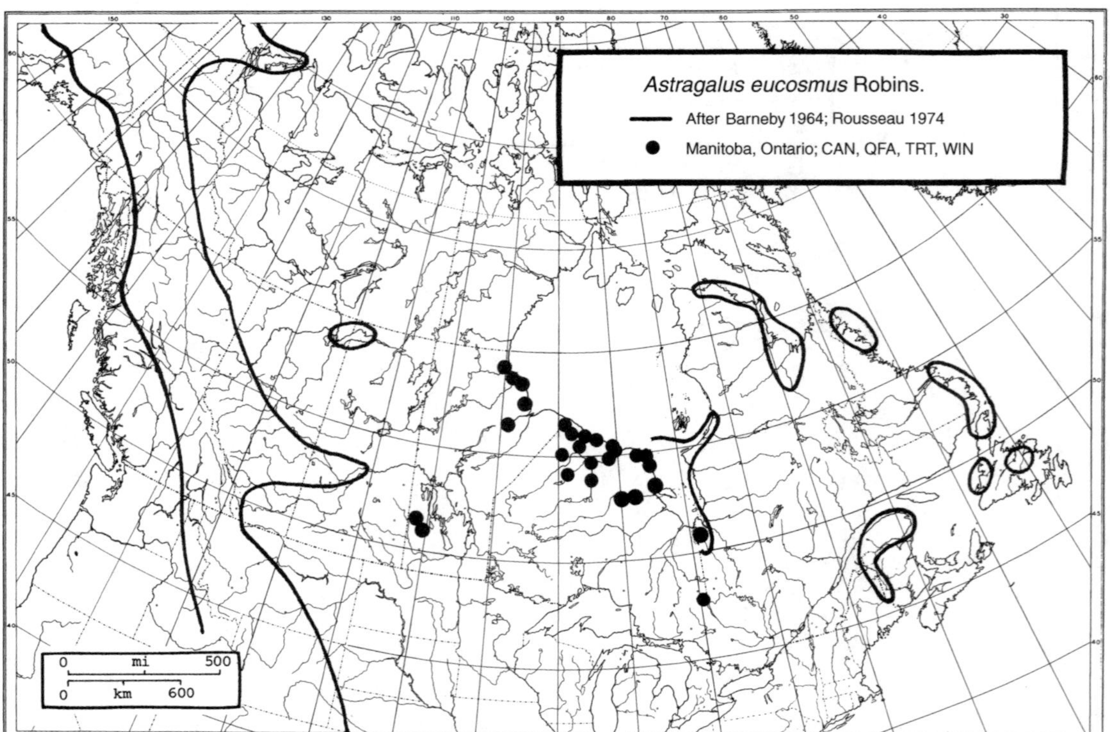

Fig. A27. Distribution of *Astragalus tenellus* in Canada.

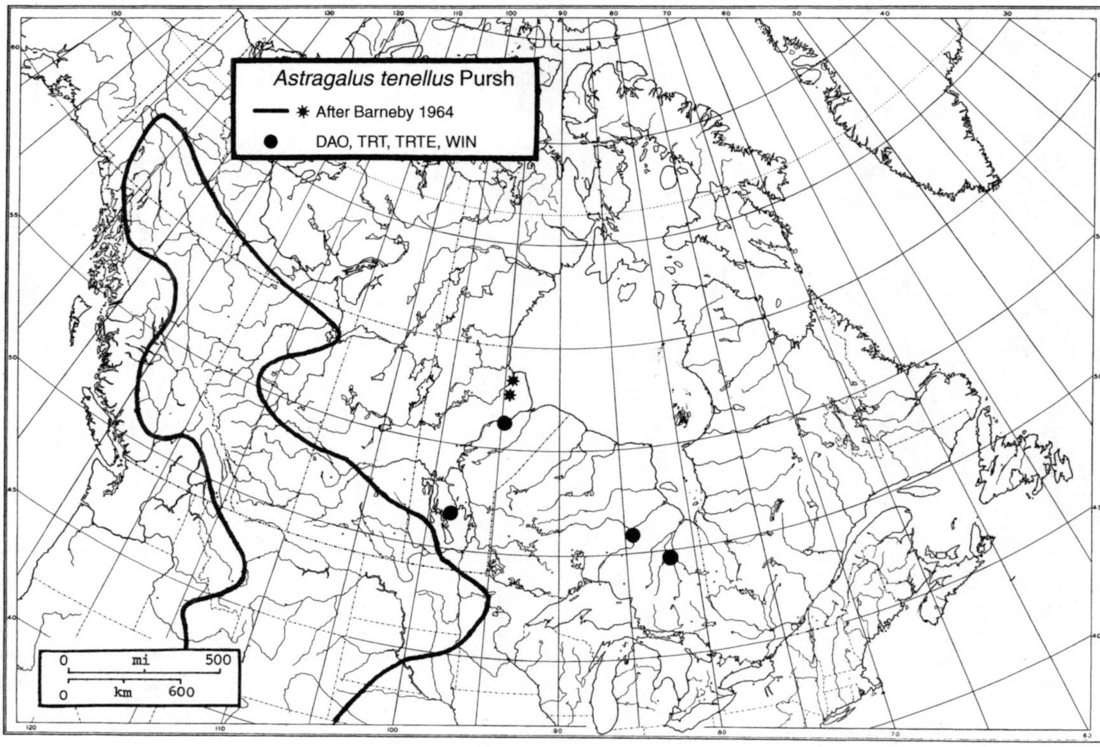

Fig. A28. Distribution of *Atriplex glabriuscula* in eastern Canada.

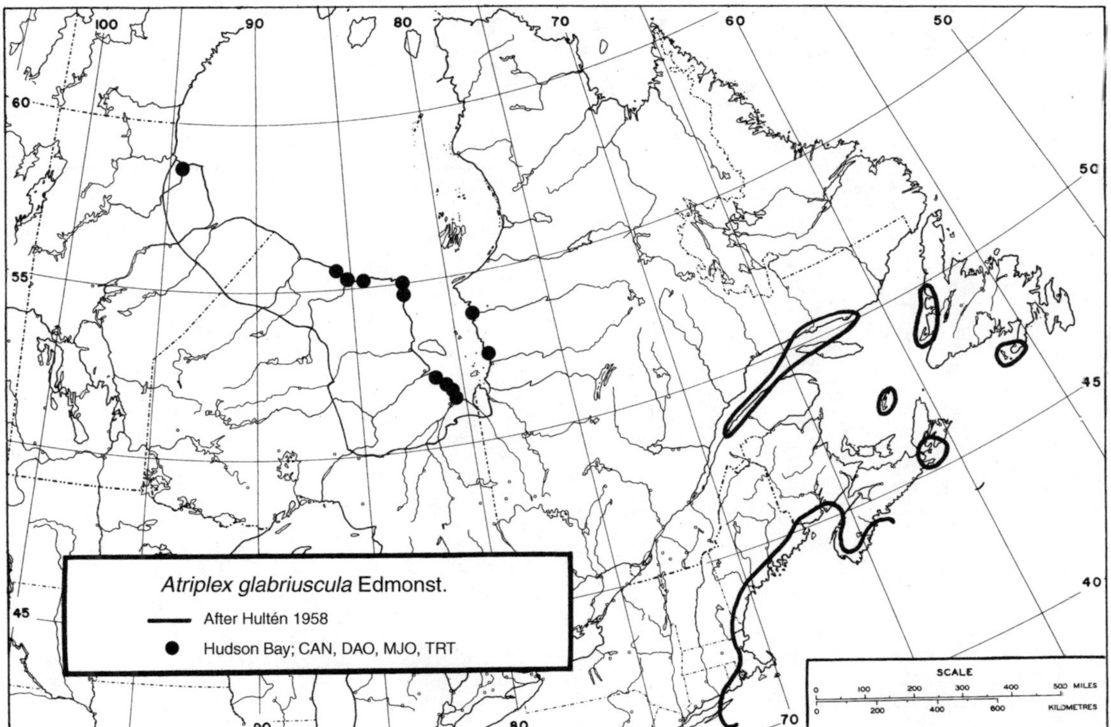

Fig. A29. Distribution of *Bartsia alpina* in eastern Canada.

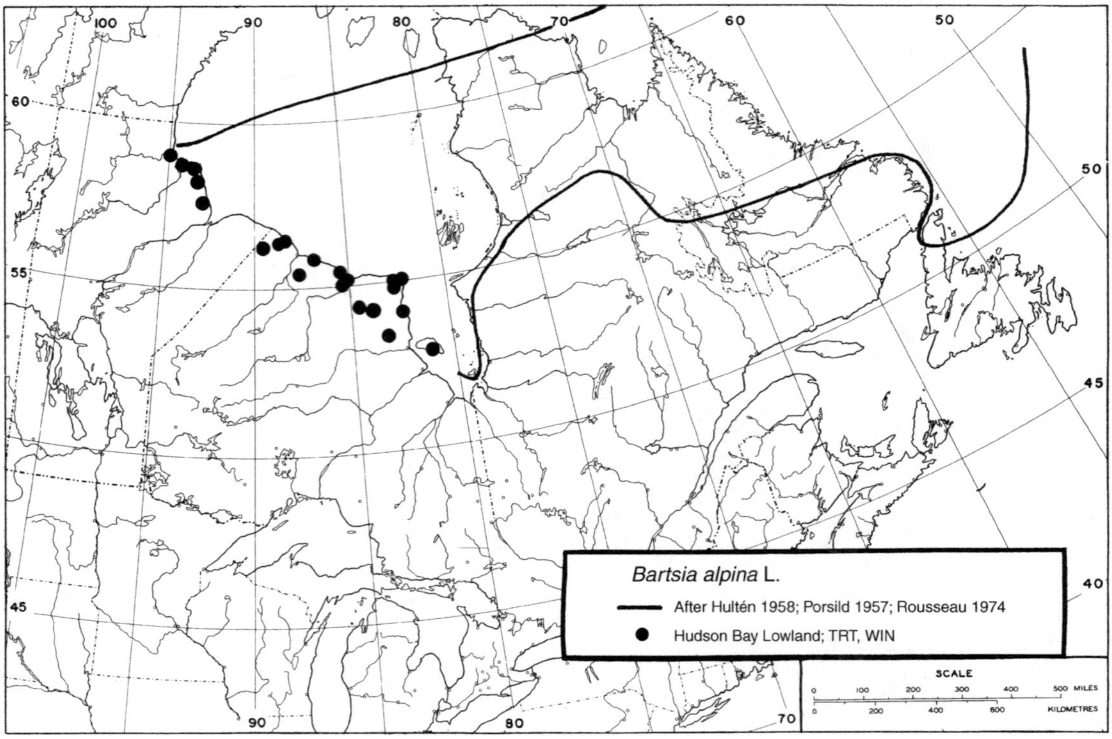

Fig. A30. Distribution of *Calamagrostis deschampsioides* in Canada.

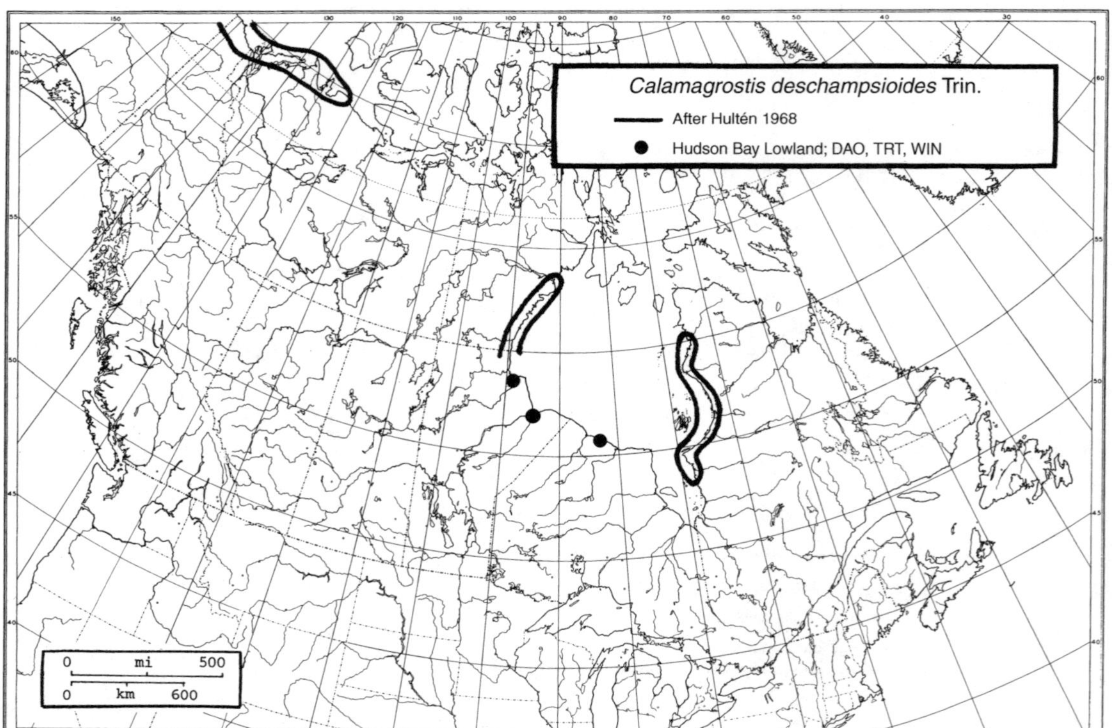

Fig. A31. Distribution of *Carex bigelowii* in Canada.

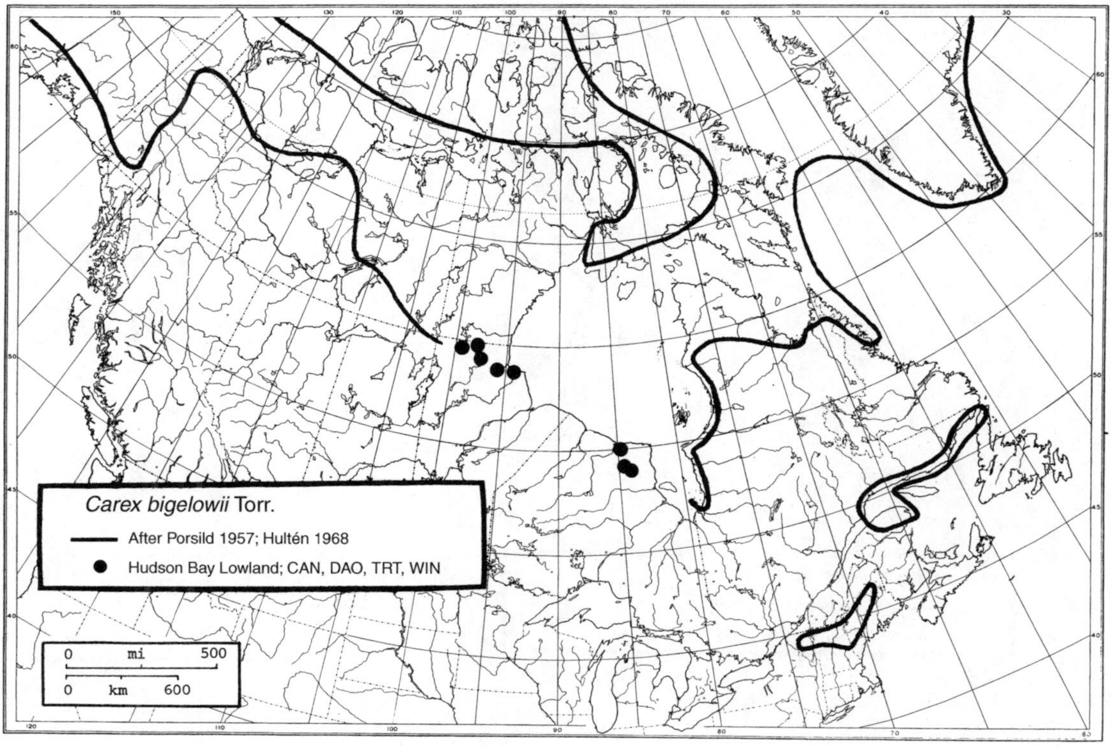

Fig. A32. Distribution of *Carex glacialis* in eastern Canada.

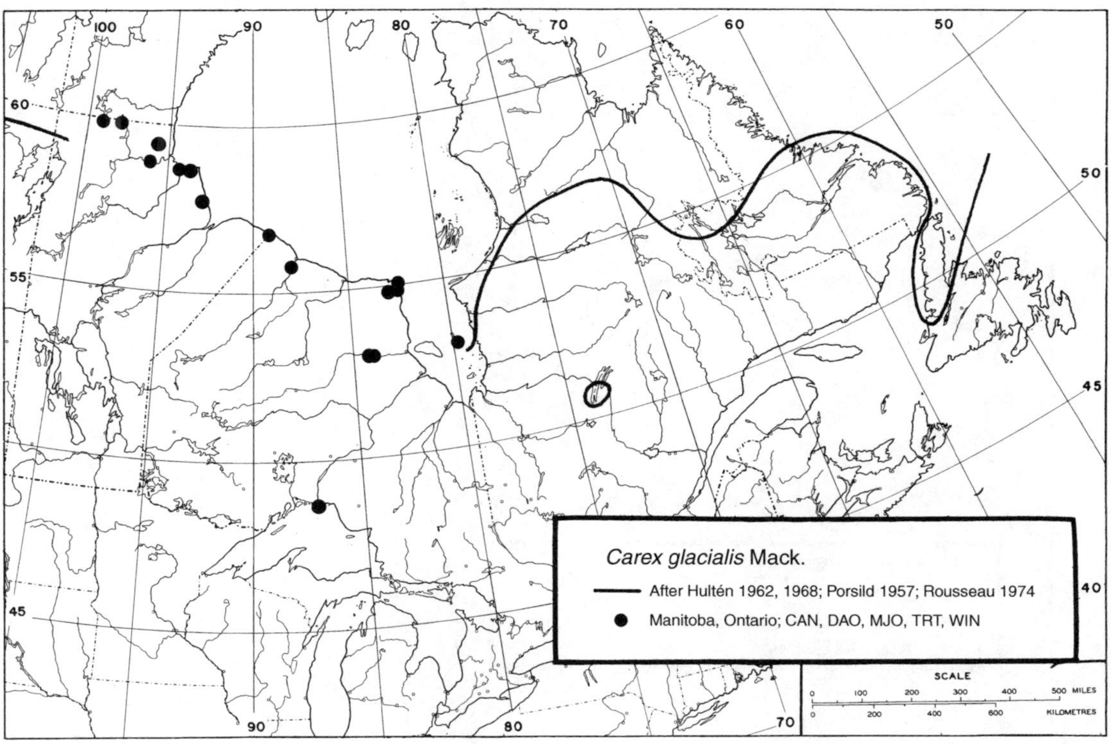

Fig. A33. Distribution of *Carex glareosa* in eastern Canada.

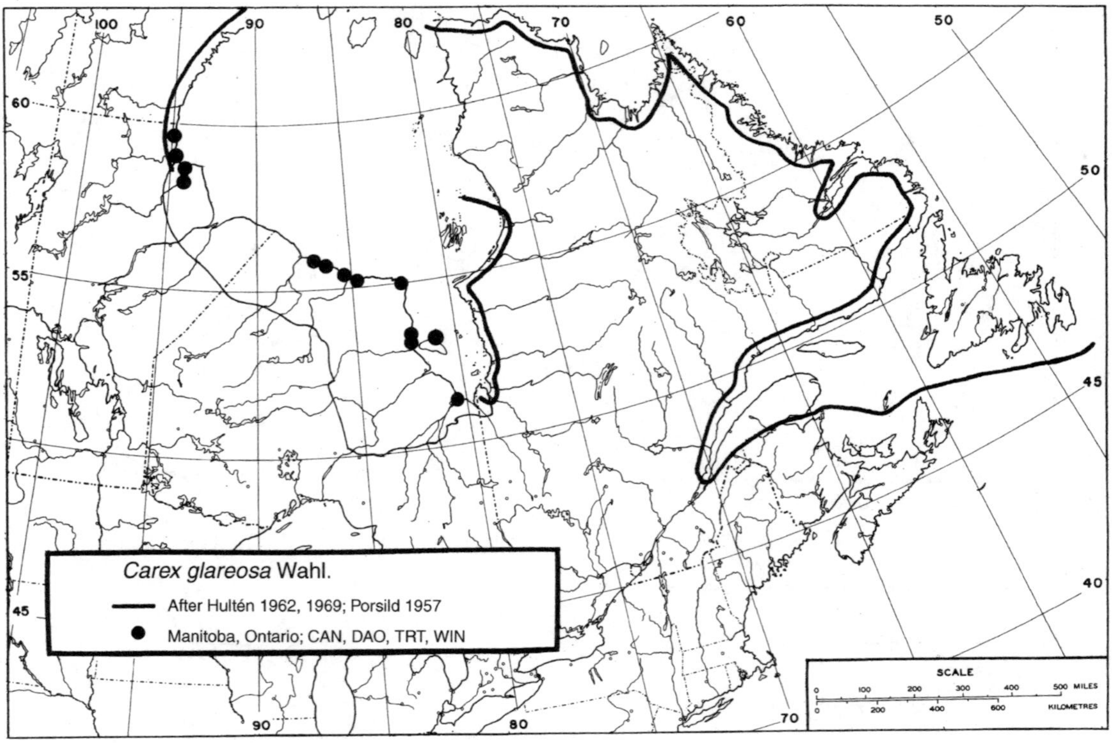

Fig. A34. Distribution of *Carex mackenziei* in eastern Canada.

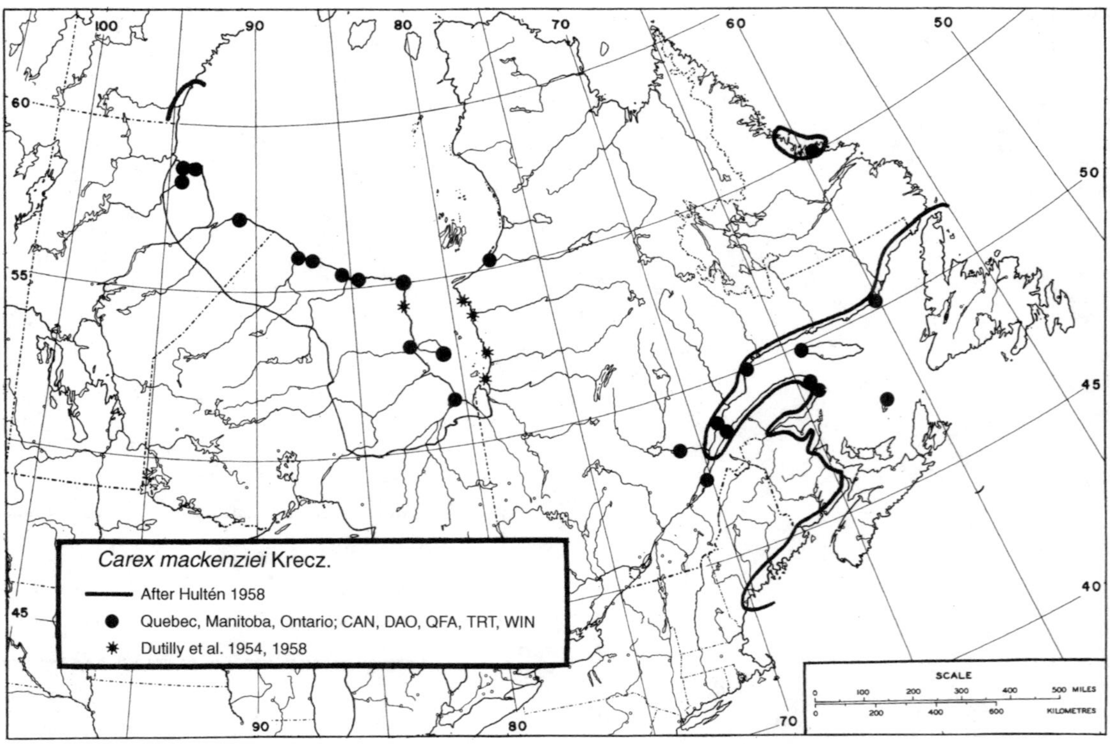

Fig. A35. Distribution of *Carex maritima* in eastern Canada.

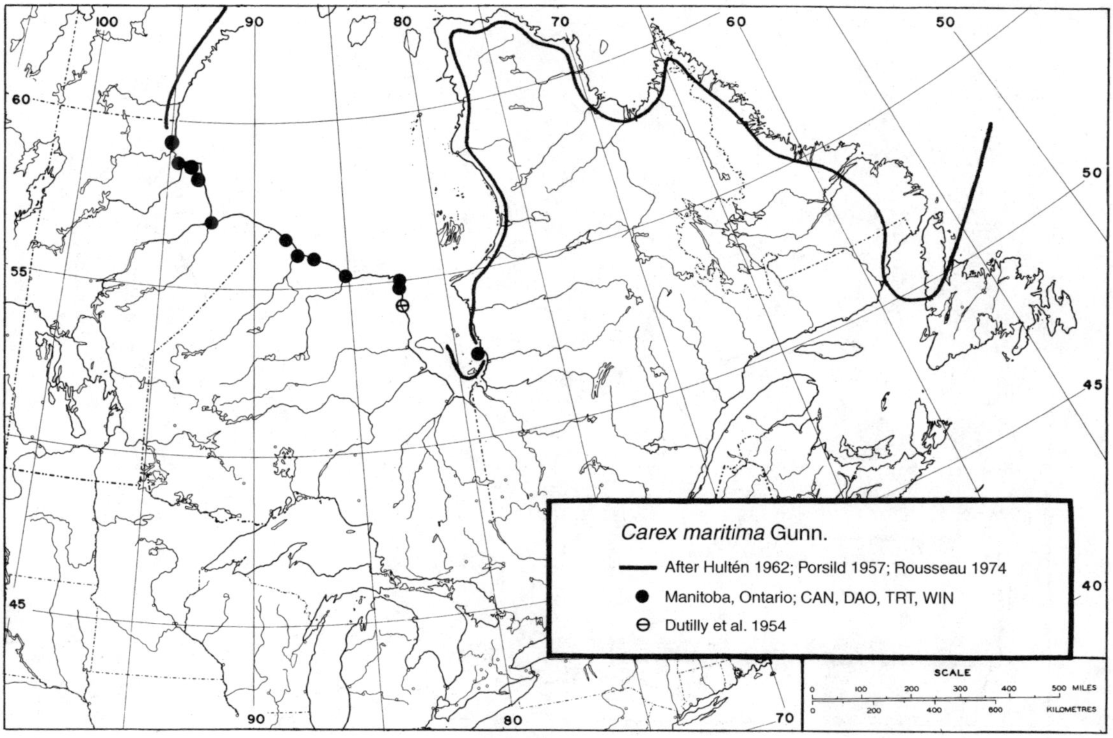

Fig. A36. Distribution of *Carex microglochin* in eastern Canada.

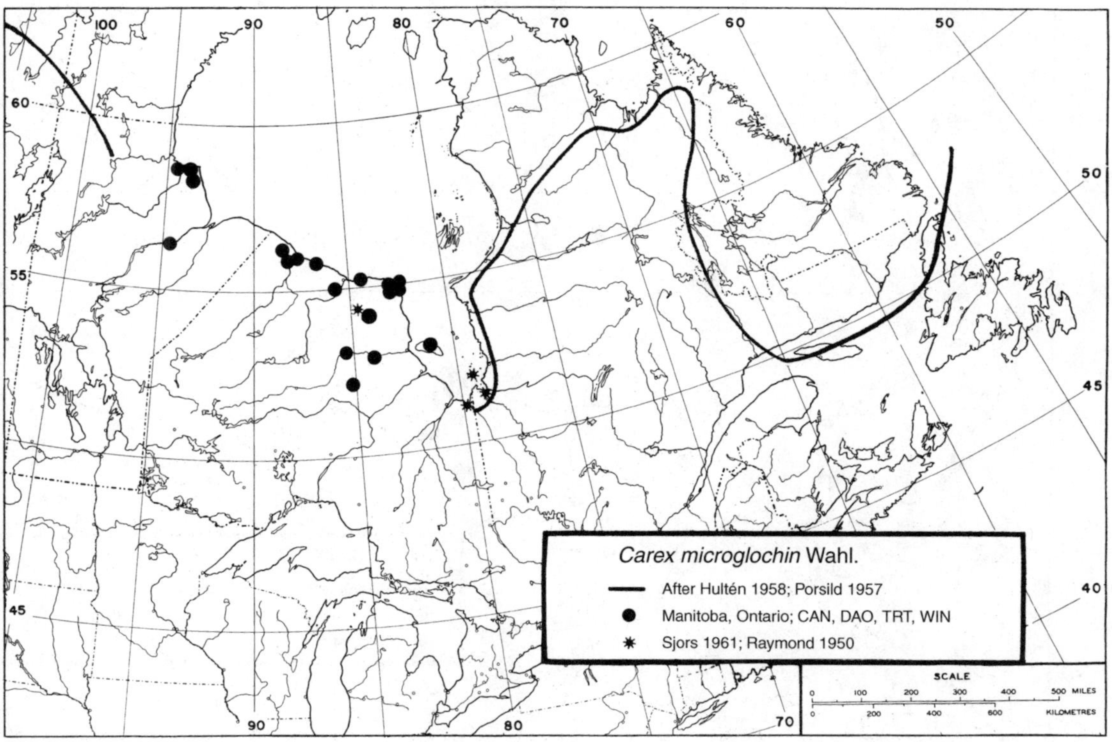

Fig. A37. Distribution of *Carex paleacea* in eastern Canada.

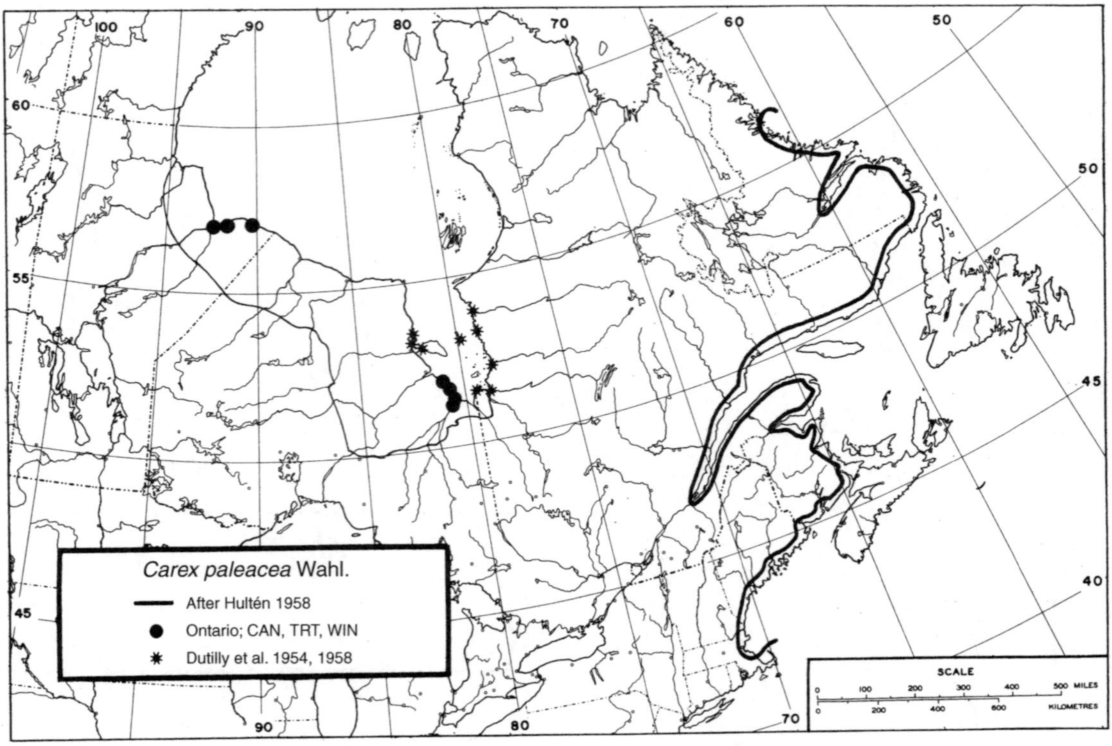

Fig. A38. Distribution of *Carex saxatilis* in eastern Canada.

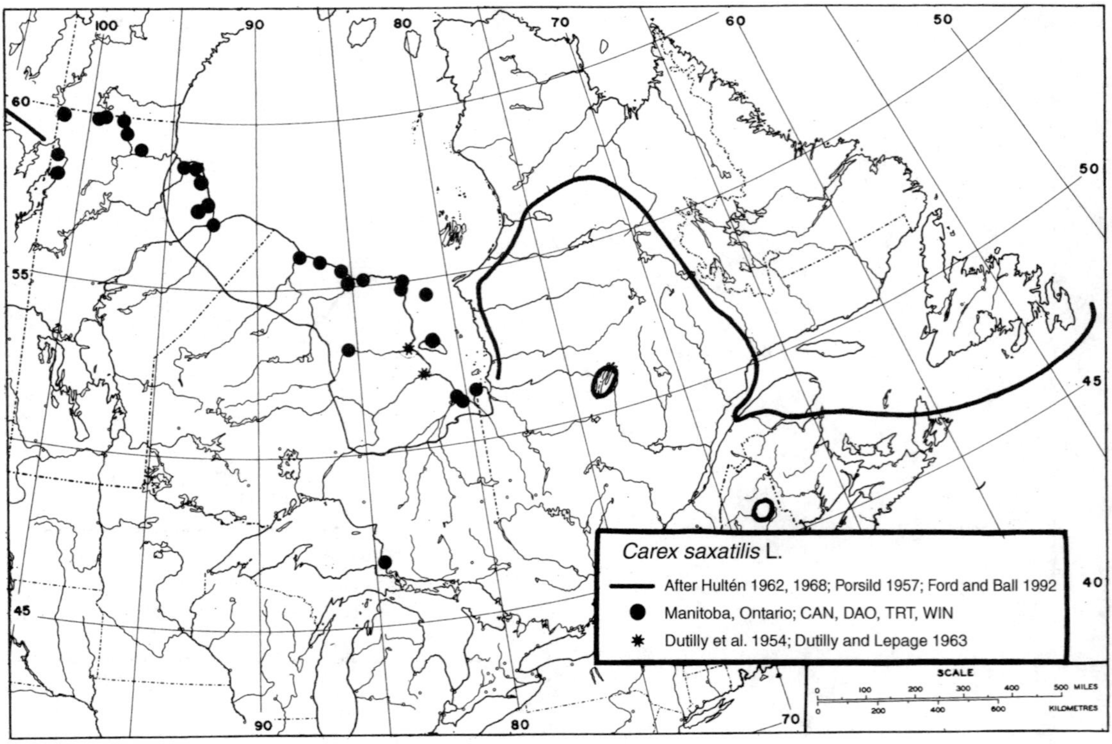

Fig. A39. Distribution of *Carex subspathacea* in eastern Canada.

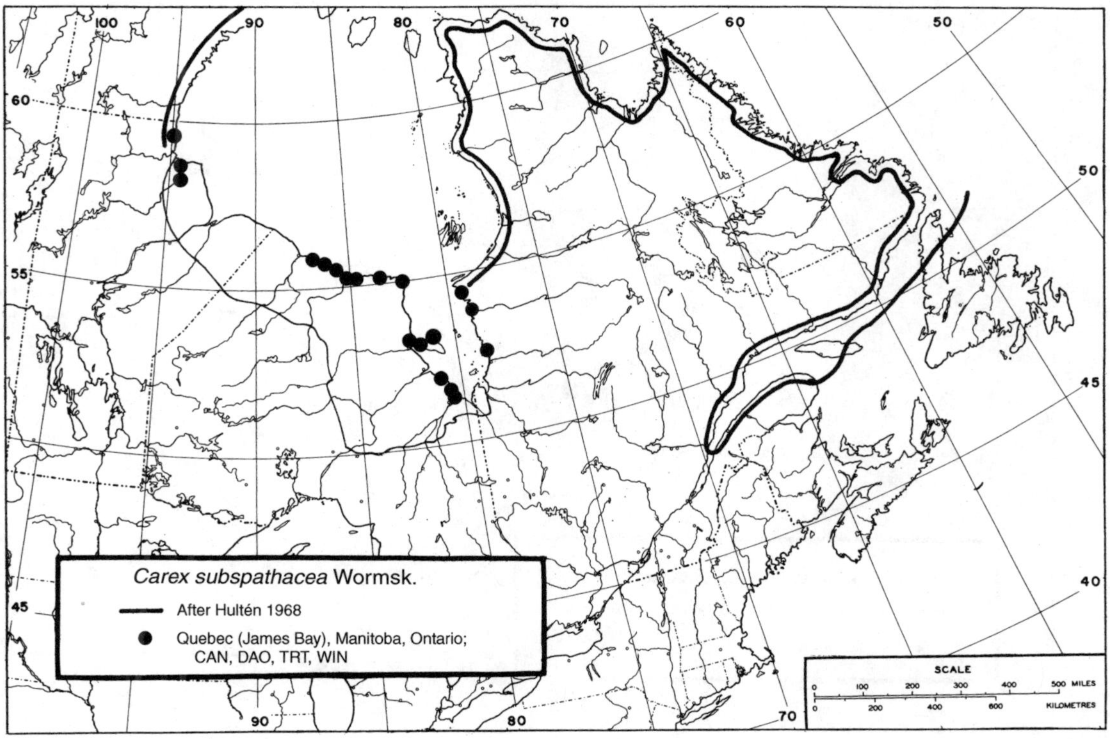

Fig. A40. Distribution of *Castilleja raupii* in eastern Canada.

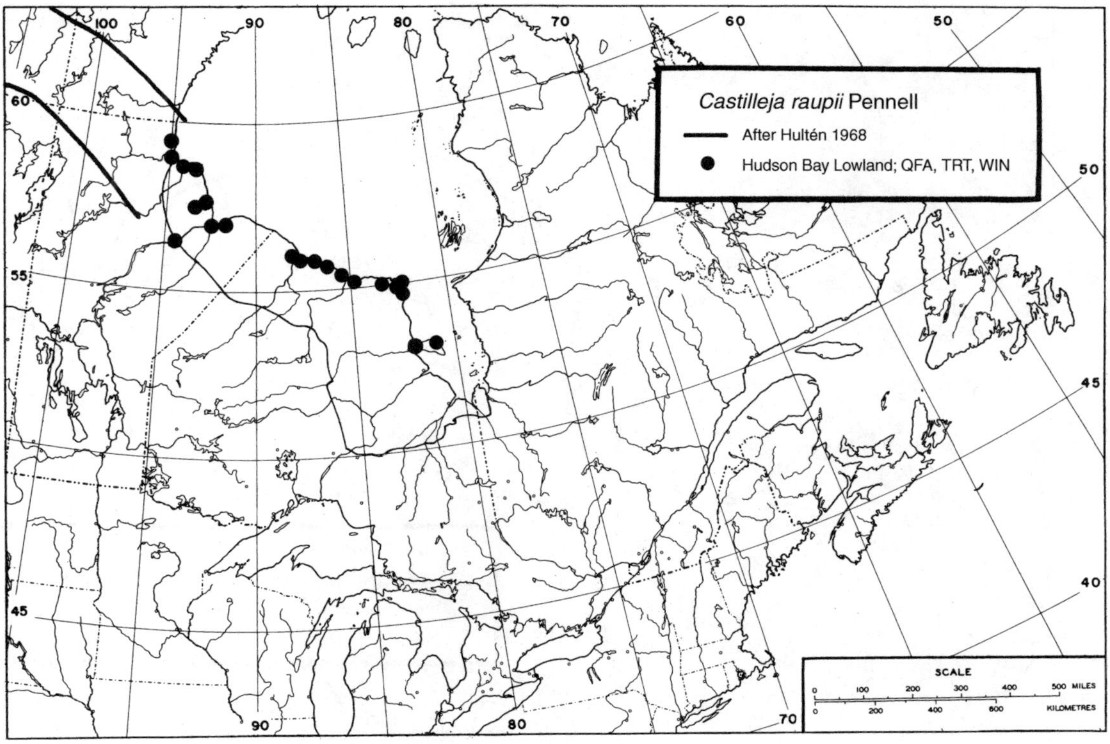

Fig. A41. Distribution of *Catabrosa aquatica* in eastern Canada.

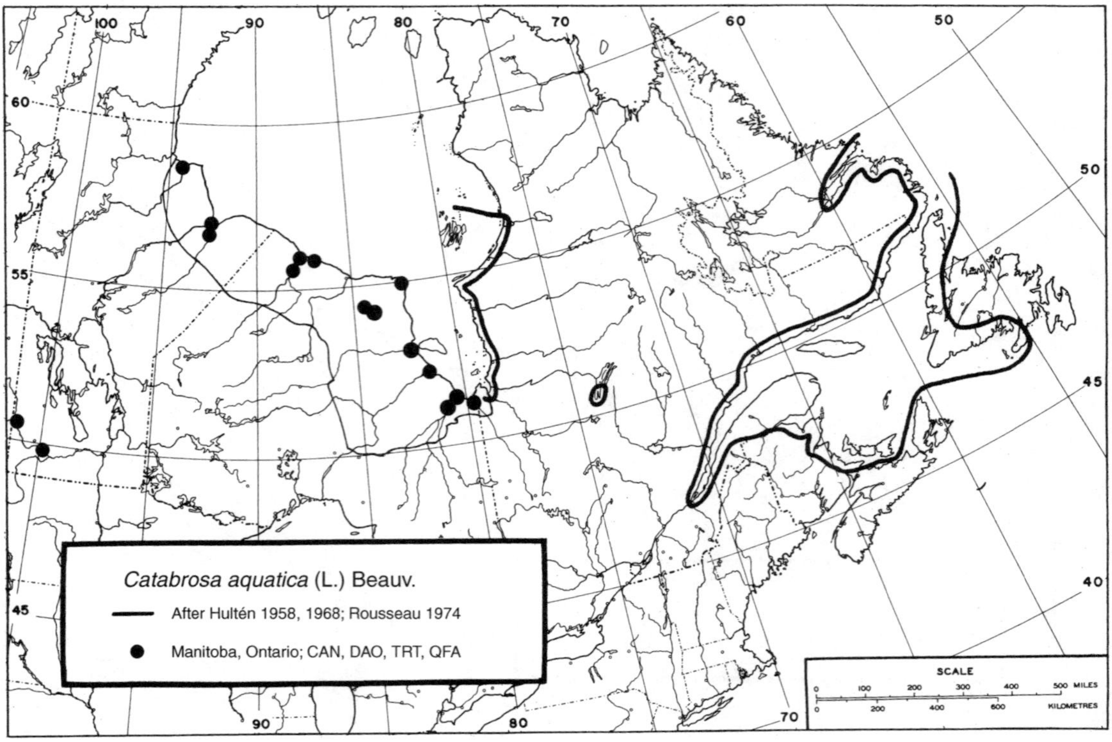

Fig. A42. Distribution of *Cerastium alpinum* in eastern Canada.

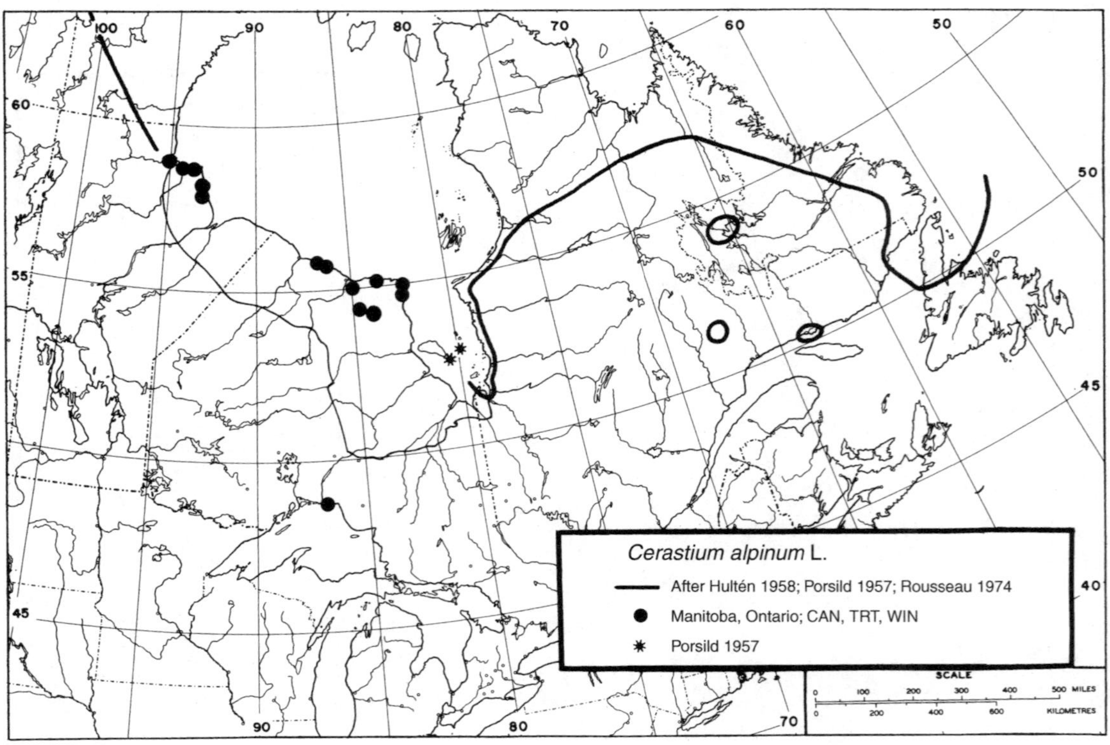

Fig. A43. Distribution of *Ceratophyllum demersum* in eastern Canada.

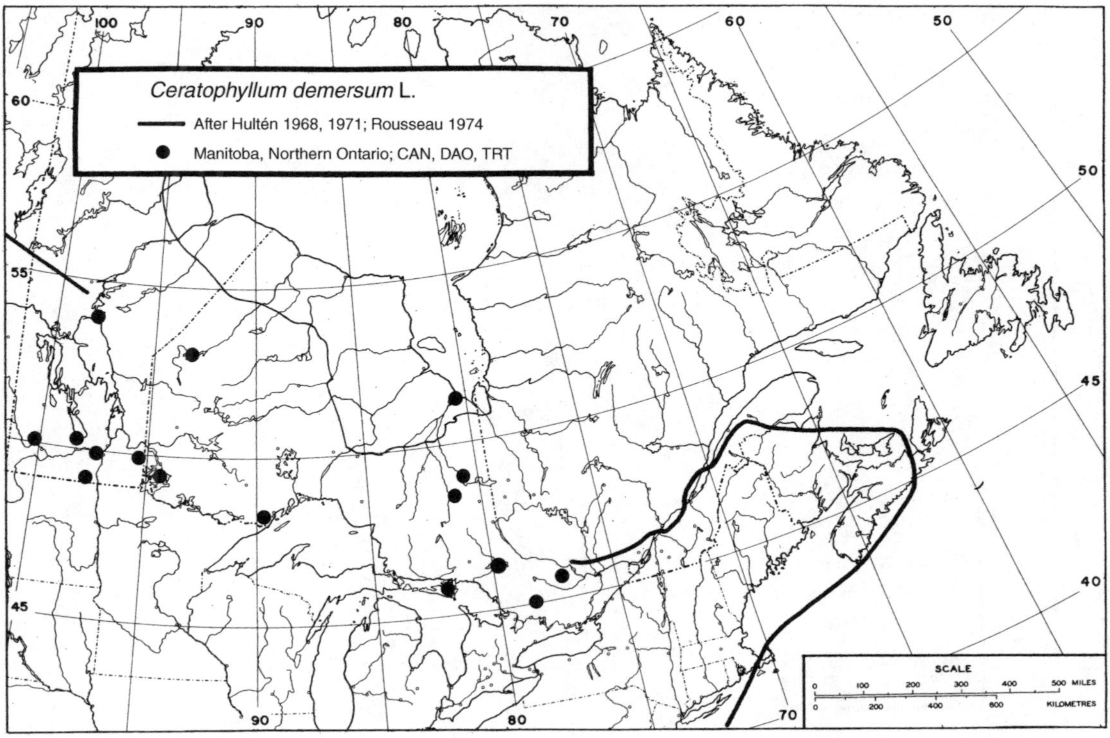

Fig. A44. Distribution of *Chenopodium glaucum* var. *salinum* in eastern Canada.

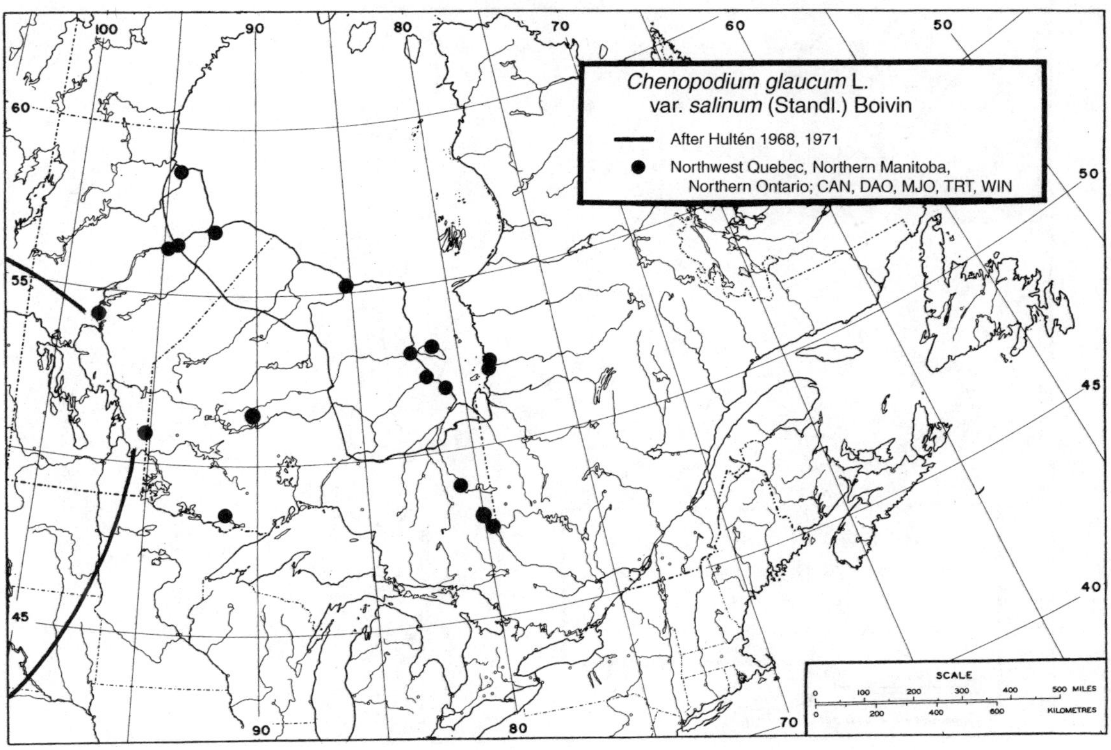

Fig. A45. Distribution of *Chrysanthemum arcticum* in Canada.

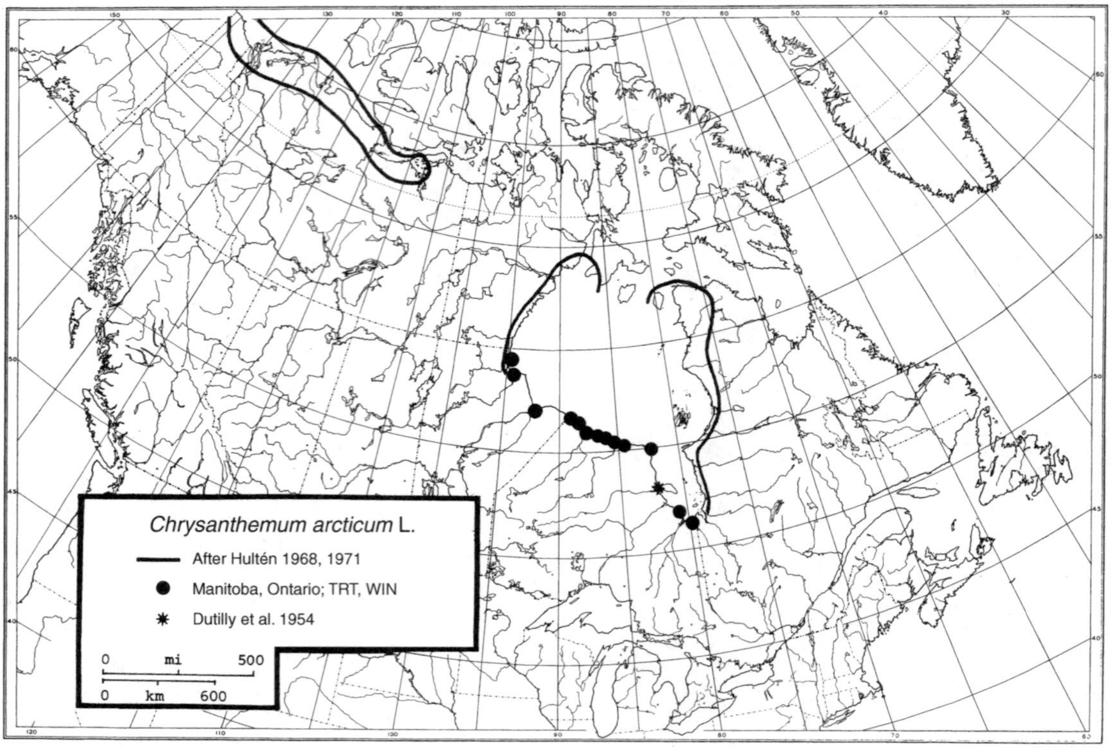

Fig. A46. Distribution of *Tanacetum huronense* in Canada.

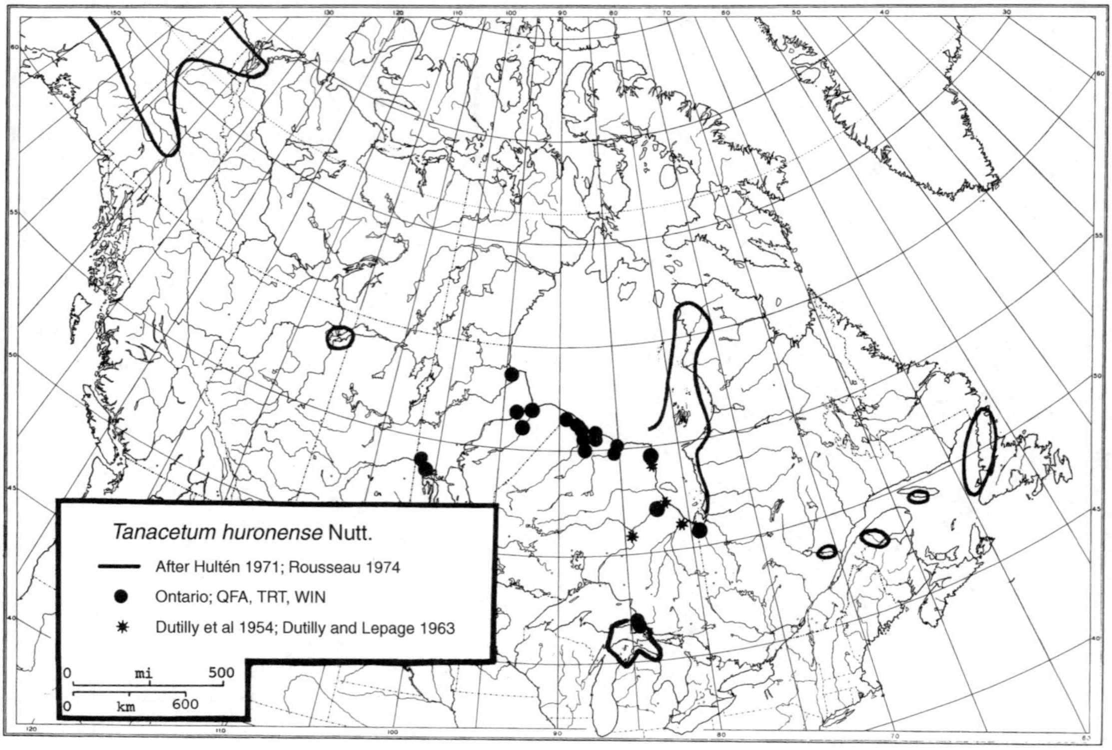

Fig. A47. Distribution of *Cicuta mackenzieana* in Canada.

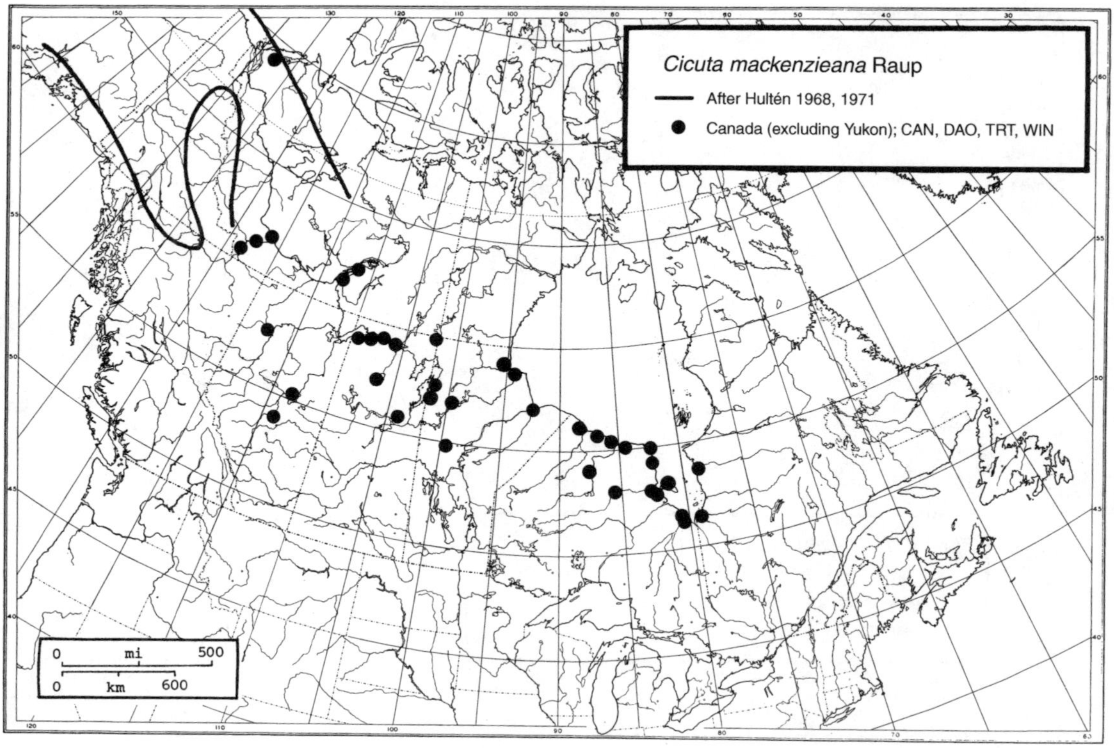

Fig. A48. Distribution of *Conioselinum chinense* in eastern Canada.

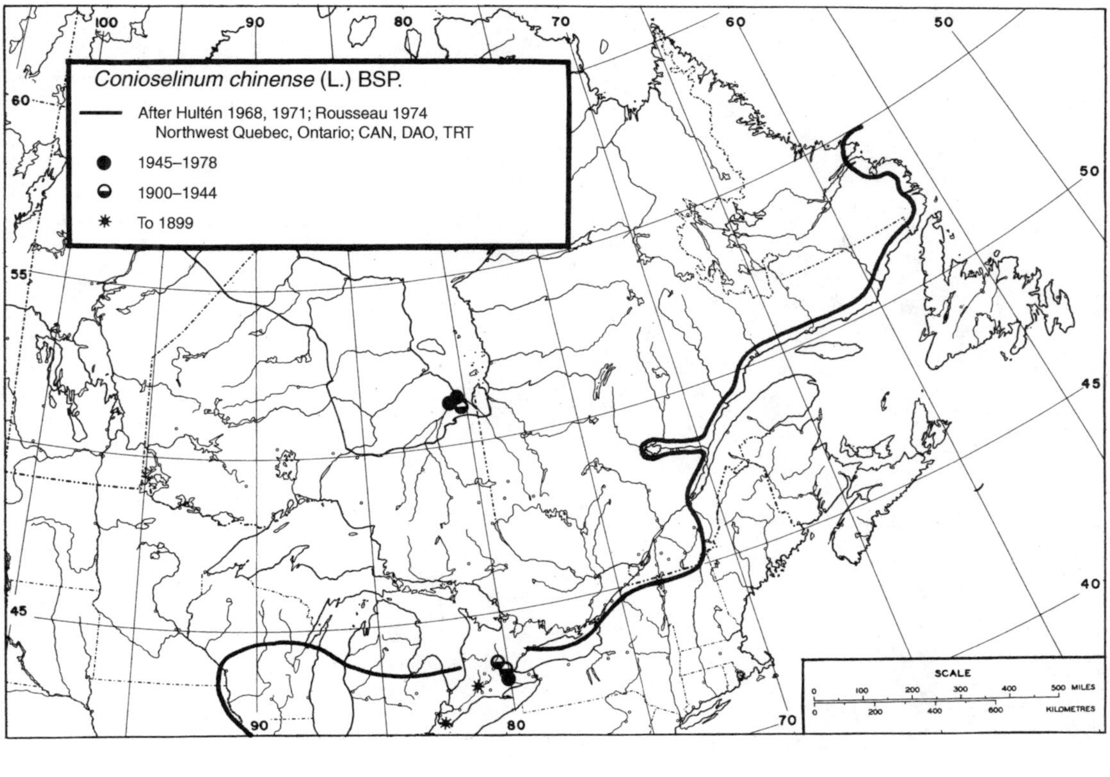

Fig. A49. Distribution of *Cypripedium parviflorum* var. *planipetalum* in eastern Canada.

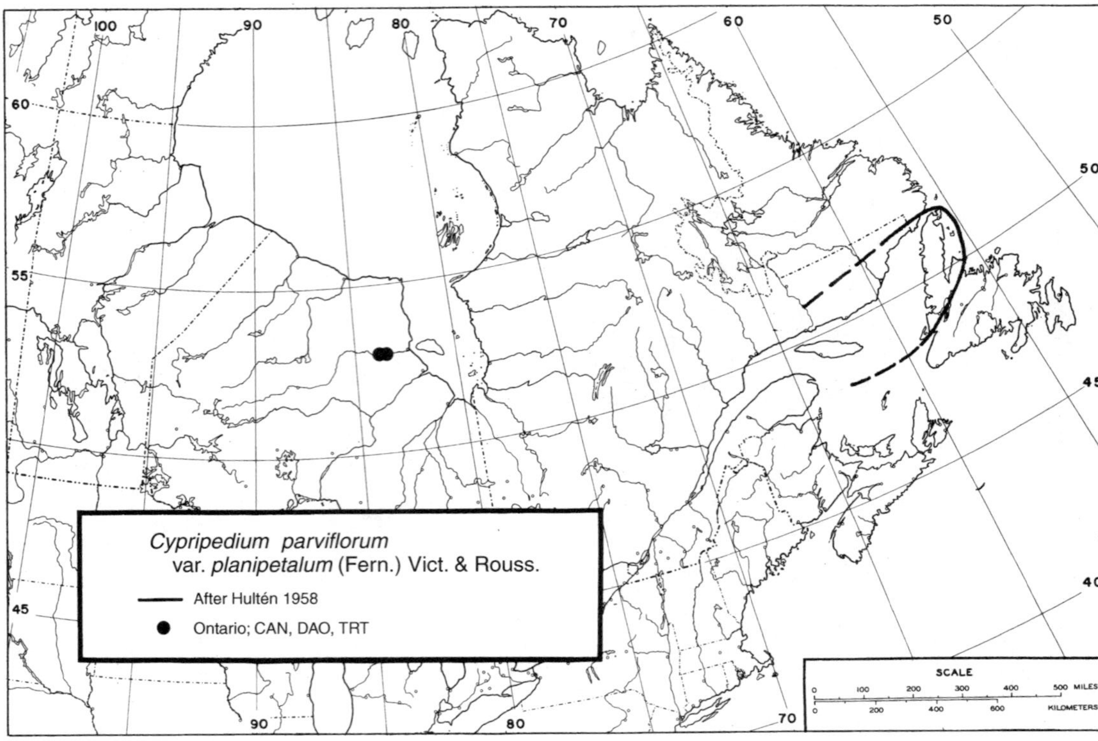

Fig. A50. Distribution of *Cypripedium passerinum* in eastern Canada.

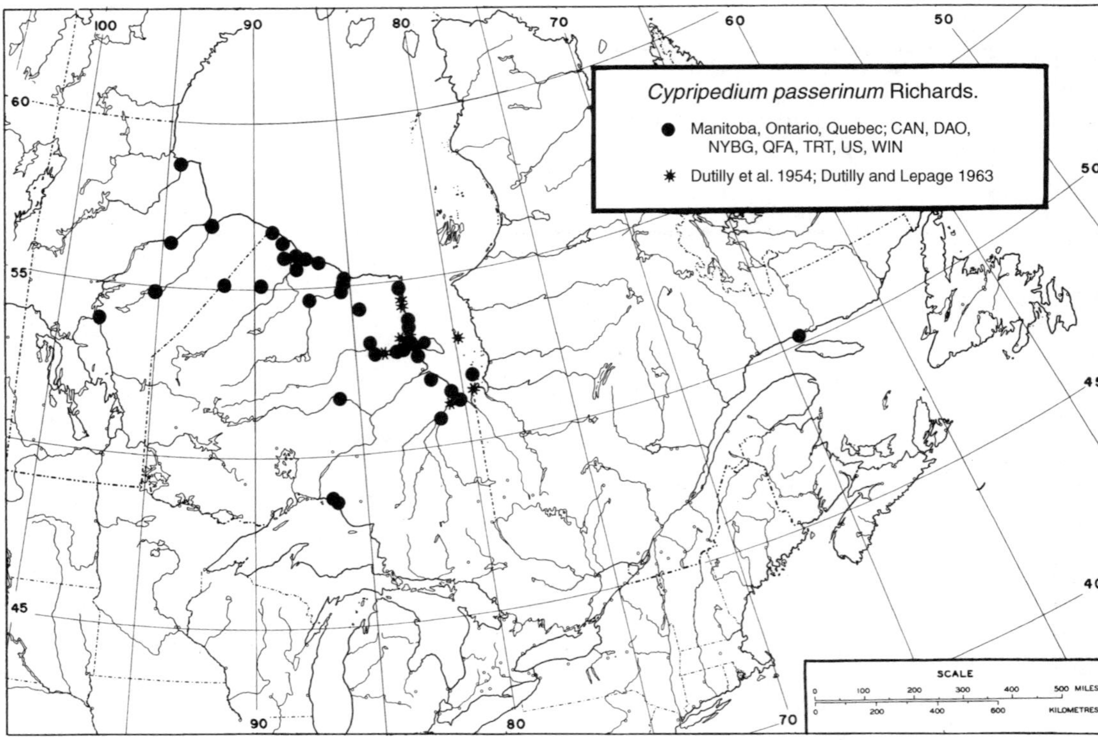

Fig. A51. Distribution of *Cypripedium passerinum* in Canada.

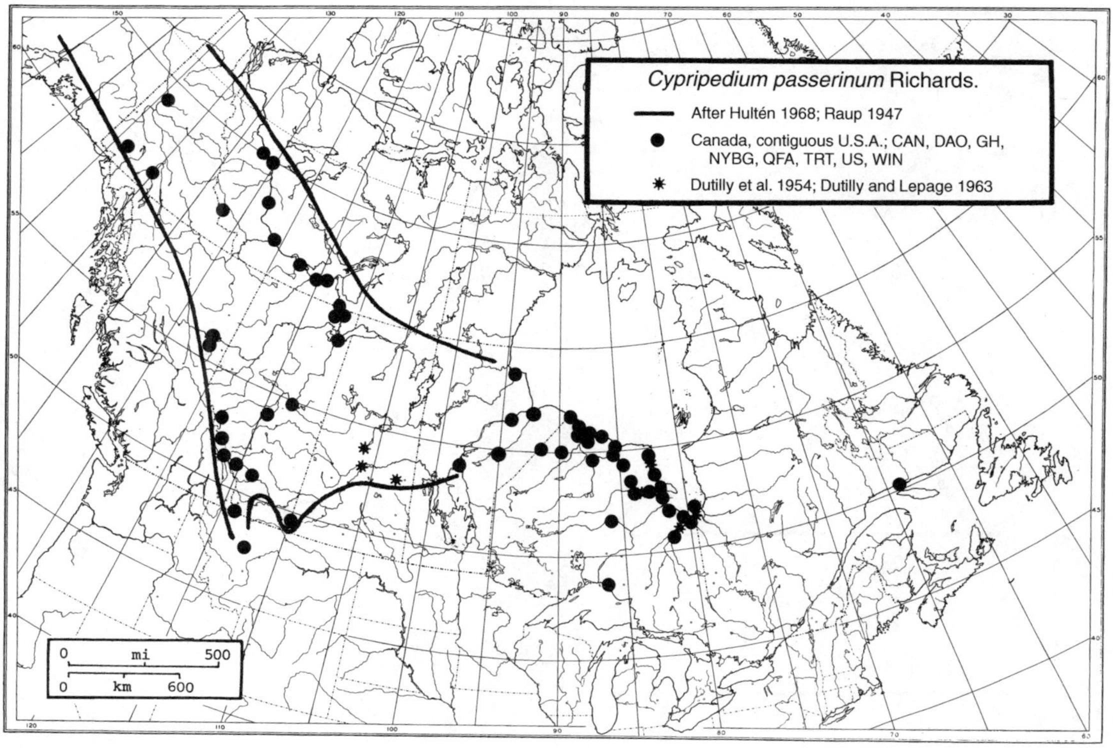

Fig. A52. Distribution of *Cypripedium reginae* in eastern Canada.

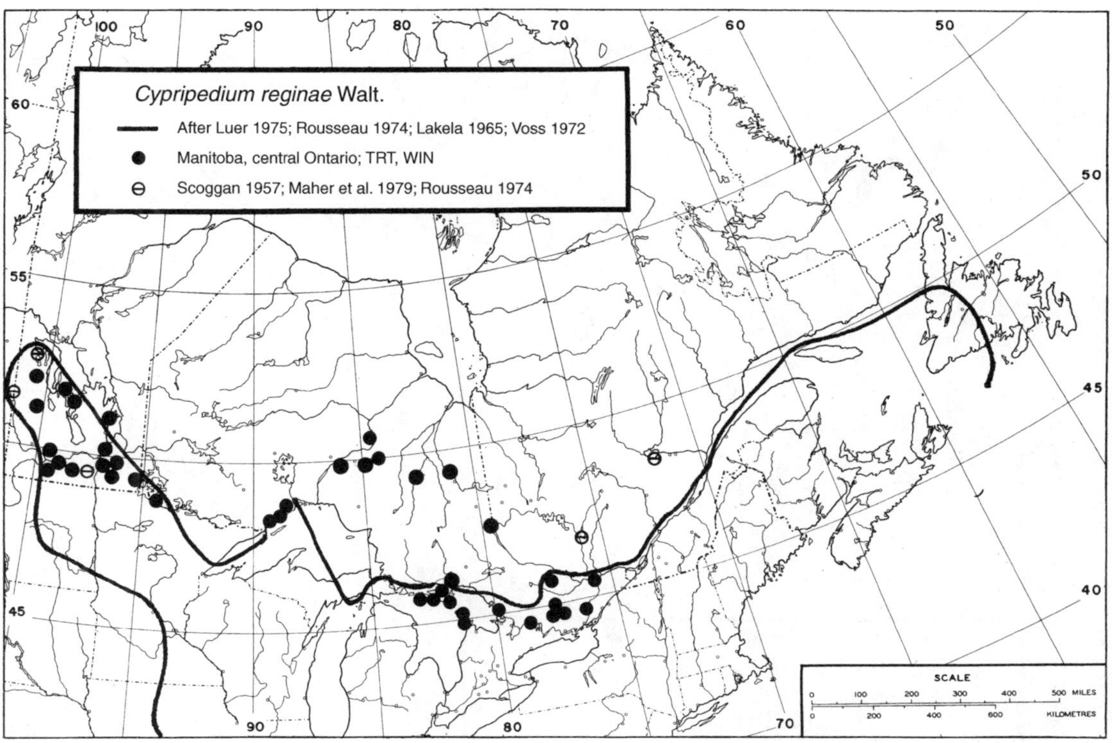

Fig. A53. Distribution of *Draba cinerea* in Canada.

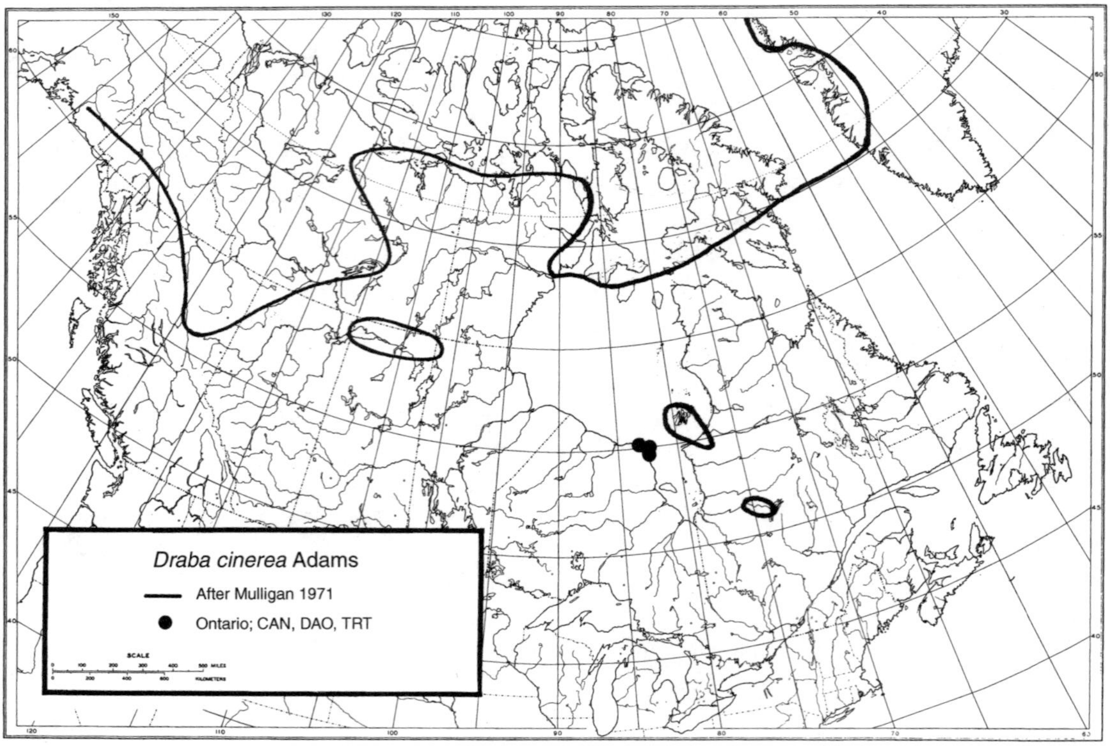

Fig. A54. Distribution of *Draba glabella* in eastern Canada.

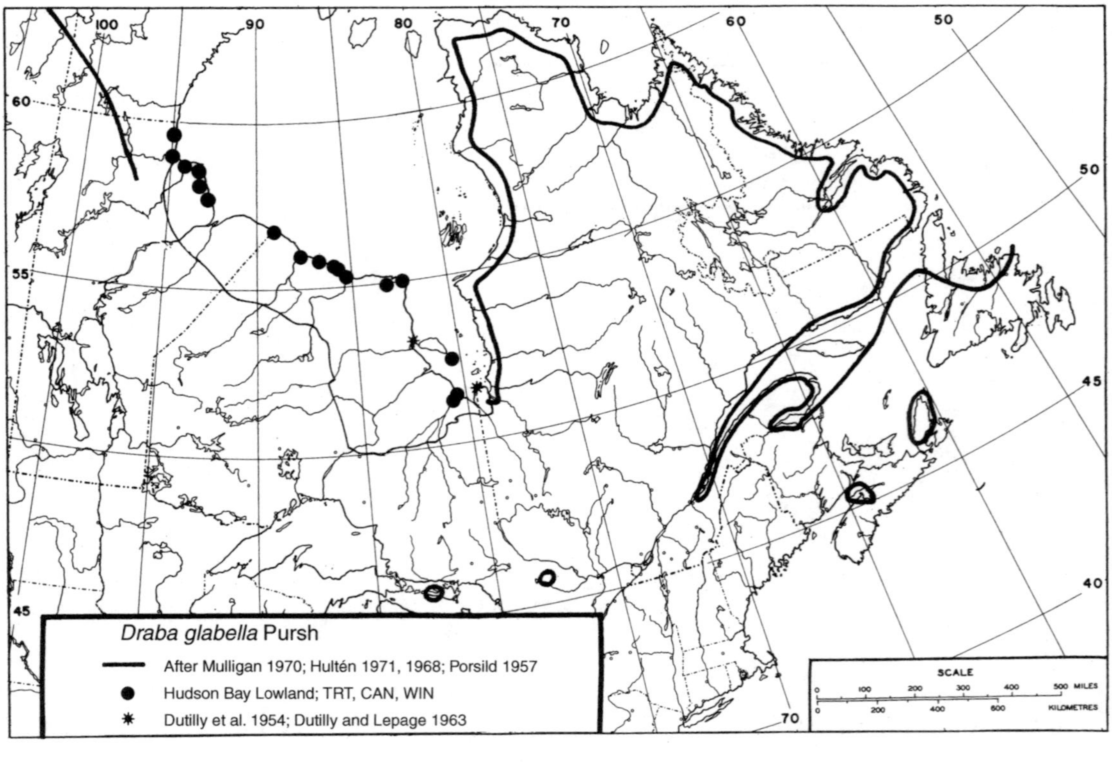

Fig. A55. Distribution of *Draba incana* in eastern Canada.

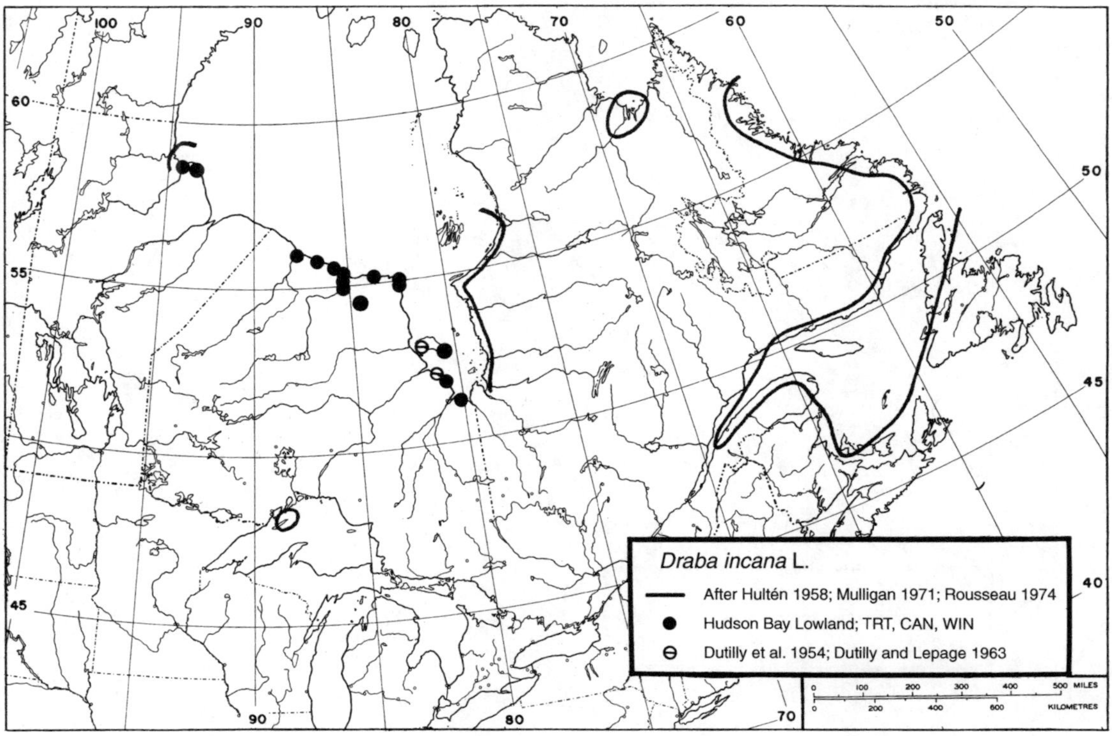

Fig. A56. Distribution of *Draba nivalis* in Canada.

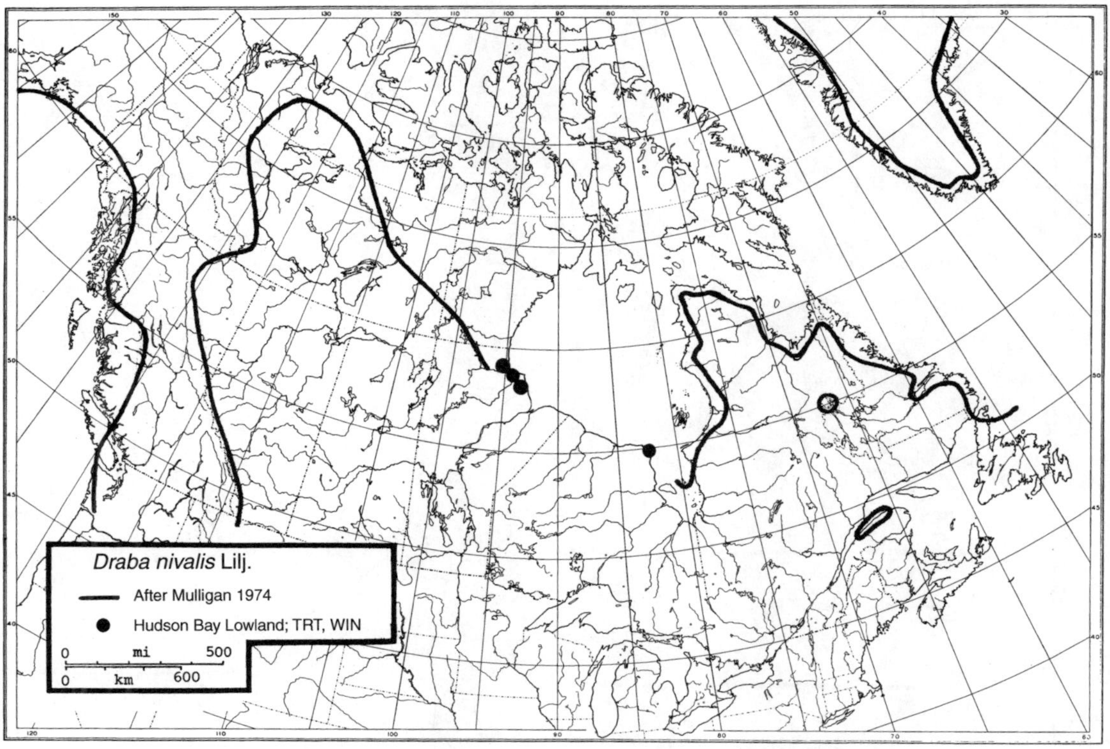

Fig. A57. Distribution of *Draba norvegica* in eastern Canada.

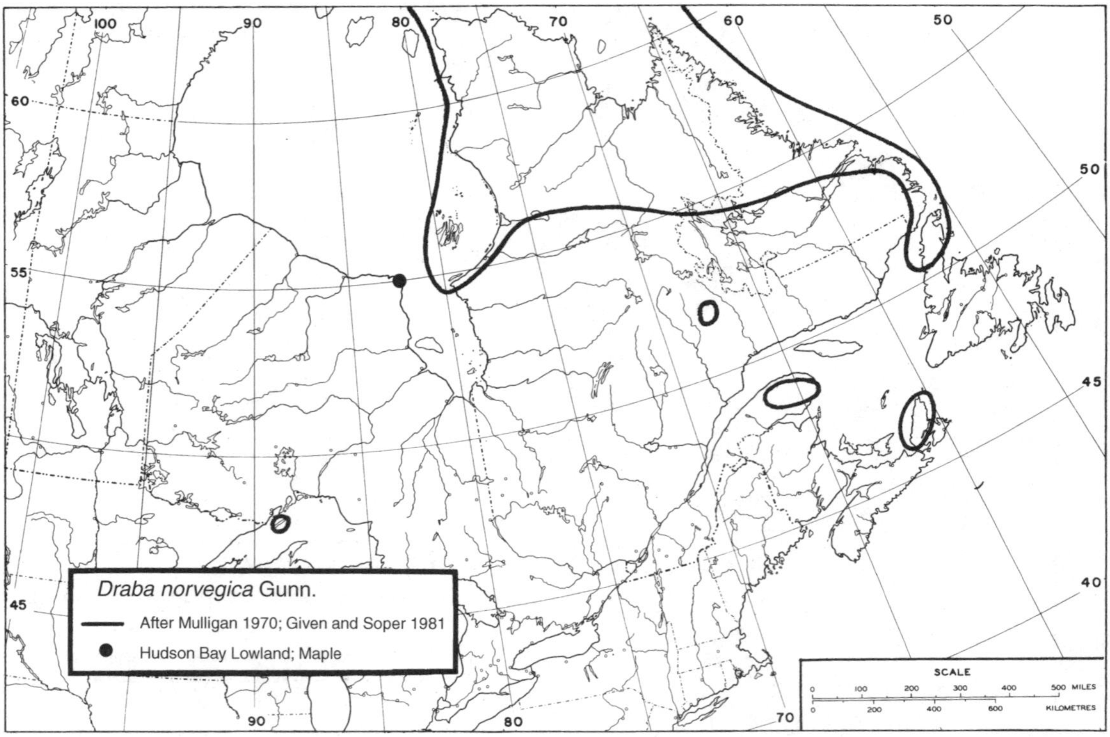

Fig. A58. Distribution of *Dupontia fisheri* in eastern Canada.

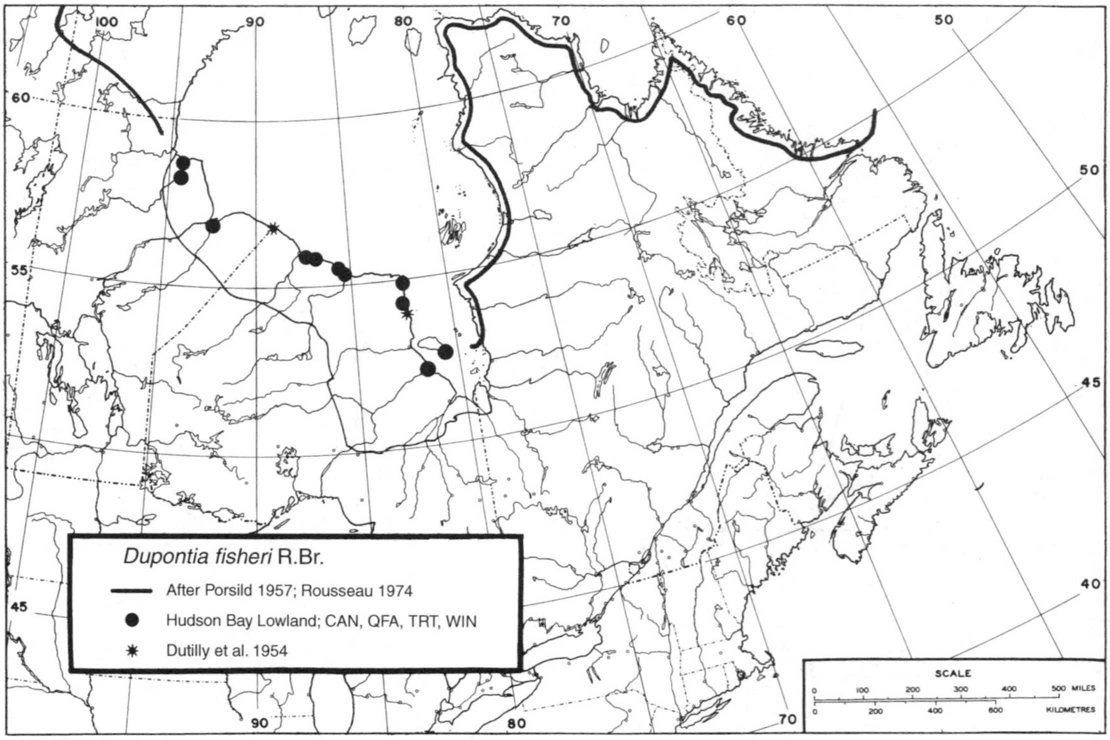

Fig. A59. Distribution of *Eleocharis kamtschatica* in Canada.

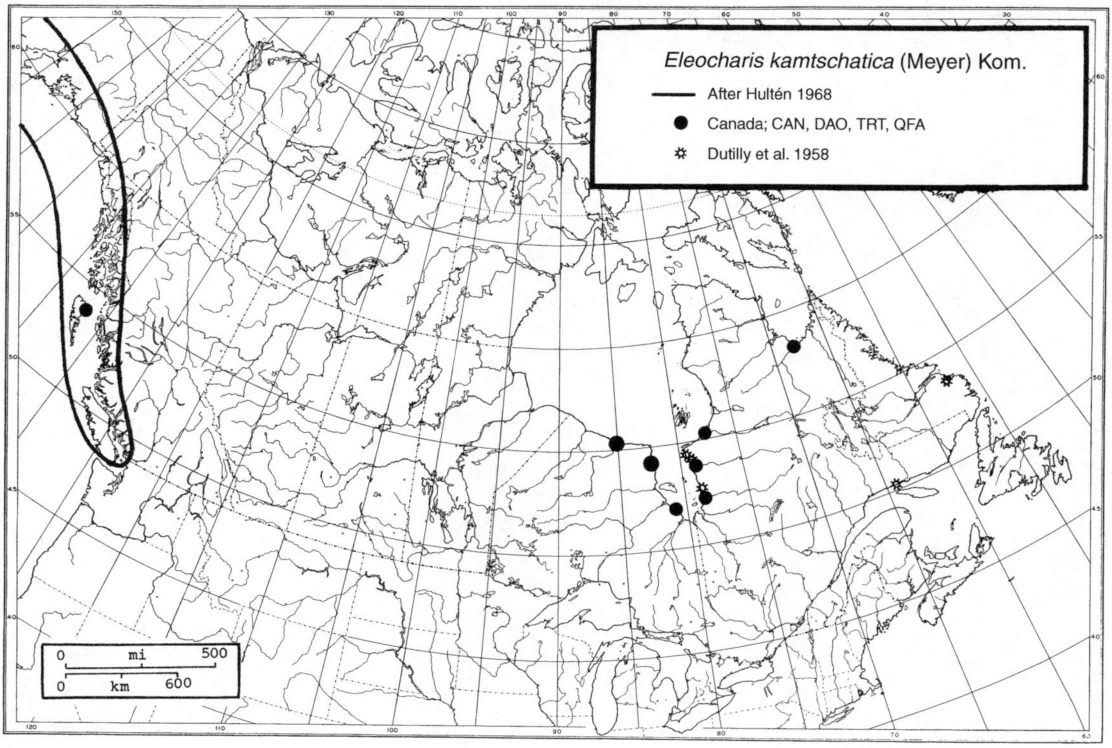

Fig. A60. Distribution of *Elymus canadensis* in eastern Canada.

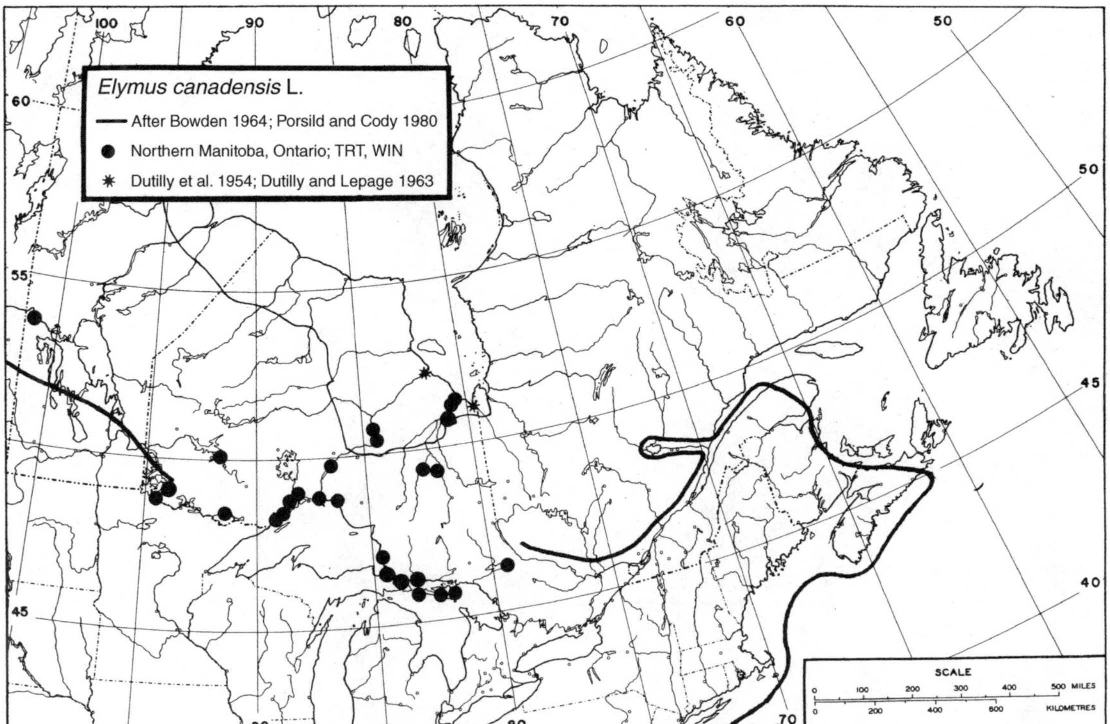

Fig. A61. Distribution of *Elymus innovatus* in eastern Canada.

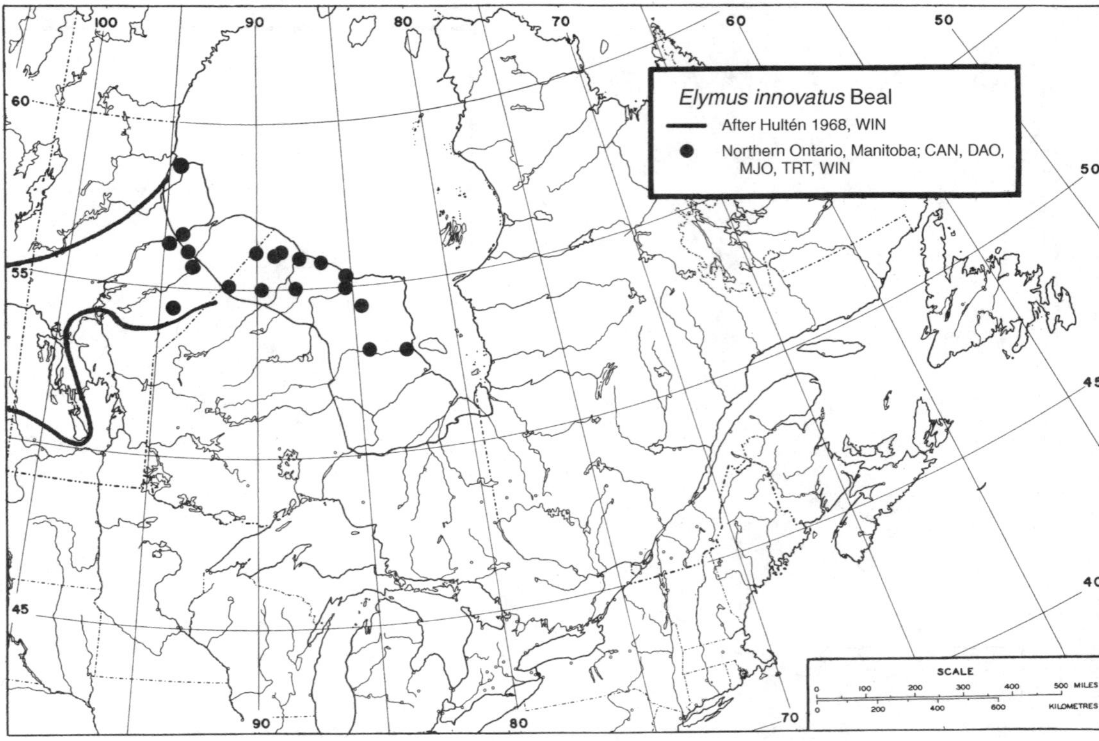

Fig. A62. Distribution of *Elymus mollis* in eastern Canada.

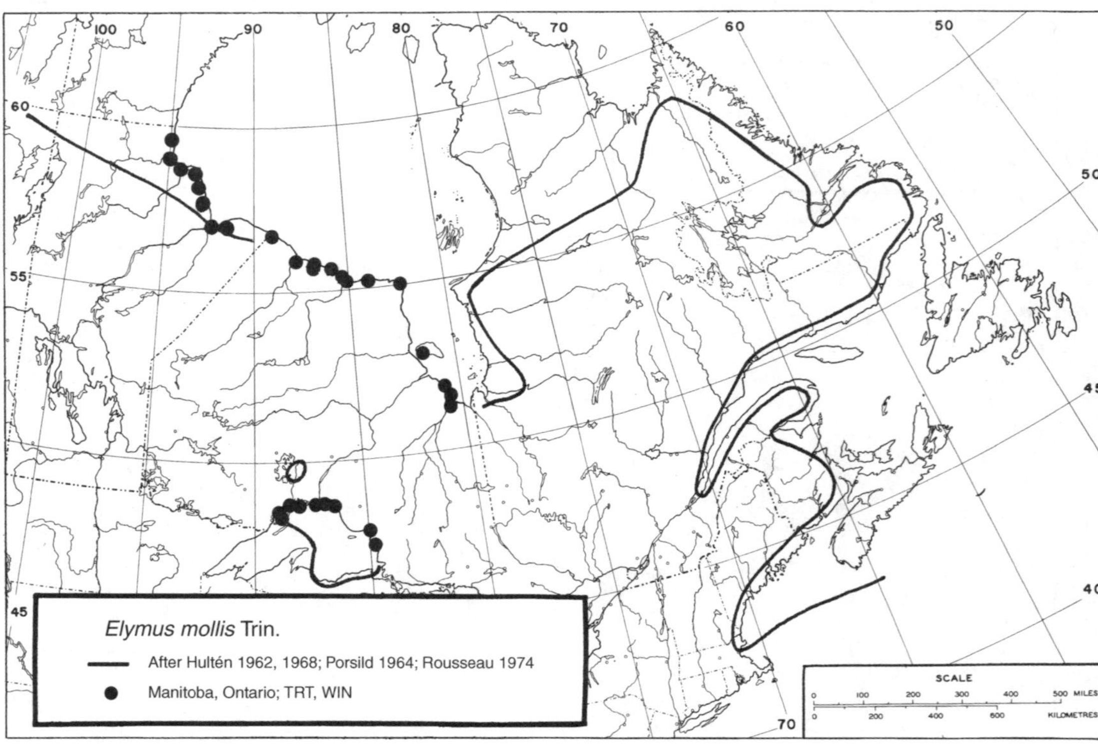

Fig. A63. Distribution of *Elymus virginicus* in eastern Canada.

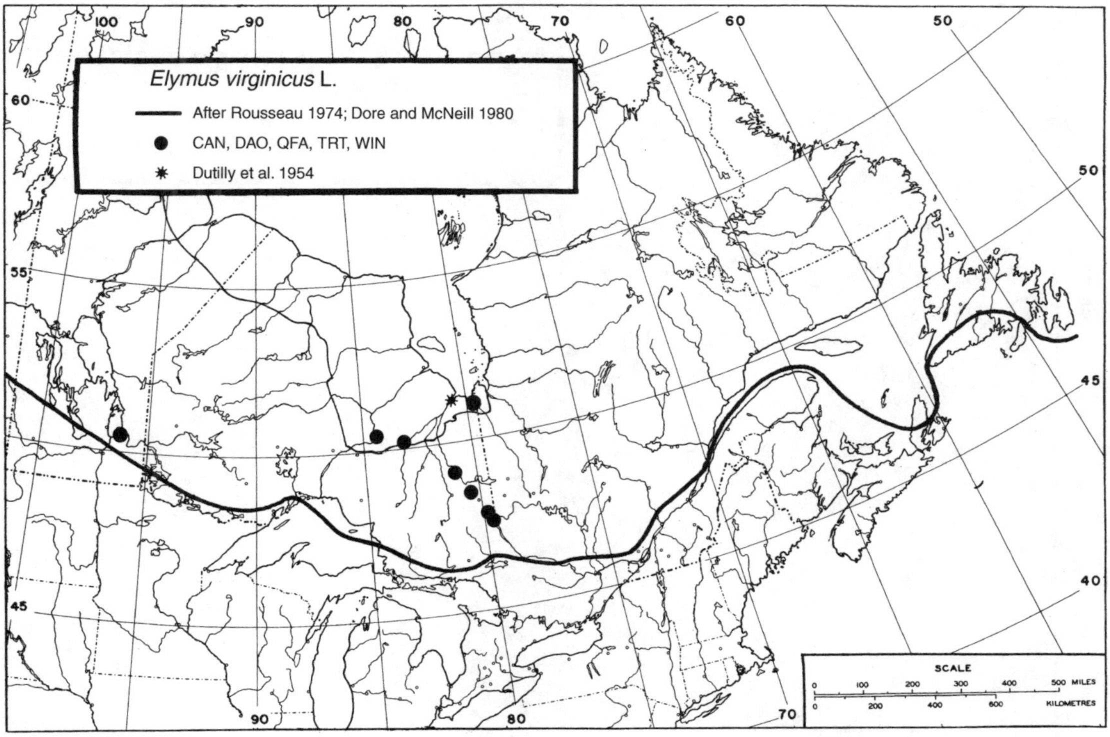

Fig. A64. Distribution of *Erigeron acris* in eastern Canada.

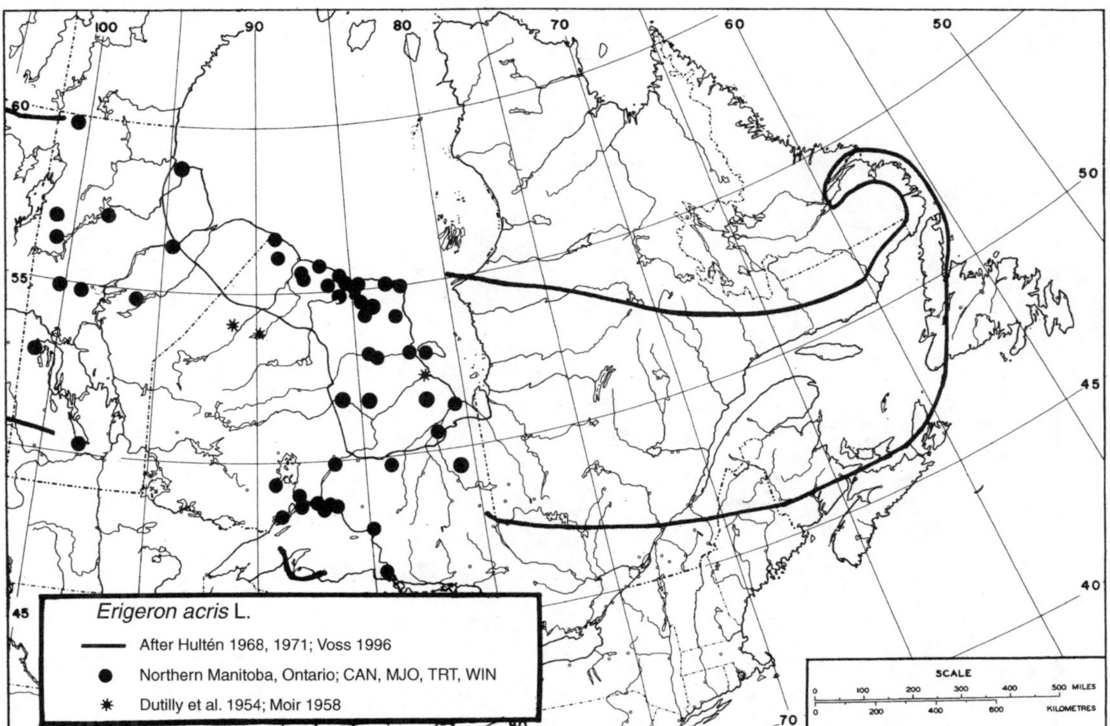

Fig. A65. Distribution of *Erigeron lonchophyllus* in eastern Canada.

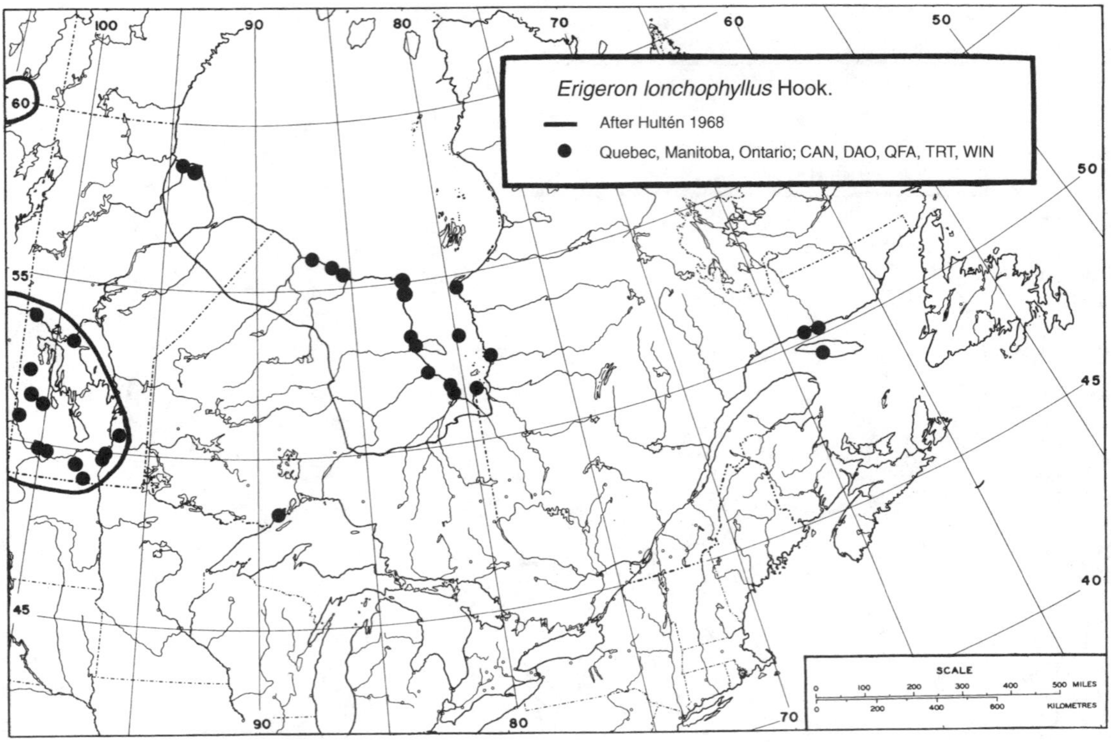

Fig. A66. Distribution of *Galium asprellum* in eastern Canada.

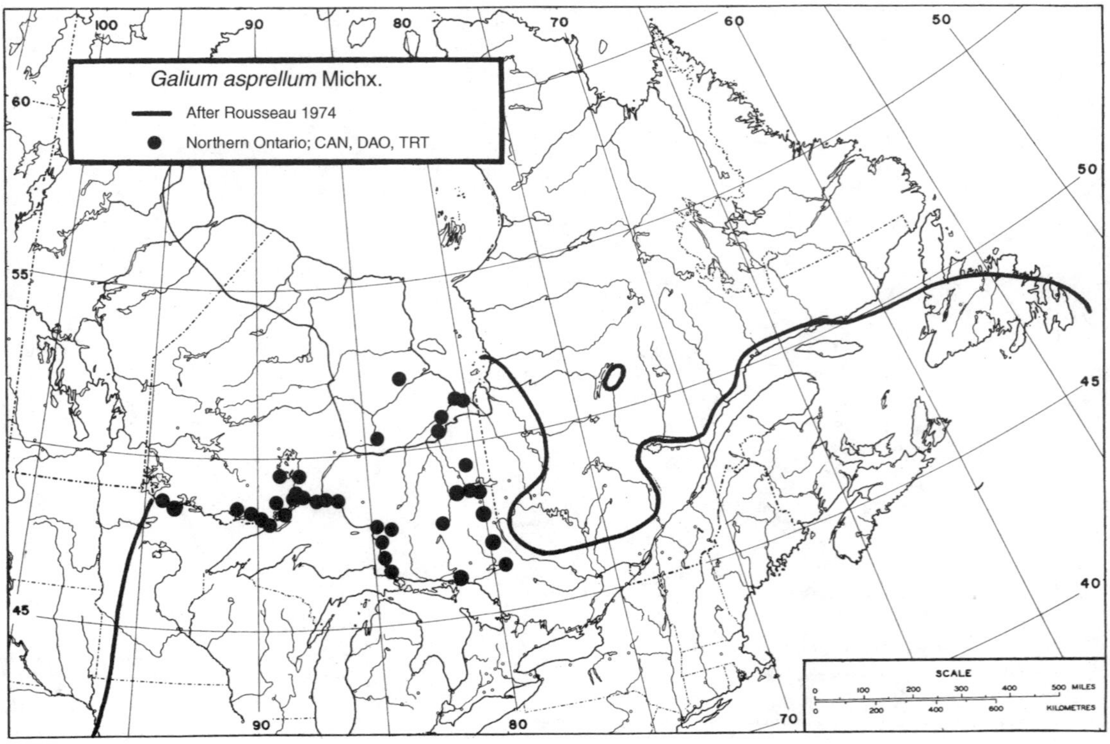

Fig. A67. Distribution of *Gentiana linearis* in eastern Canada.

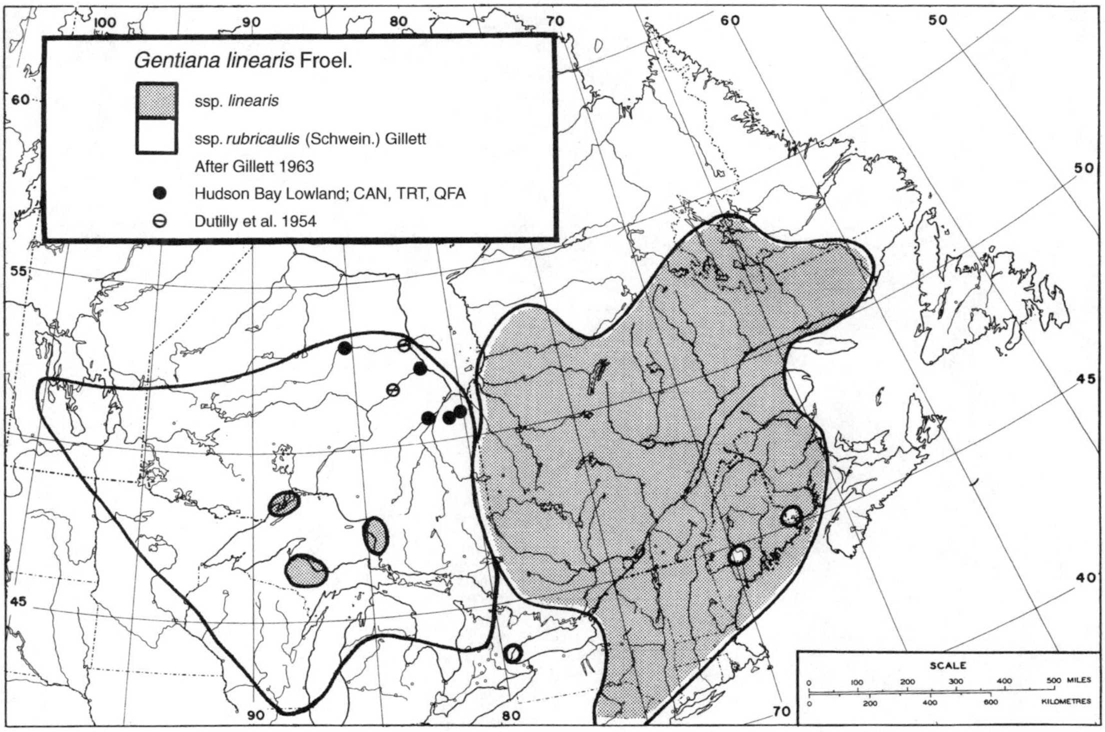

Fig. A68. Distribution of *Gentianopsis macounii* in eastern Canada.

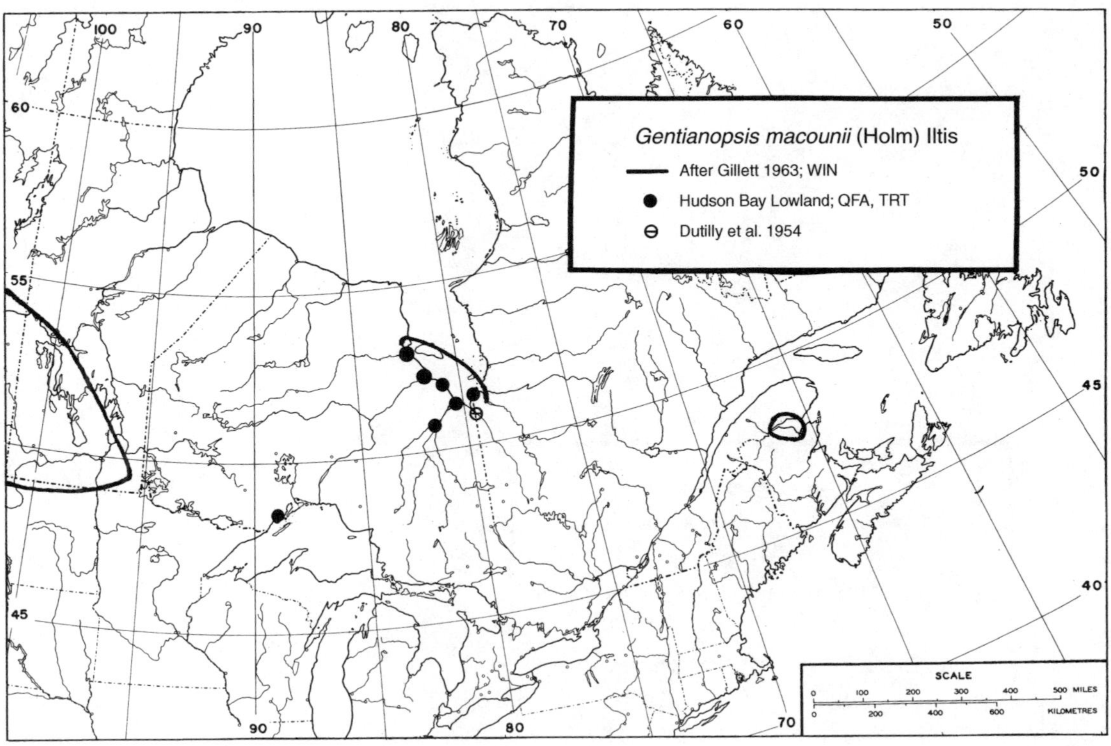

Fig. A69. Distribution of *Gentianopsis crinita*, *G. virgata*, *G. virtorinii* and *G. macounii* in Canada.

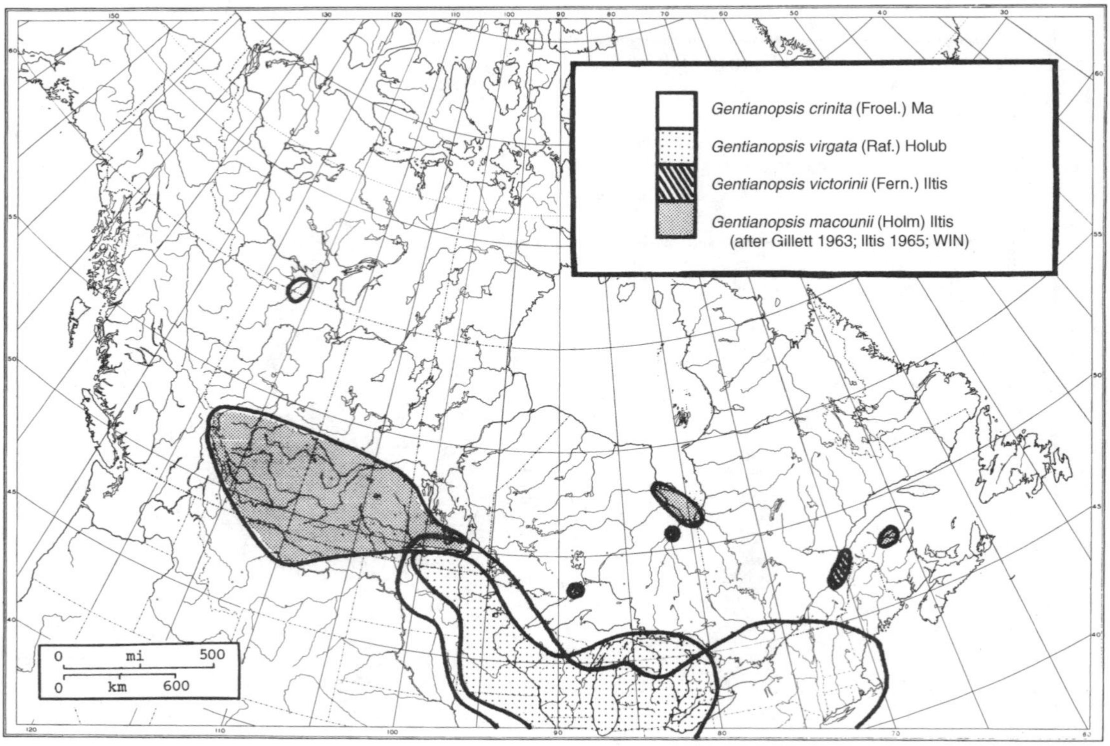

Fig. A70. Distribution of *Gentianopsis detonsa* and *G. nesophila* in Canada.

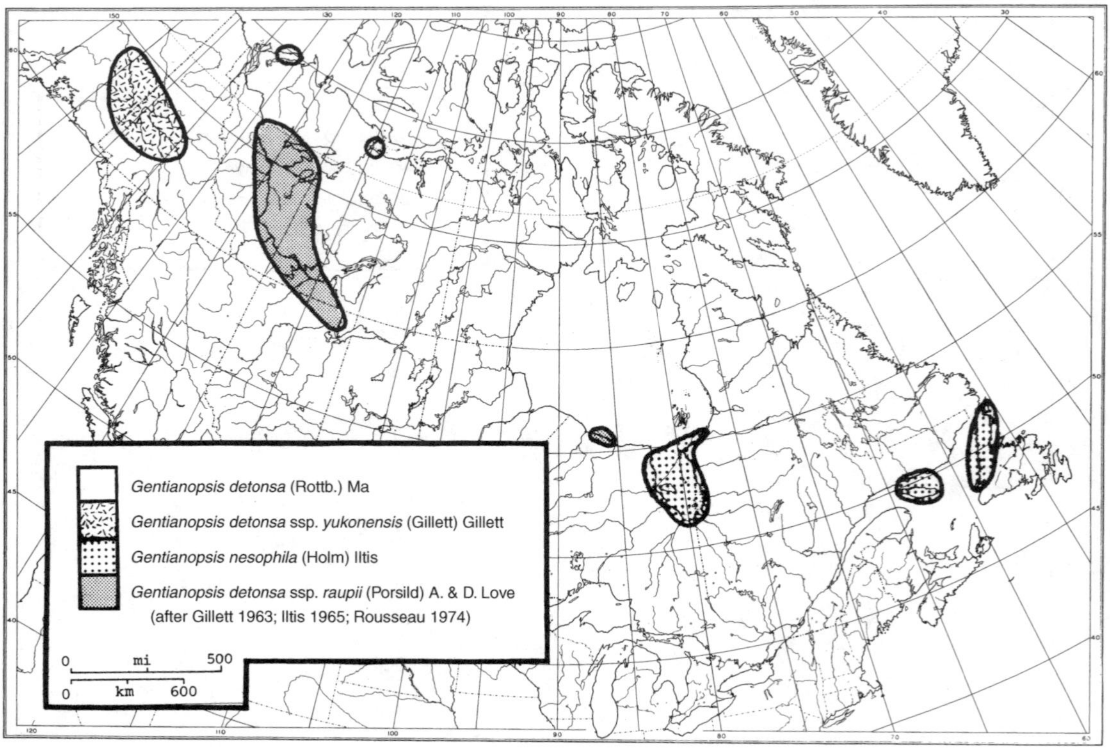

Fig. A71. Distribution of *Gentianella propinqua* in Canada.

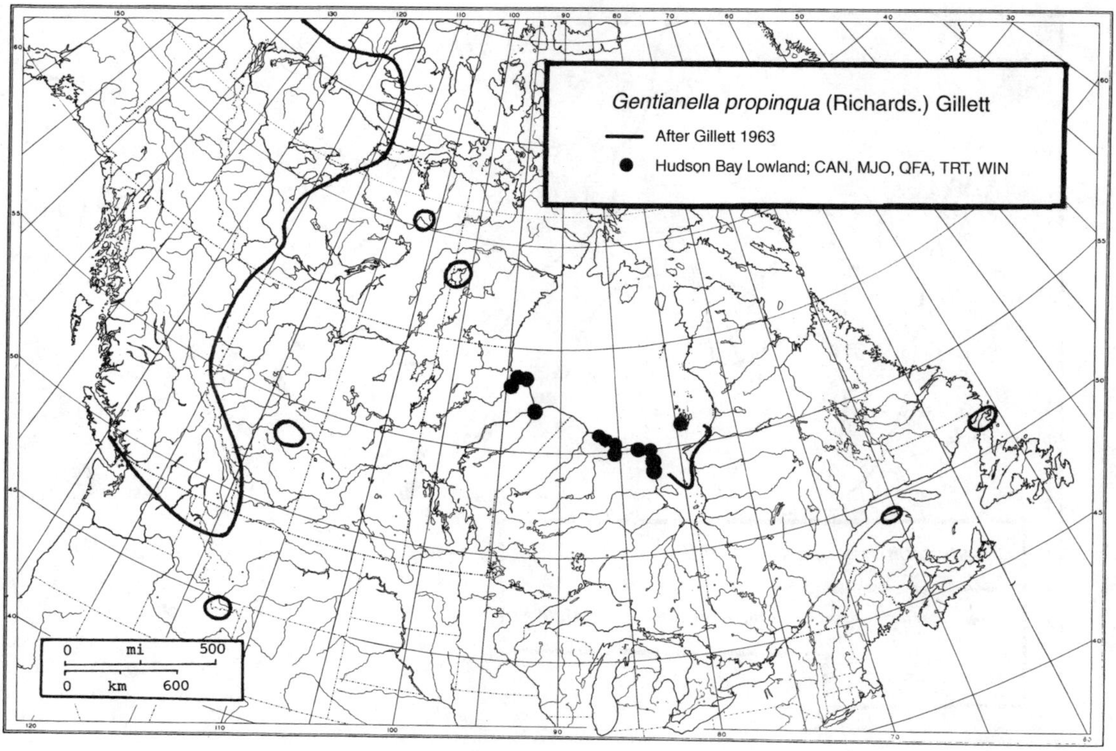

Fig. A72. Distribution of *Glaux maritima* in eastern Canada.

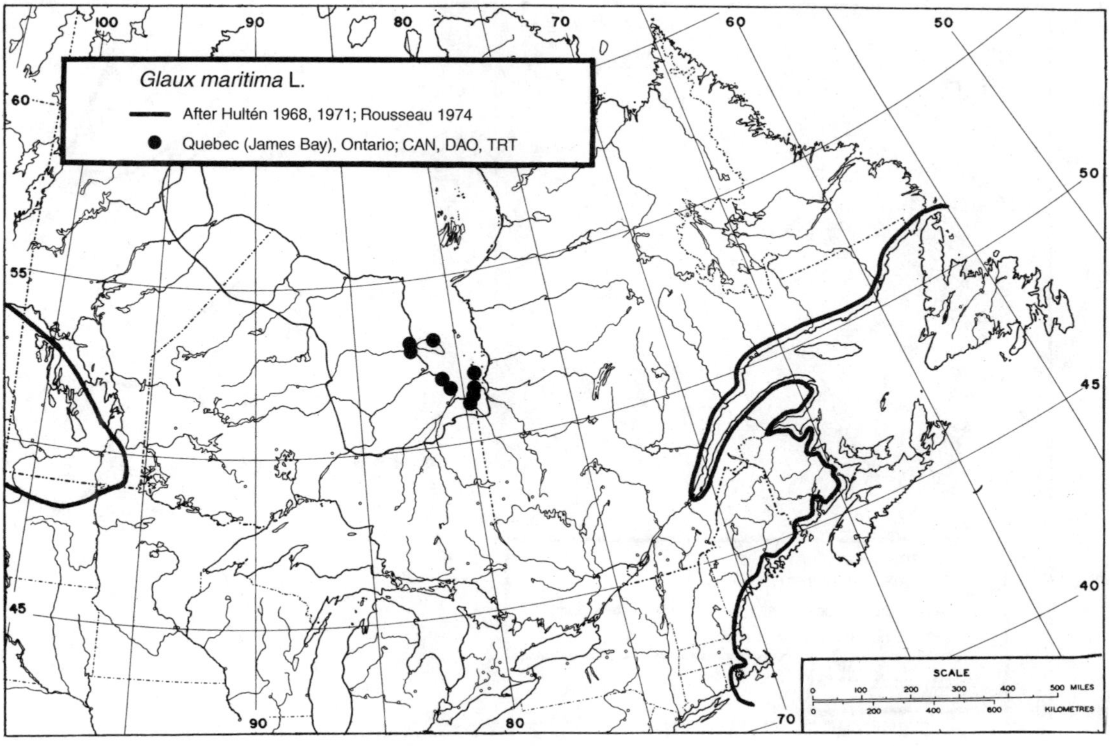

Fig. A73. Distribution of *Hedysarum mackenzii* in Canada.

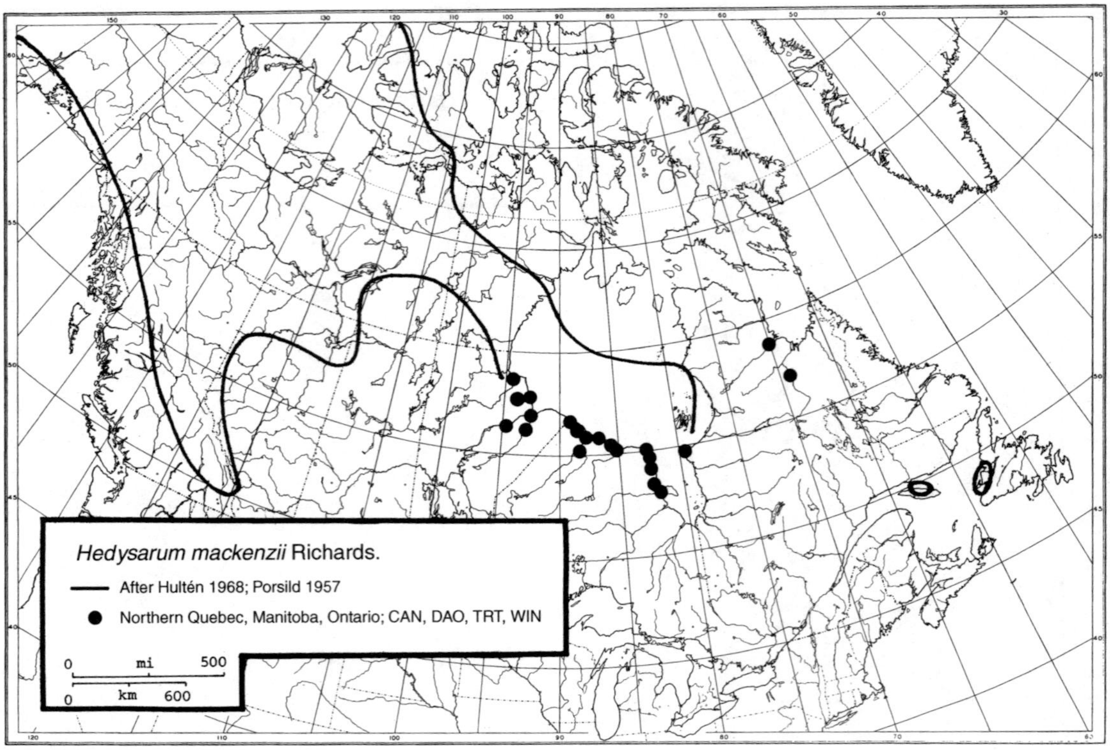

Fig. A74. Distribution of *Hierochloe alpina* var. *orthantha* in Canada.

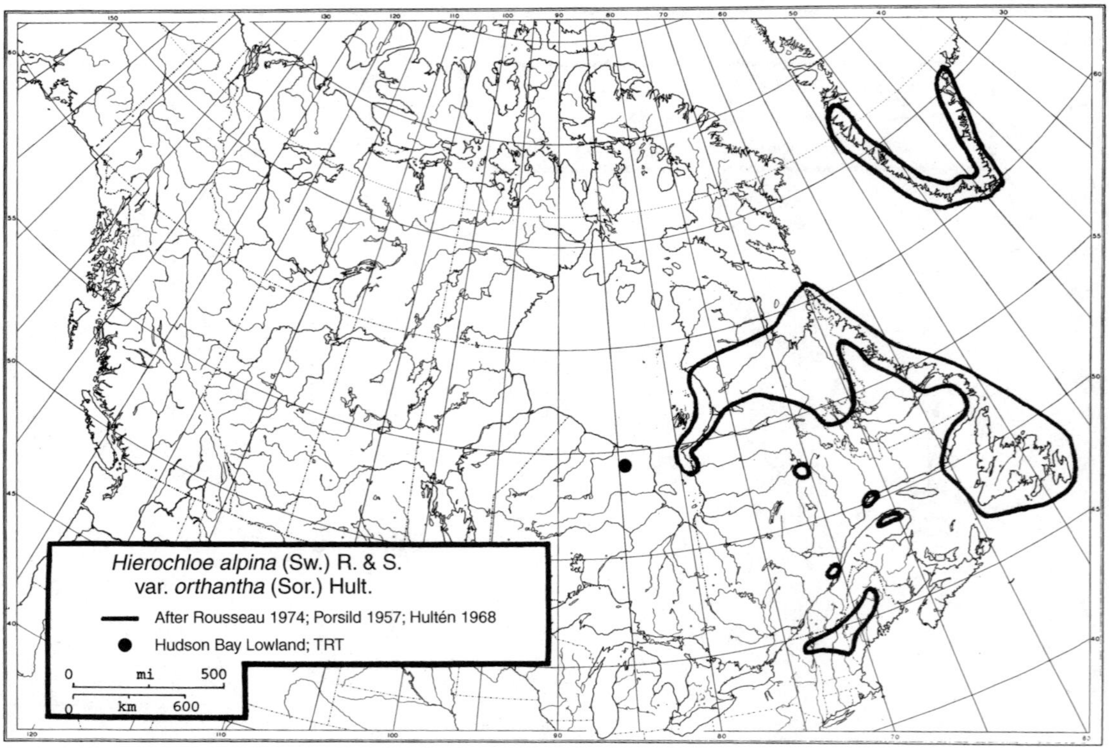

Fig. A75. Distribution of *Honkenya peploides* in eastern Canada.

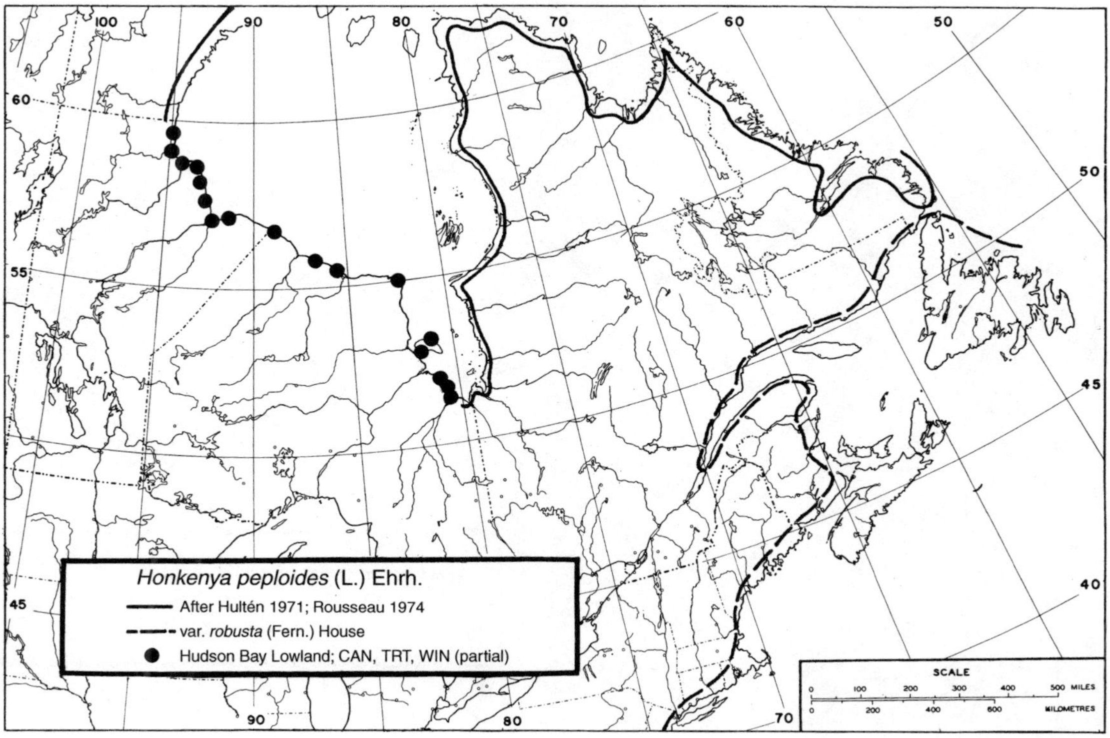

Fig. A76. Distribution of *Hutchinsia procumbens* in Canada.

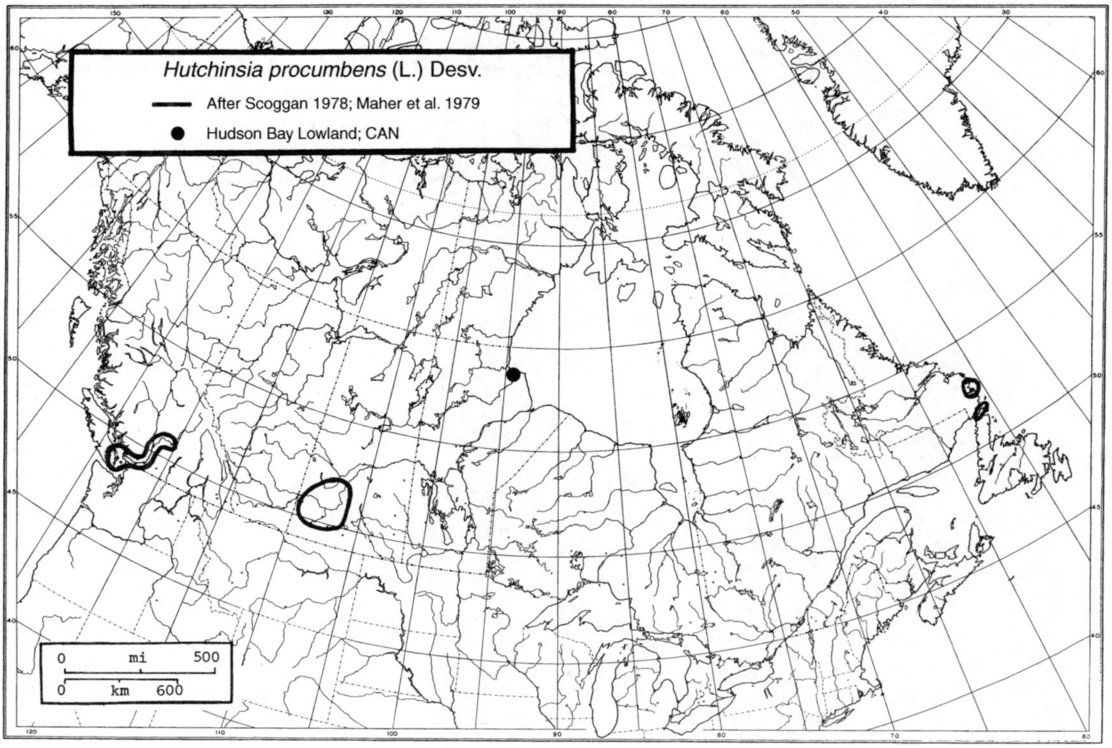

Fig. A77. Distribution of *Juncus castaneus* in eastern Canada.

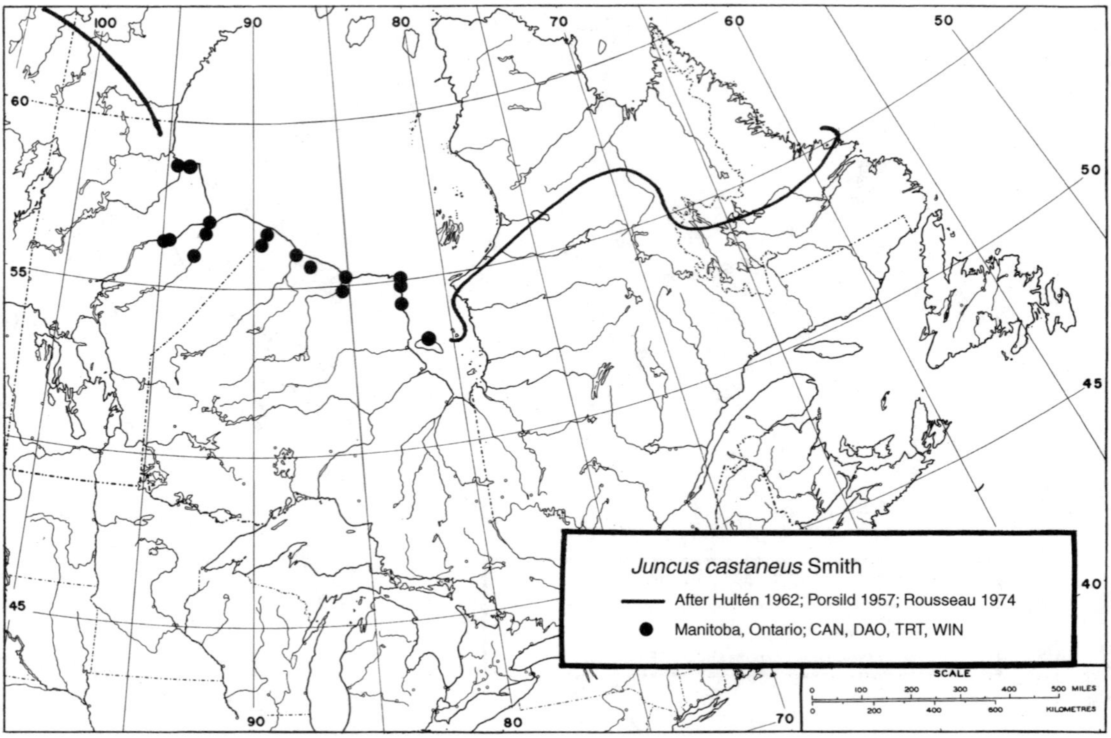

Fig. A78. Distribution of *Juncus ensifolius* in Canada.

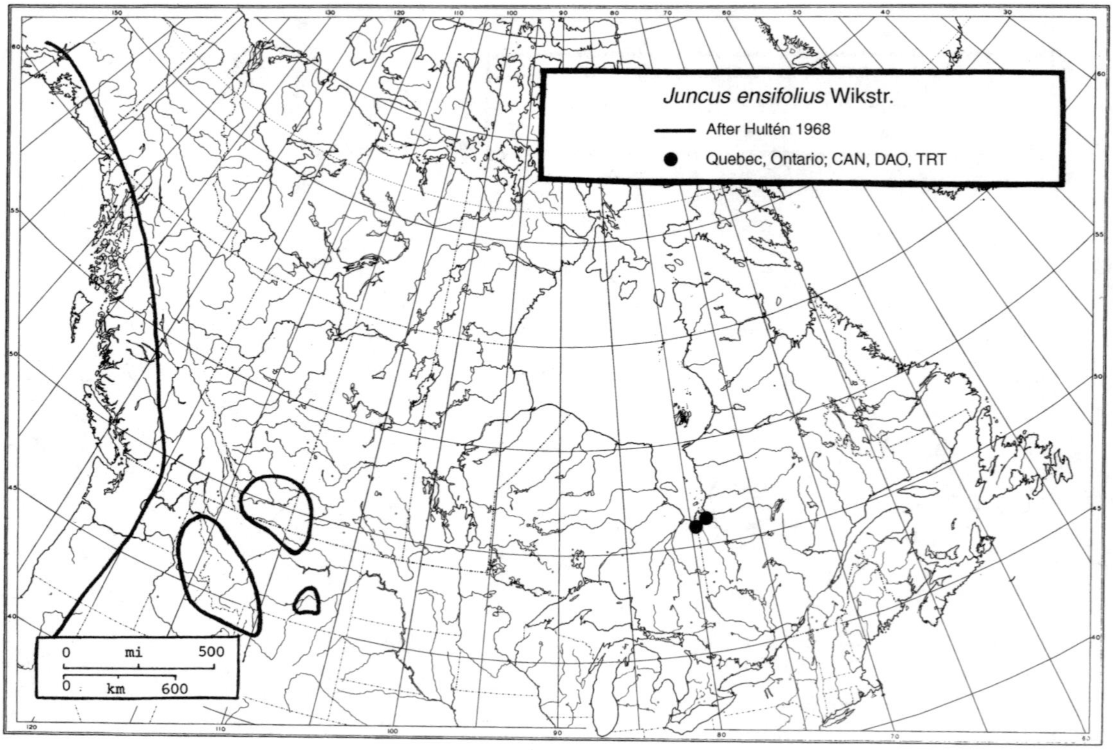

Fig. A79. Distribution of *Kobresia myosuroides* in Canada.

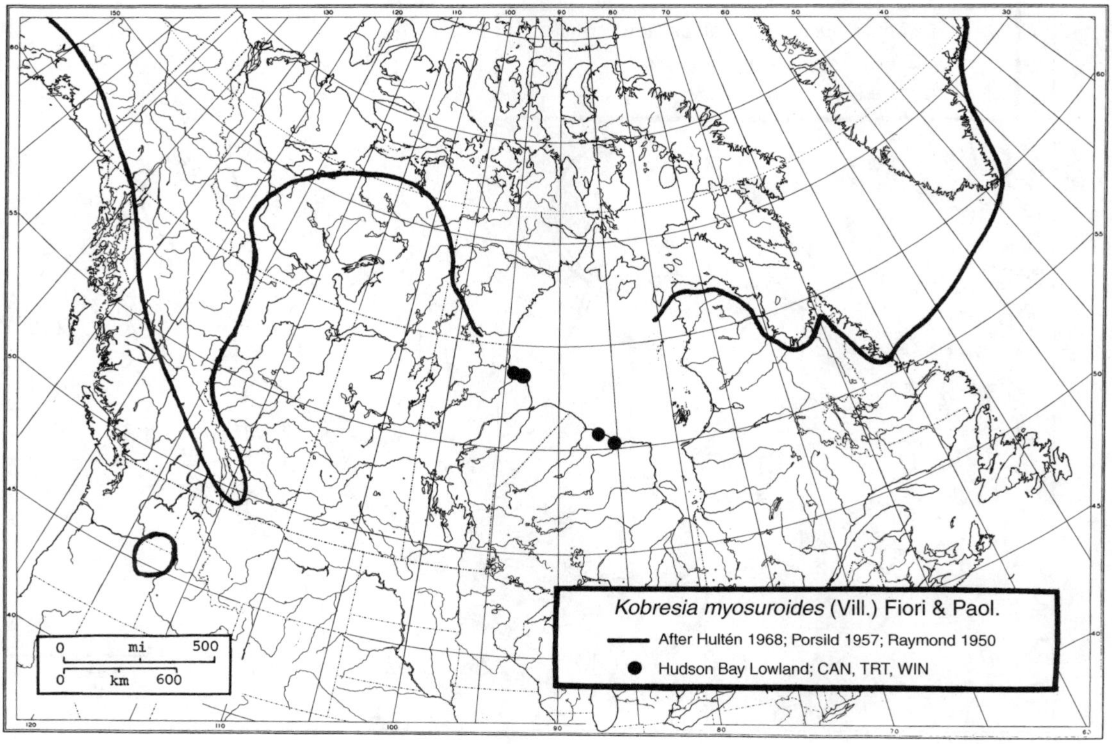

Fig. A80. Distribution of *Kobresia simpliuscula* in Canada.

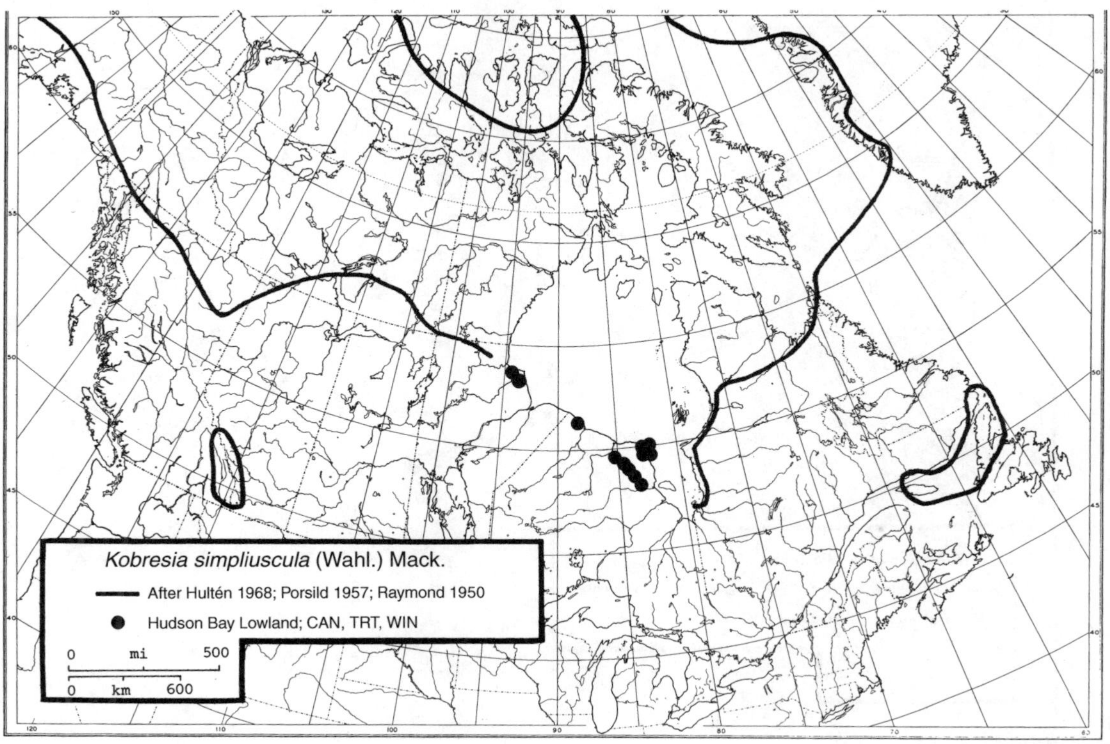

Fig. A81. Distribution of *Lathyrus japonicus* var. *pellitus* in eastern Canada.

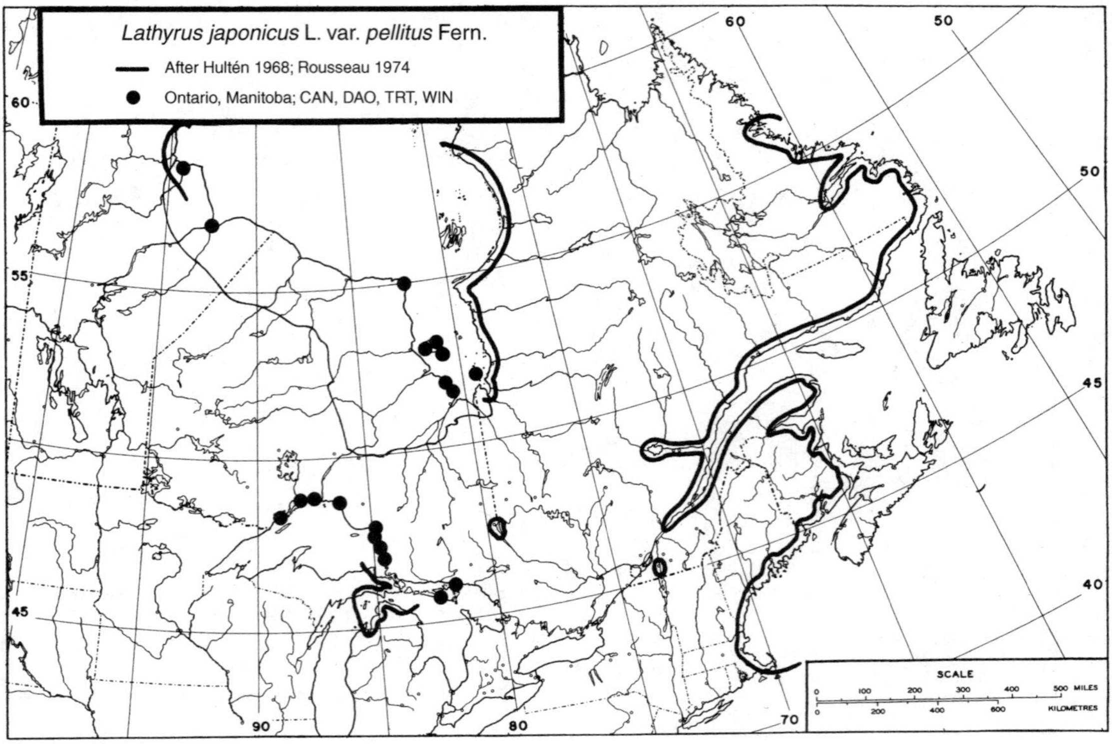

Fig. A82. Distribution of *Ledum decumbens* in Canada.

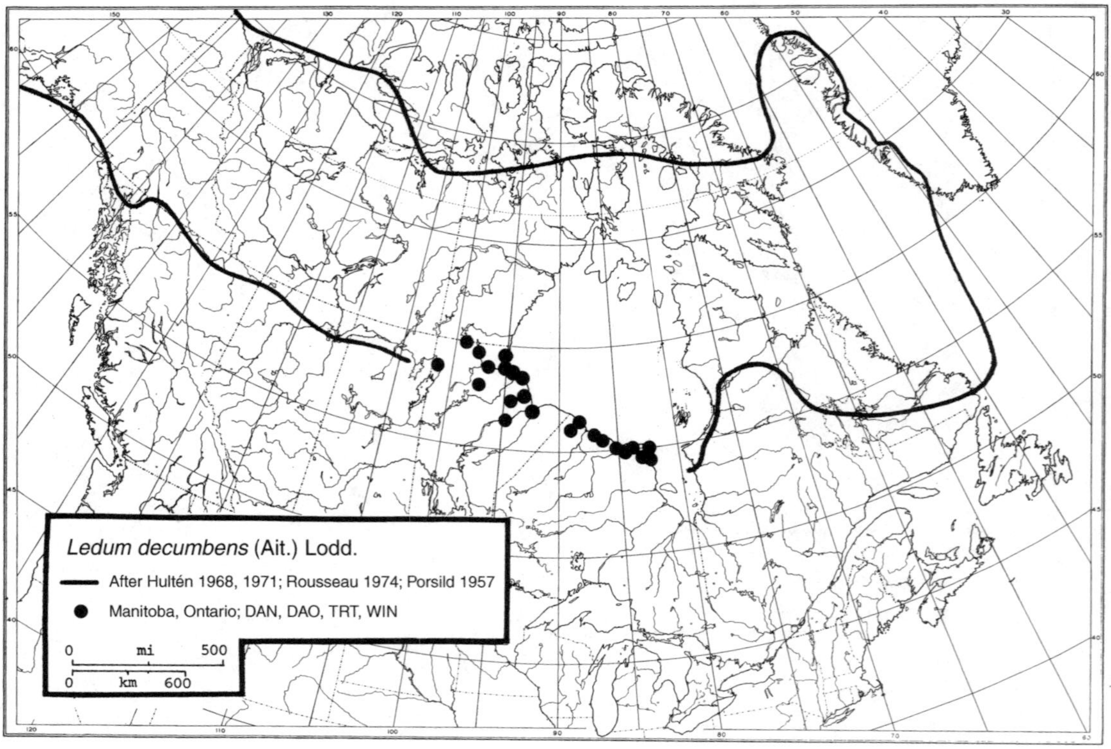

Fig. A83. Distribution of *Ledum decumbens* in eastern Canada.

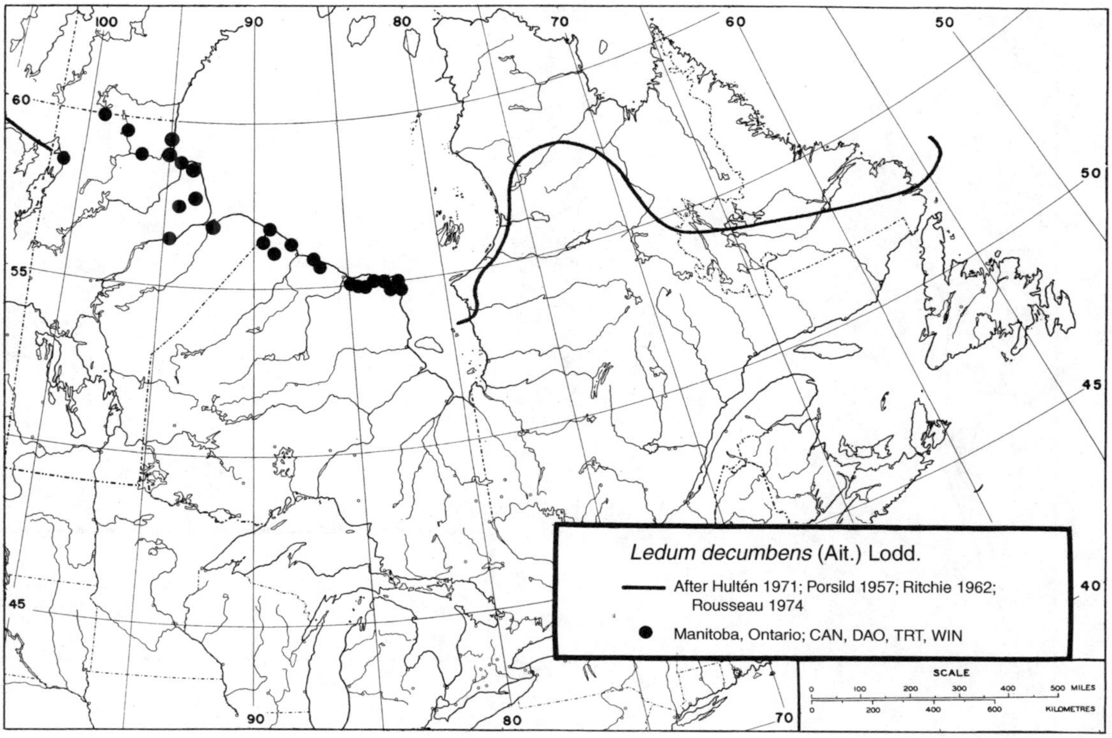

Fig. A84. Distribution of *Ligusticum scothicum* in eastern Canada.

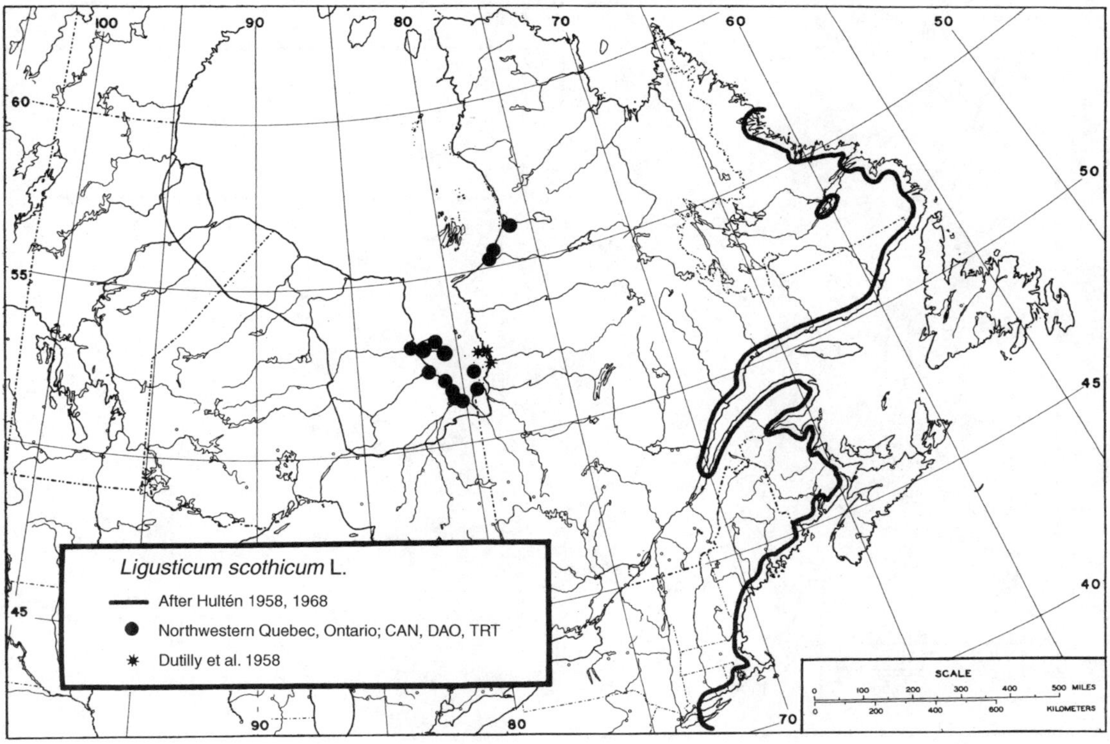

Fig. A85. Distribution of *Limosella aquatica* in eastern Canada.

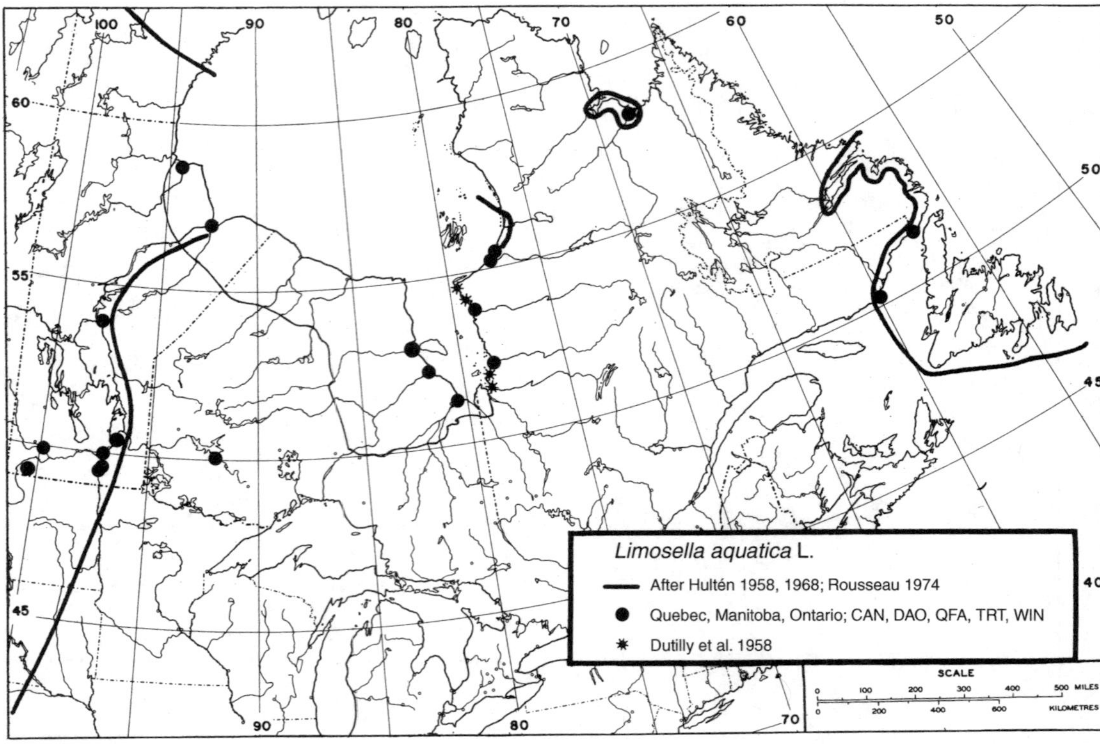

Fig. A86. Distribution of *Linum lewisii* in Canada.

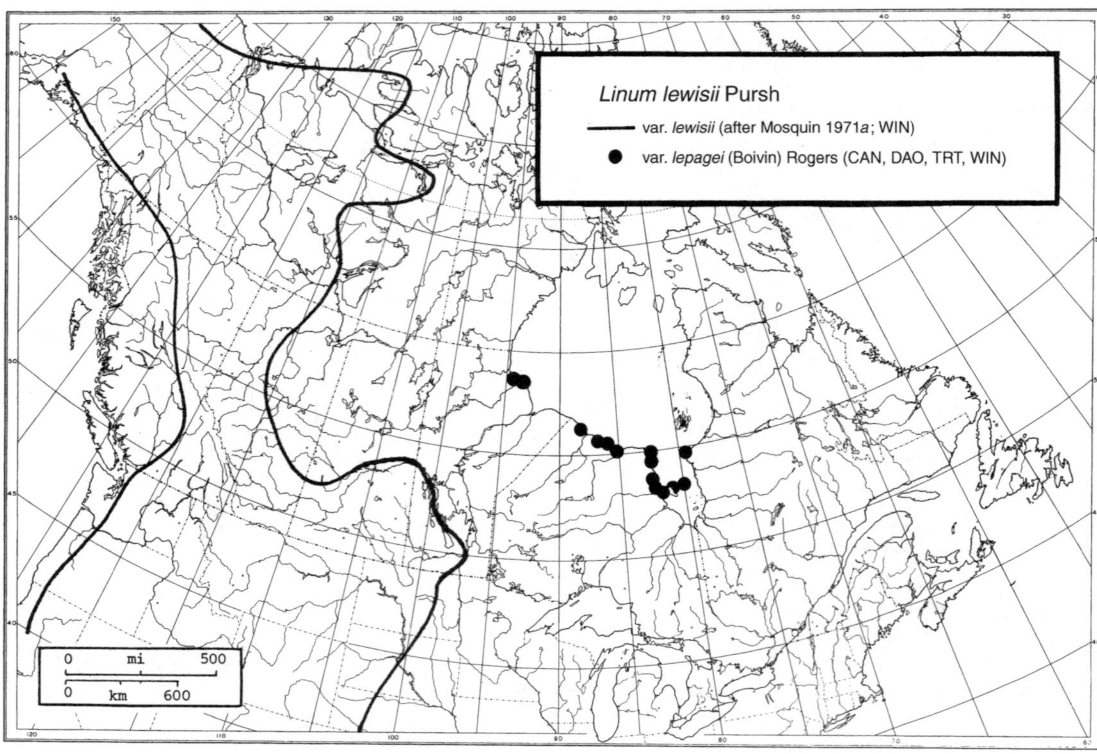

Fig. A87. Distribution of *Loiseleuria procumbens* in eastern Canada.

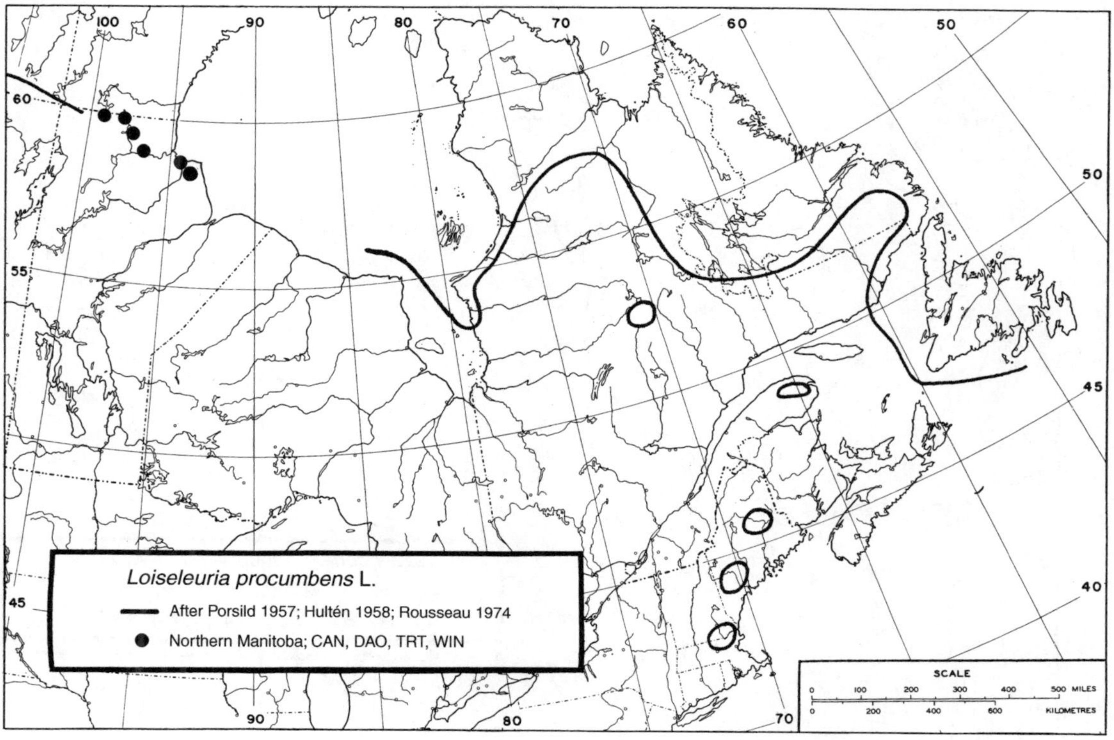

Fig. A88. Distribution of *Lomatogonium rotatum* in eastern Canada.

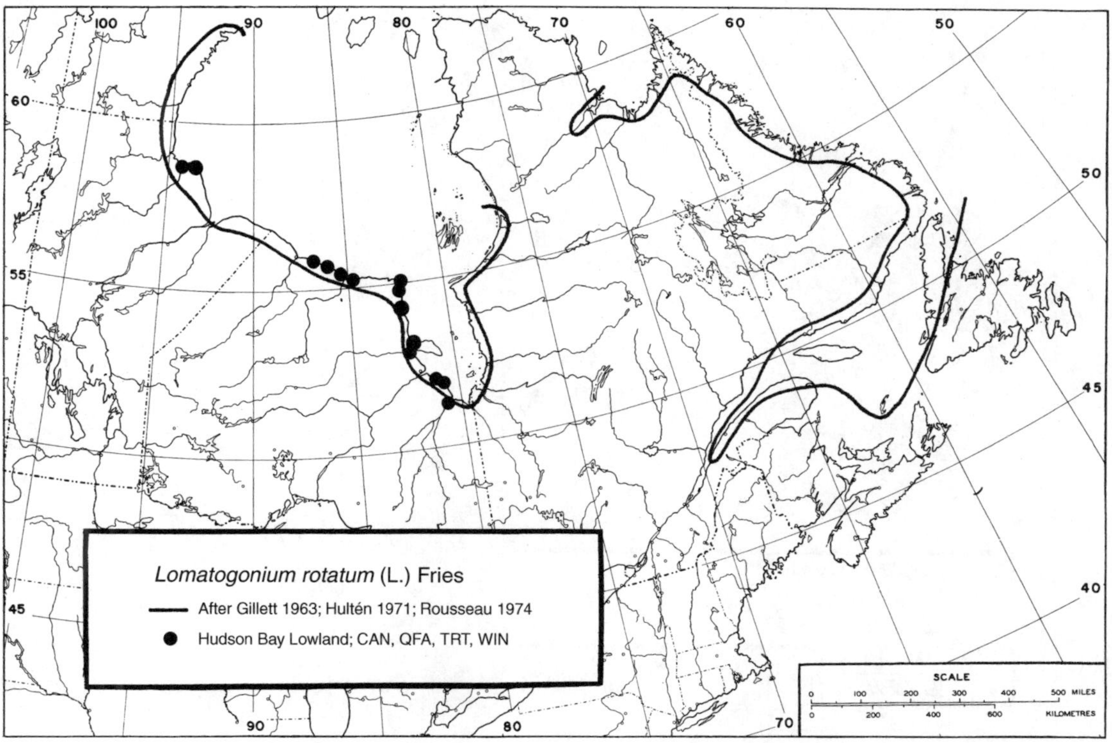

Fig. A89. Distribution of *Luzula confusa* in eastern Canada.

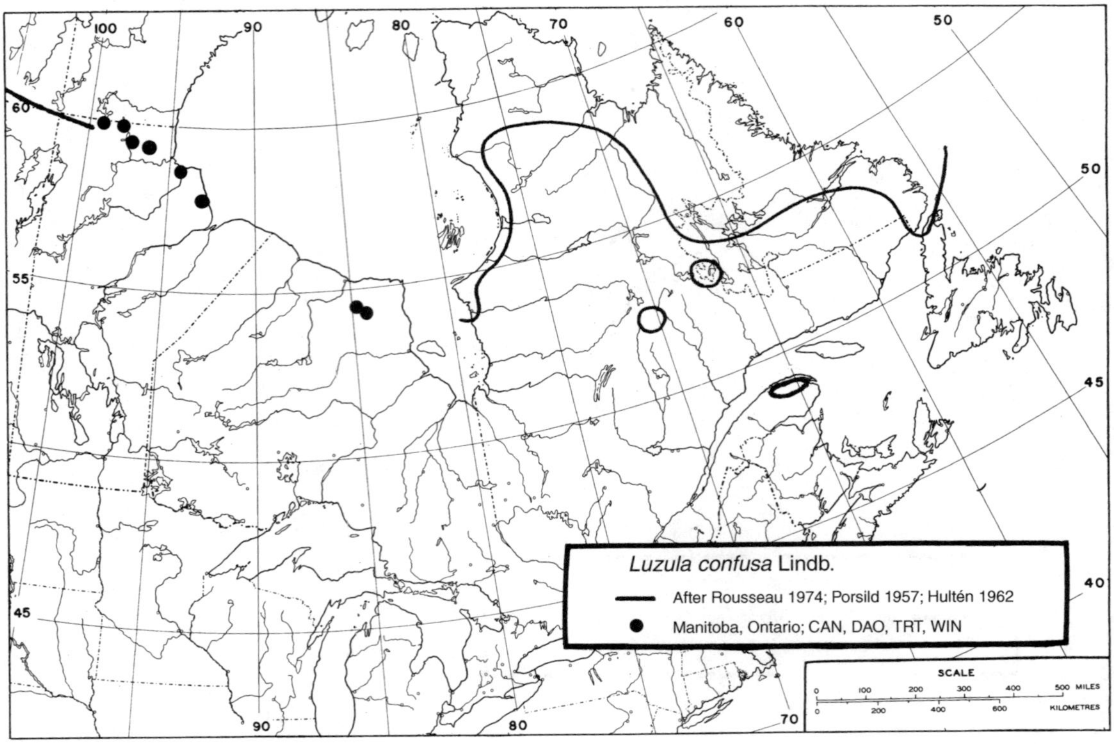

Fig. A90. Distribution of *Luzula confusa* in Canada.

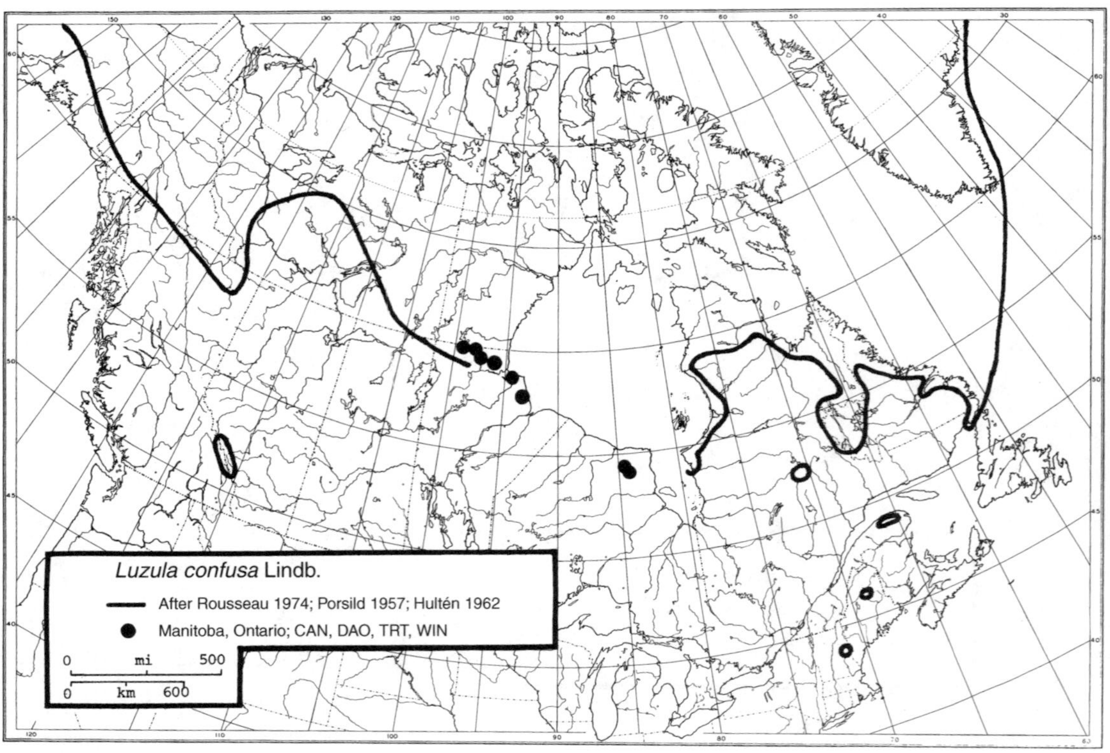

Fig. A91. Distribution of *Luzula multiflora* in eastern Canada.

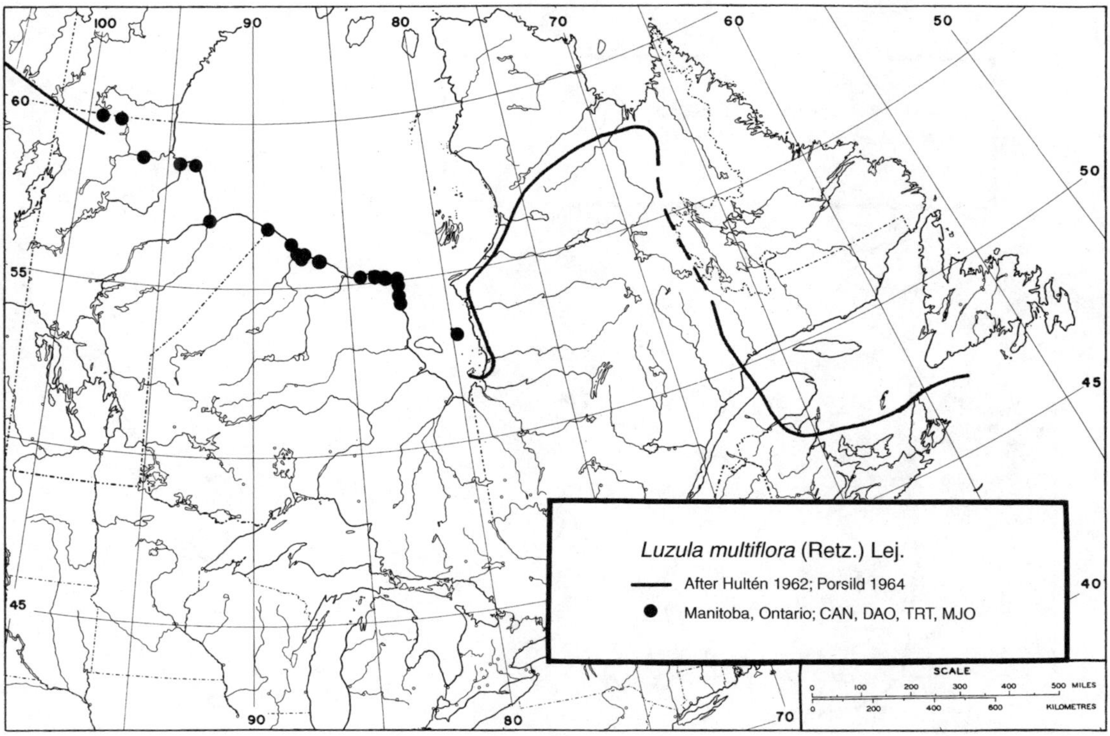

Fig. A92. Distribution of *Luzula wahlenbergii* in eastern Canada.

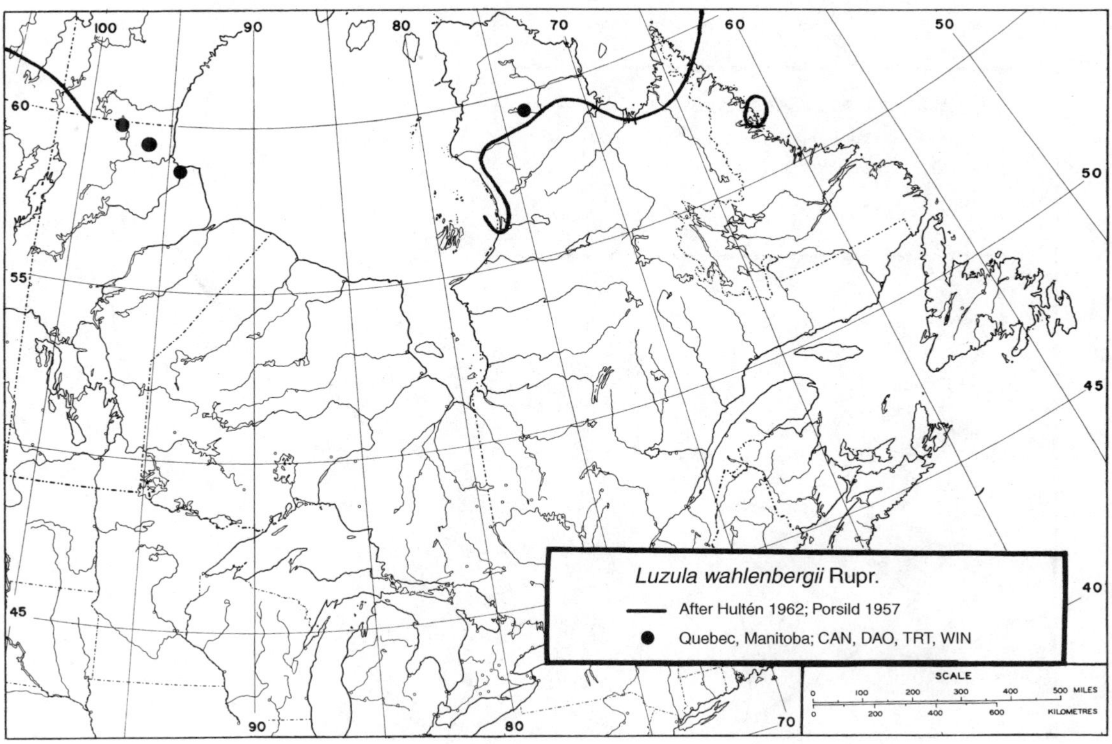

Fig. A93. Distribution of *Lycopus asper* in eastern Canada.

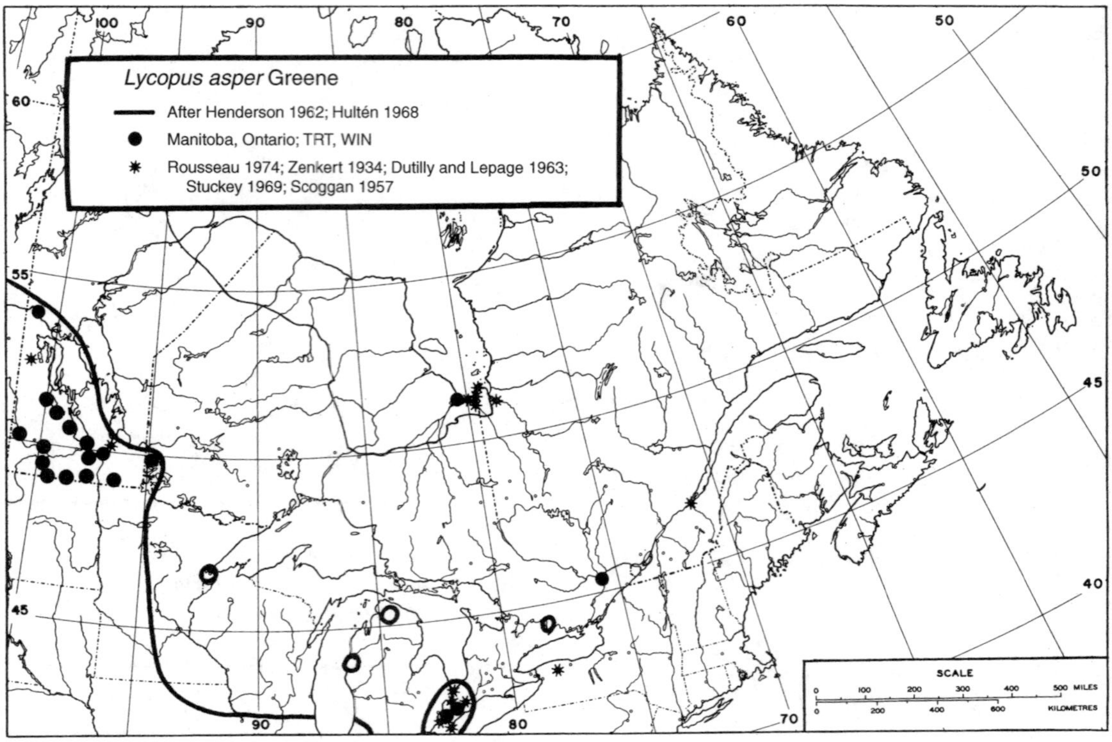

Fig. A94. Distribution of *Malaxis monophyllos* in eastern Canada.

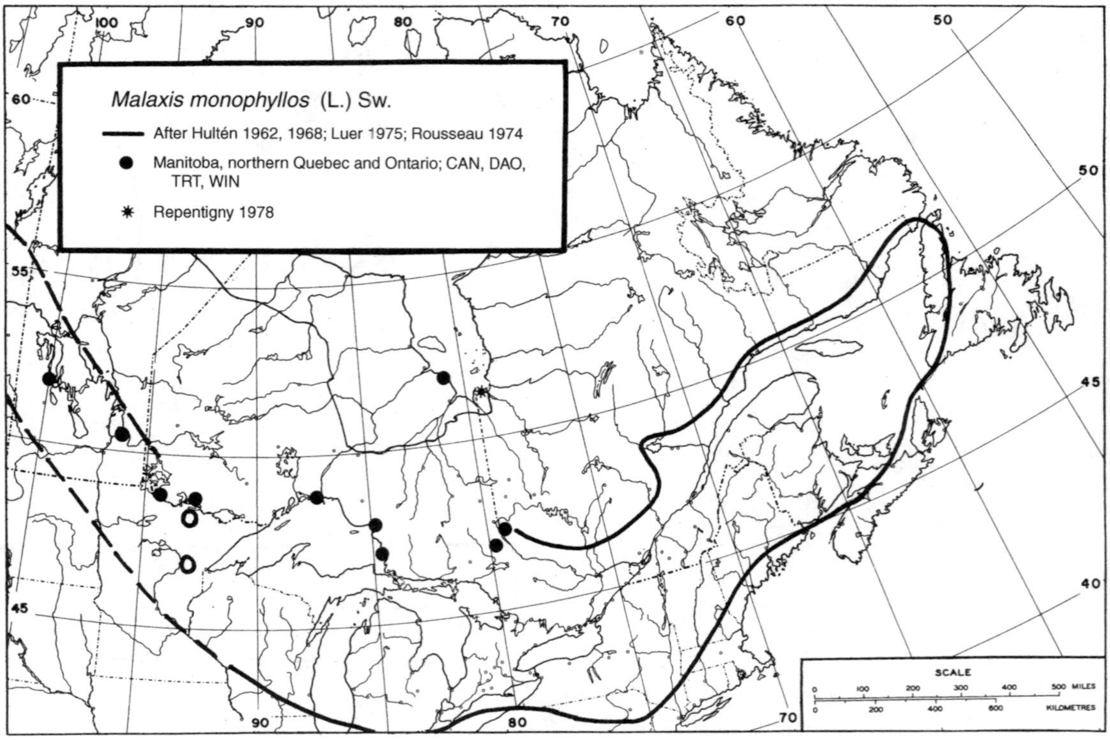

Fig. A95. Distribution of *Malaxis paludosa* in Canada.

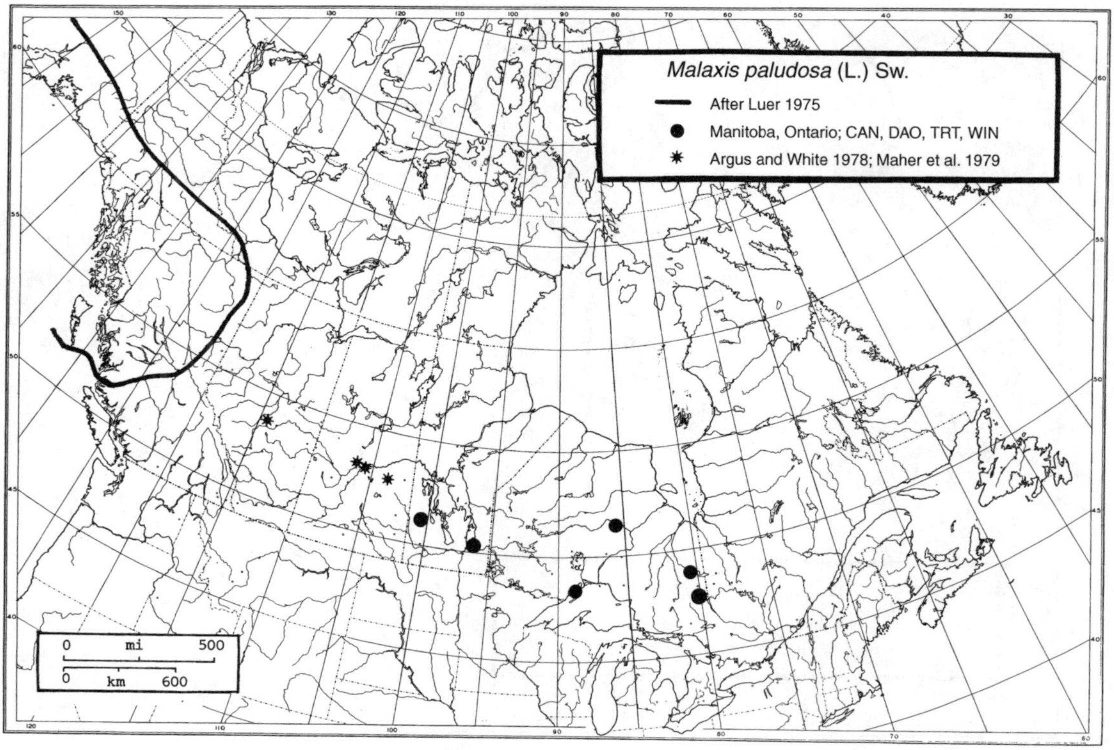

Fig. A96. Distribution of *Malaxis unifolia* in eastern Canada.

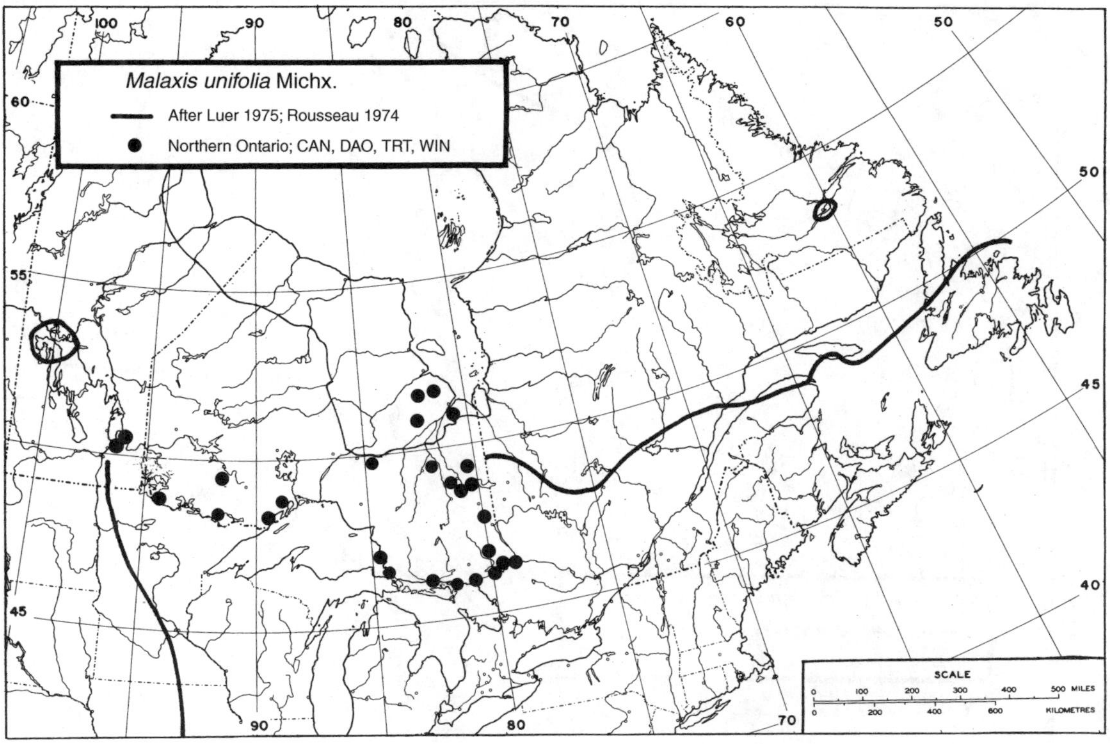

Fig. A97. Distribution of *Mertensia maritima* in eastern Canada.

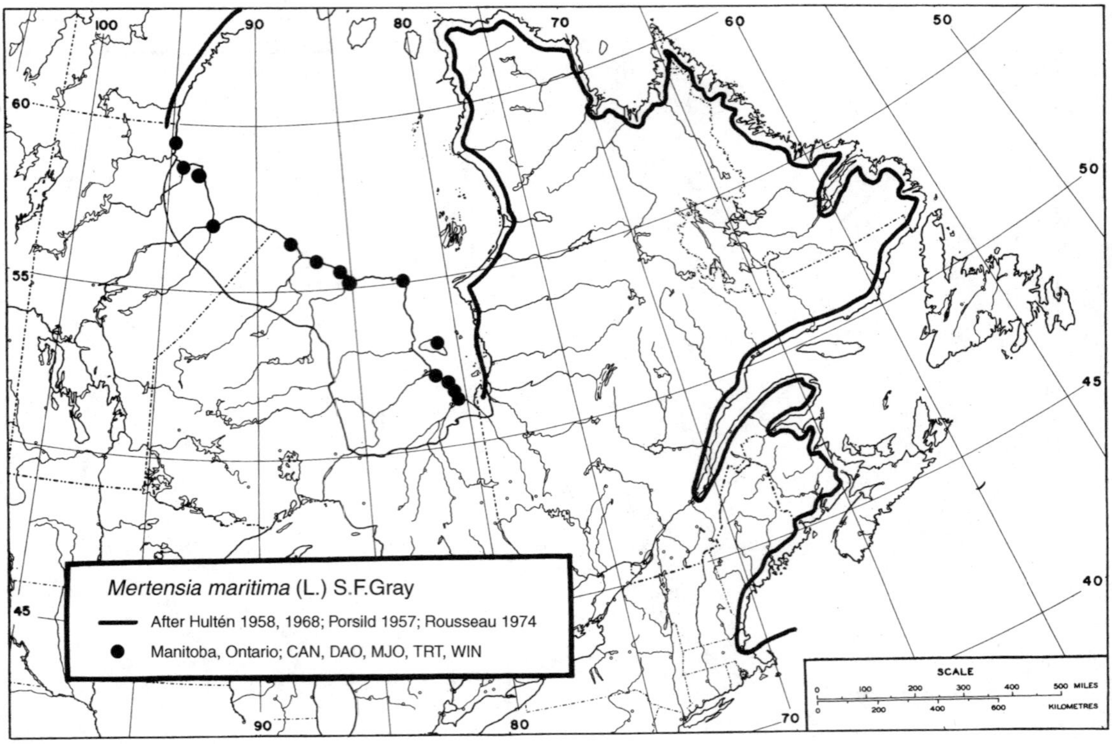

Fig. A98. Distribution of *Minuartia groenlandica* in eastern Canada.

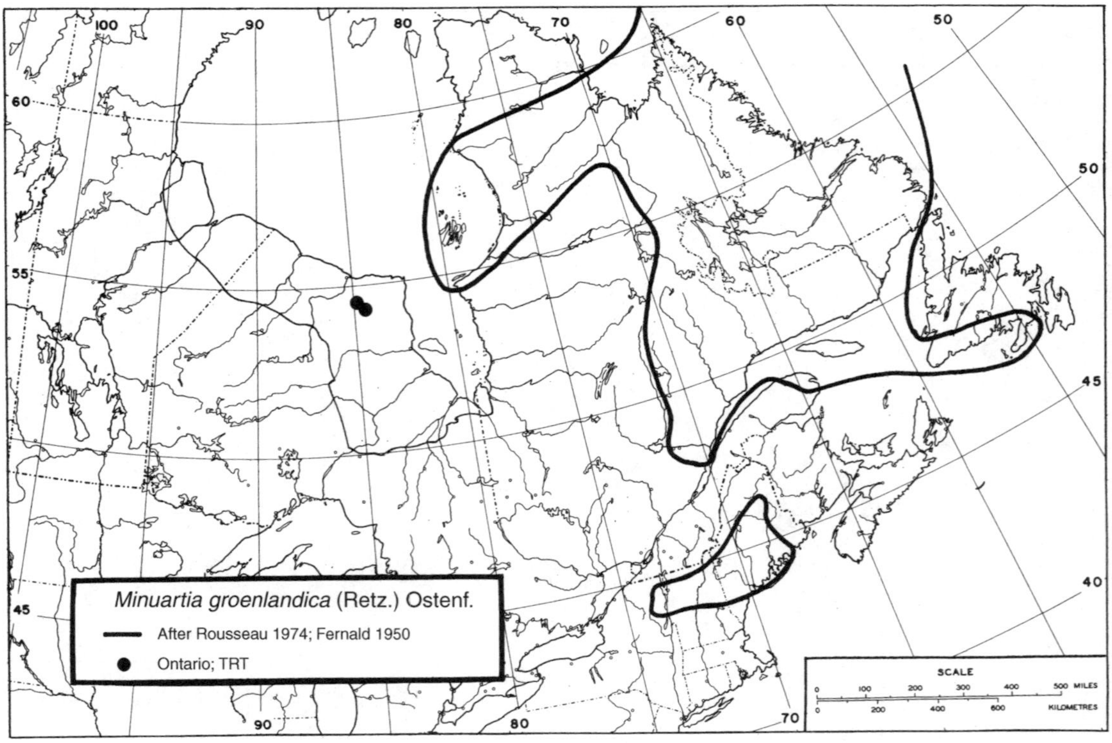

Fig. A99. Distribution of *Minuartia rubella* in eastern Canada.

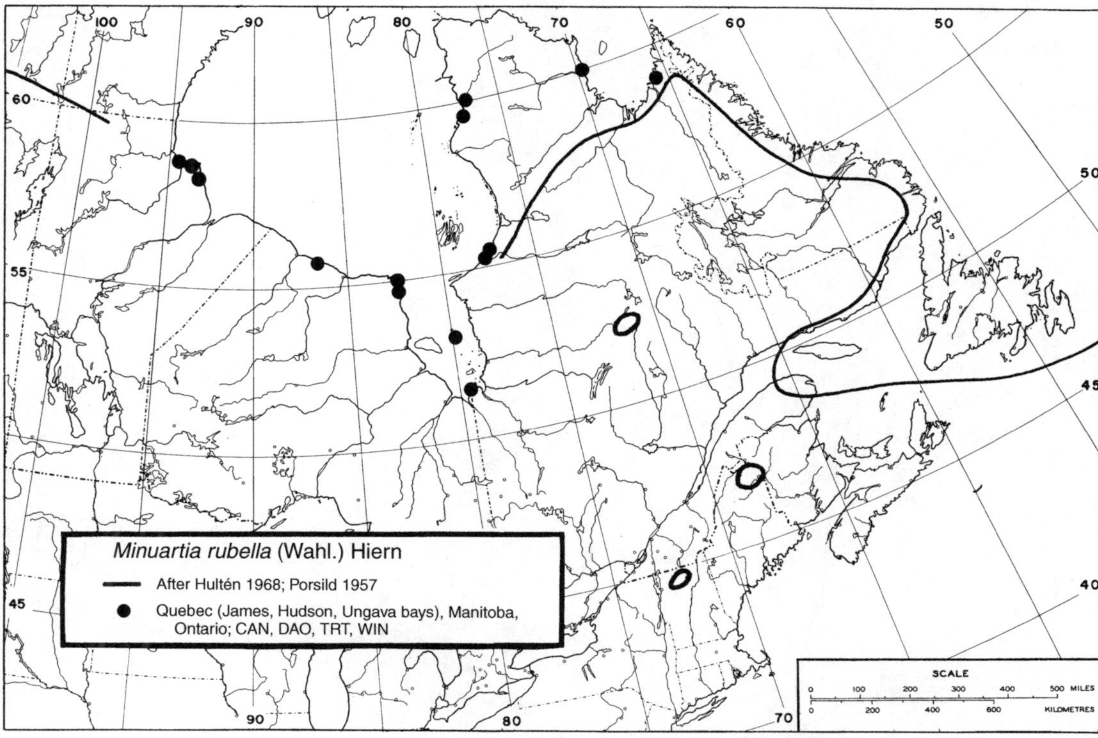

Fig. A100. Distribution of *Oxytropis campestris* var. *johannensis* in eastern Canada.

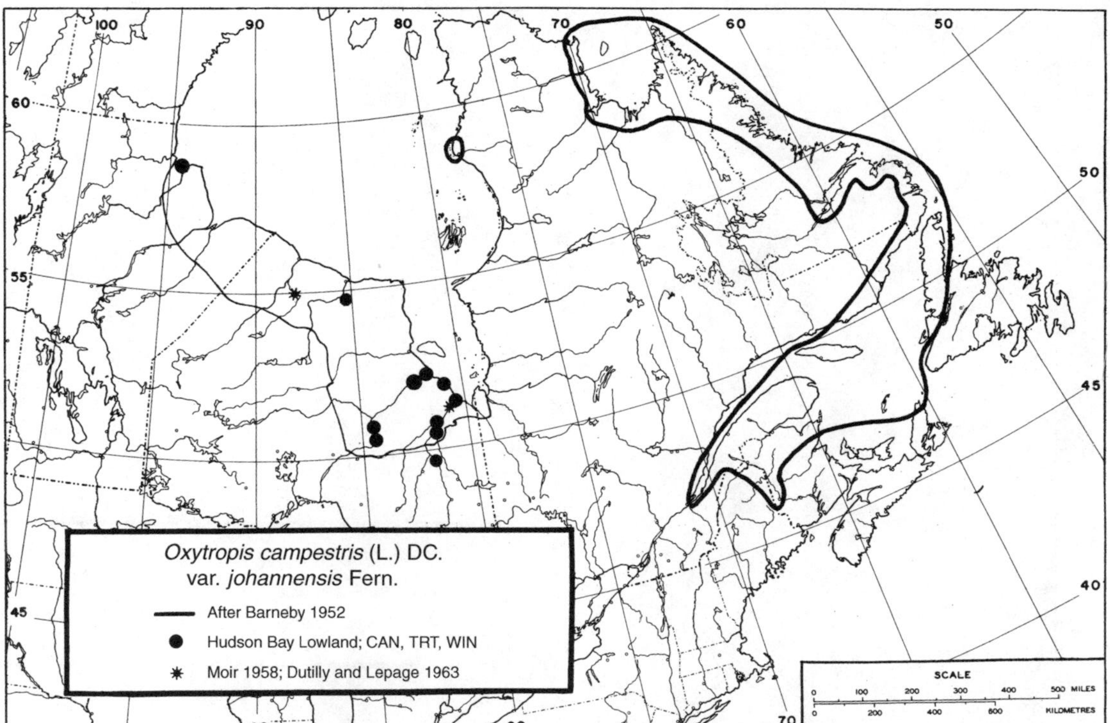

Fig. A101. Distribution of *Oxytropis splendens* in Canada.

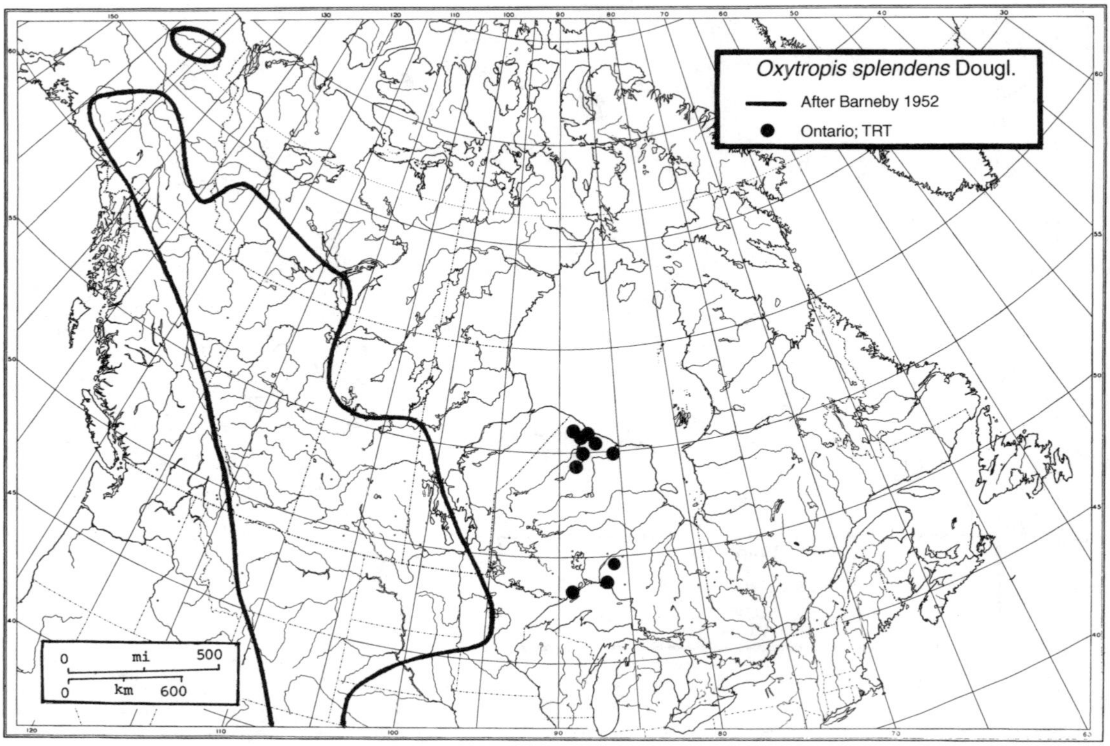

Fig. A102. Distribution of *Pedicularis flammea* in eastern Canada.

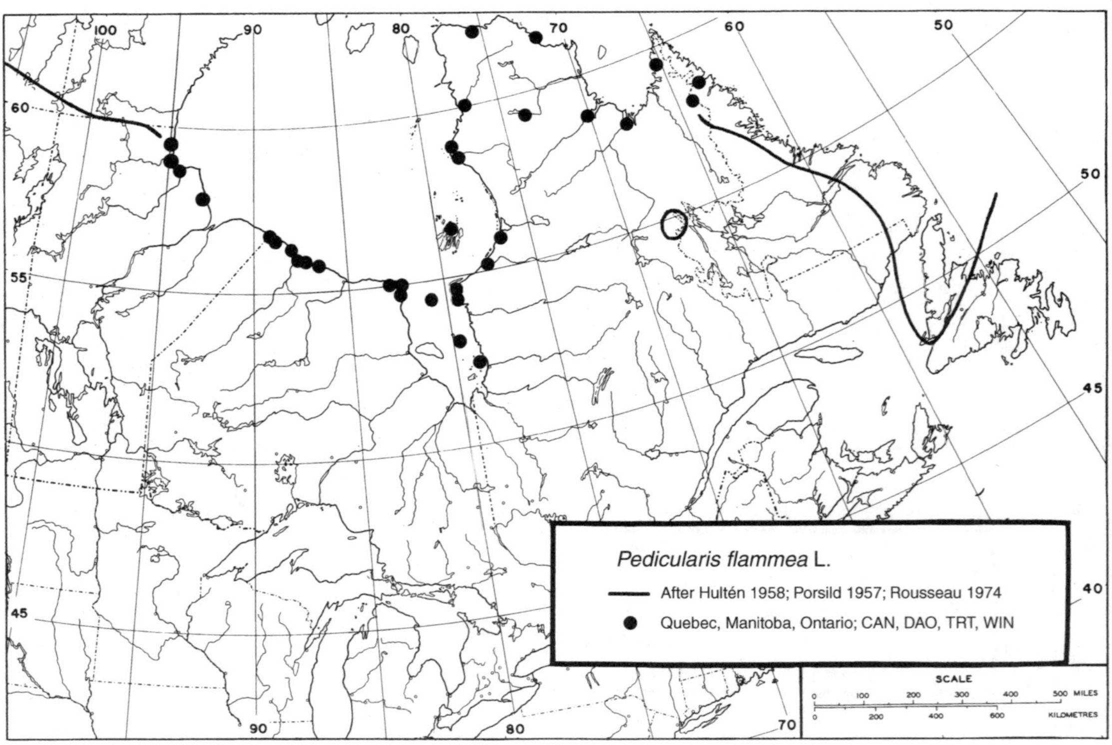

Fig. A103. Distribution of *Pedicularis flammea* in Canada.

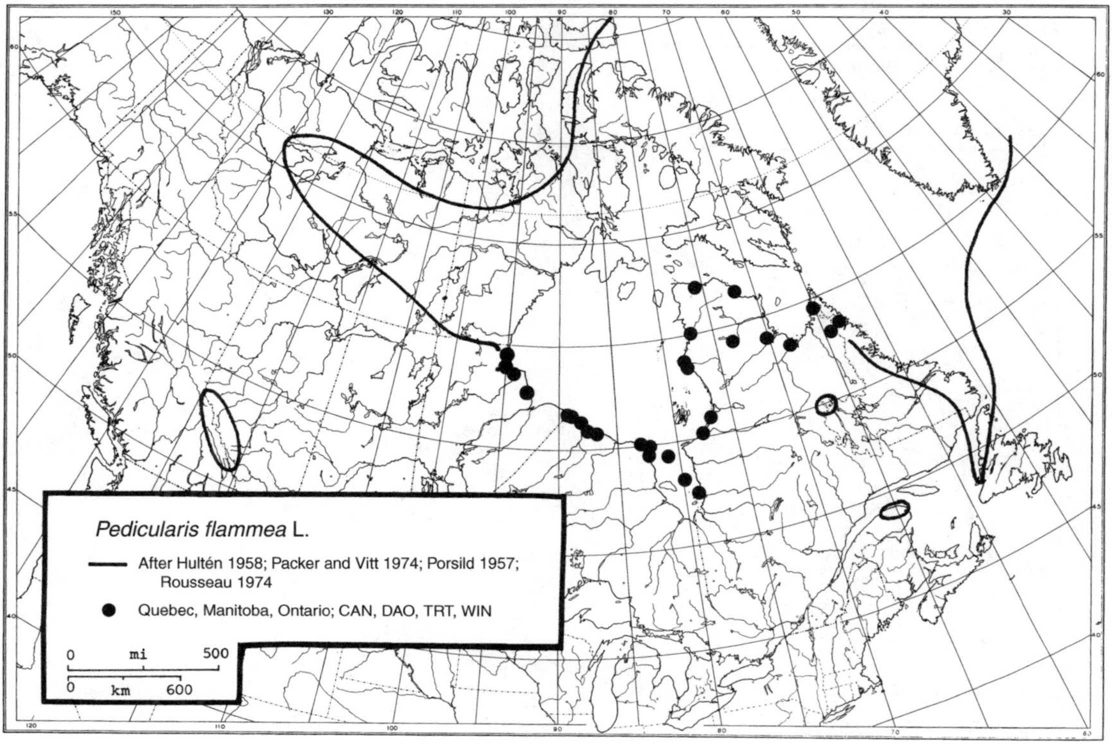

Fig. A104. Distribution of *Pedicularis groenlandica* in eastern Canada.

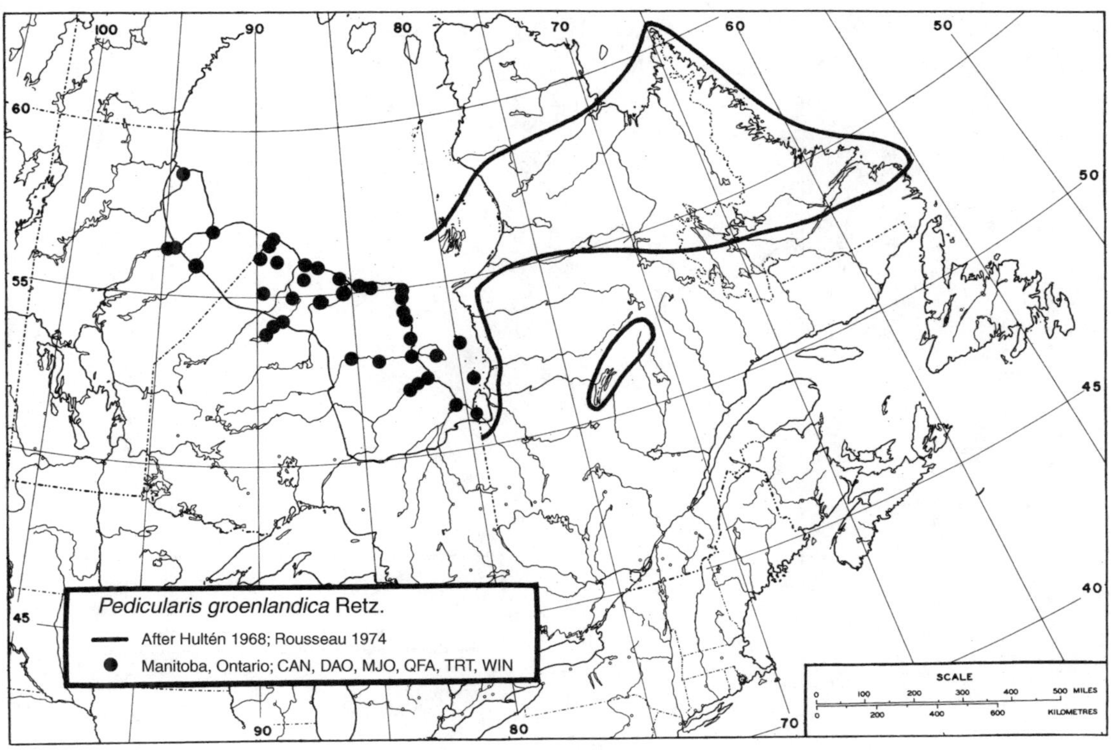

Fig. A105. Distribution of *Pedicularis groenlandica* in Canada.

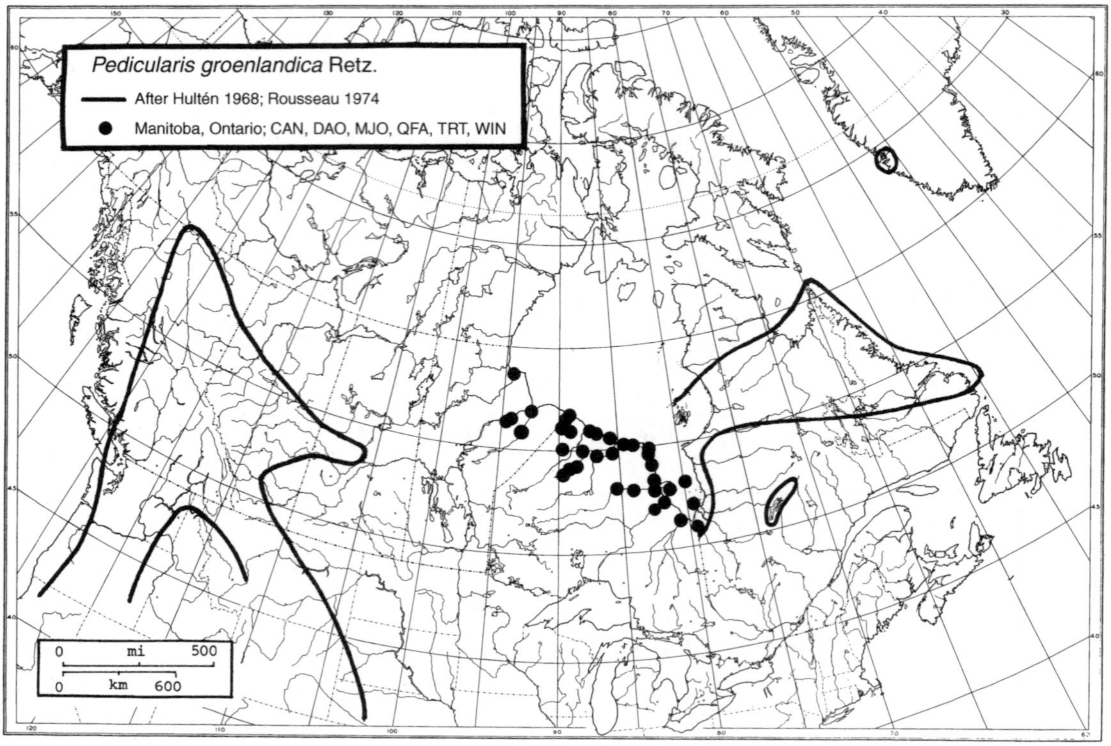

Fig. A106. Distribution of *Pedicularis labradorica* in eastern Canada.

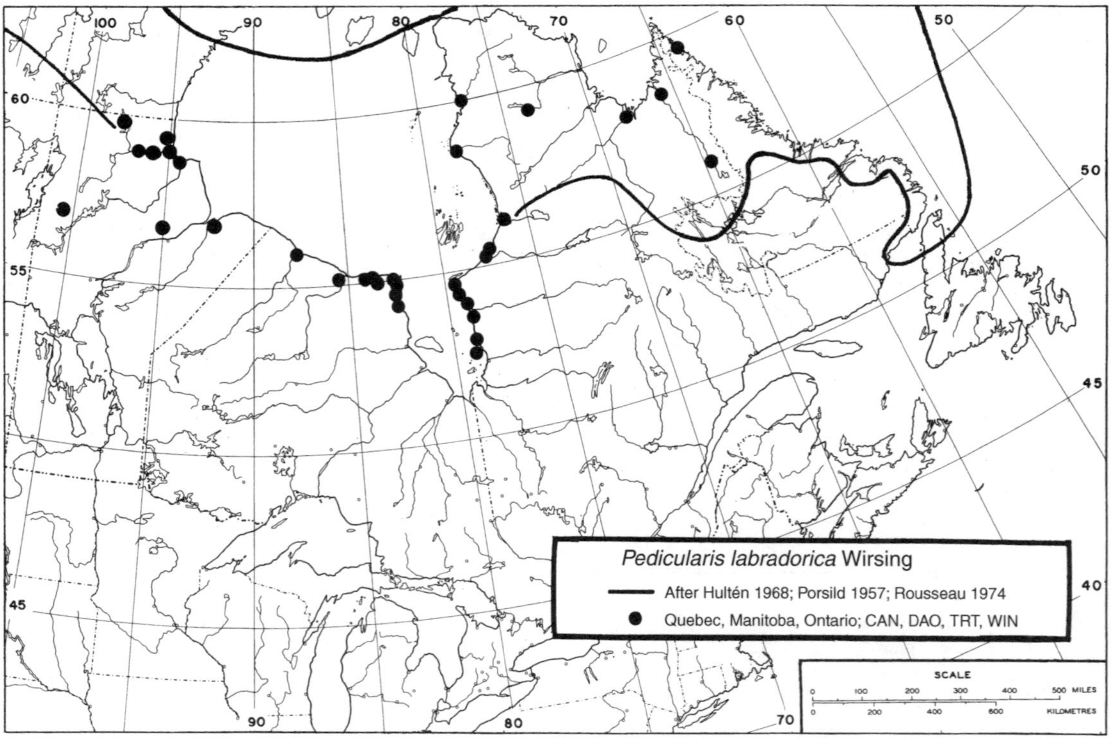

Fig. A107. Distribution of *Pedicularis lapponica* in eastern Canada.

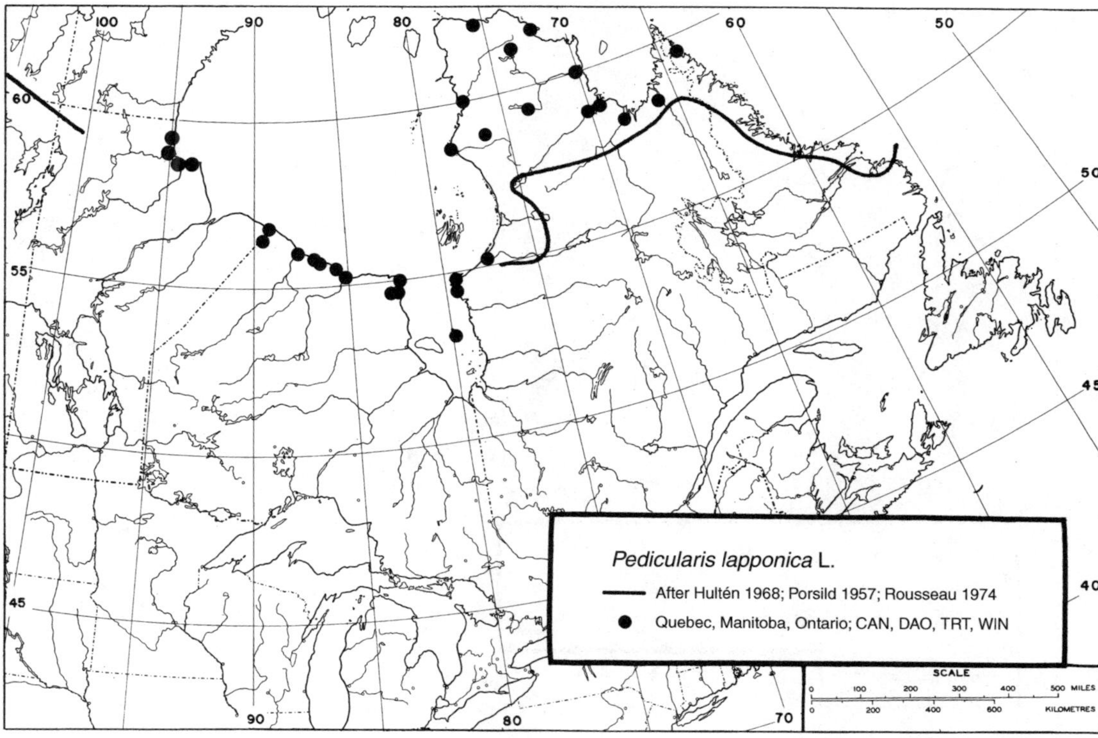

Fig. A108. Distribution of *Pedicularis parviflora* in eastern Canada.

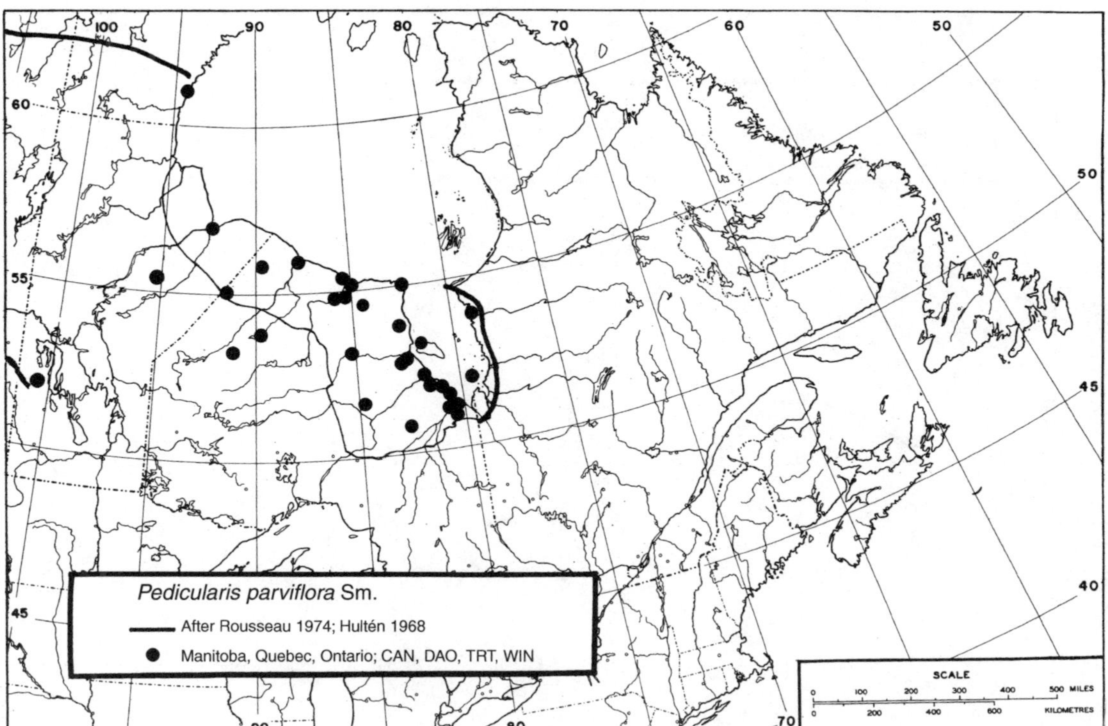

Fig. A109. Distribution of *Pedicularis sudetica* in Canada.

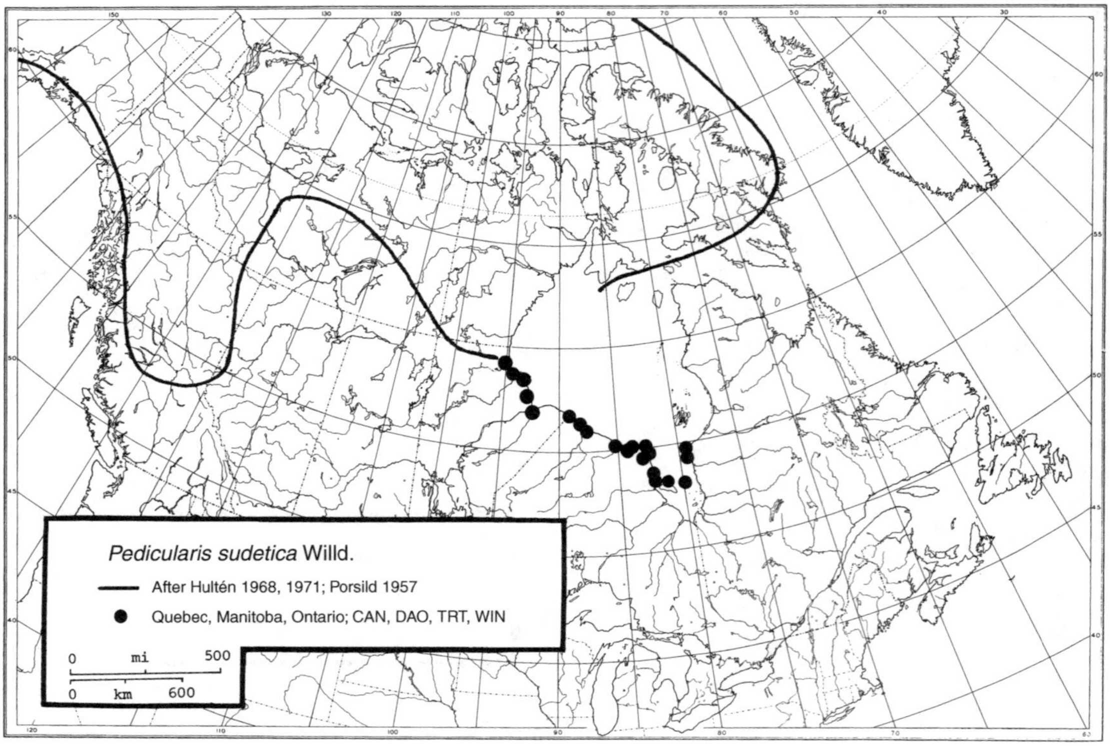

Fig. A110. Distribution of *Phragmites australis* in eastern Canada.

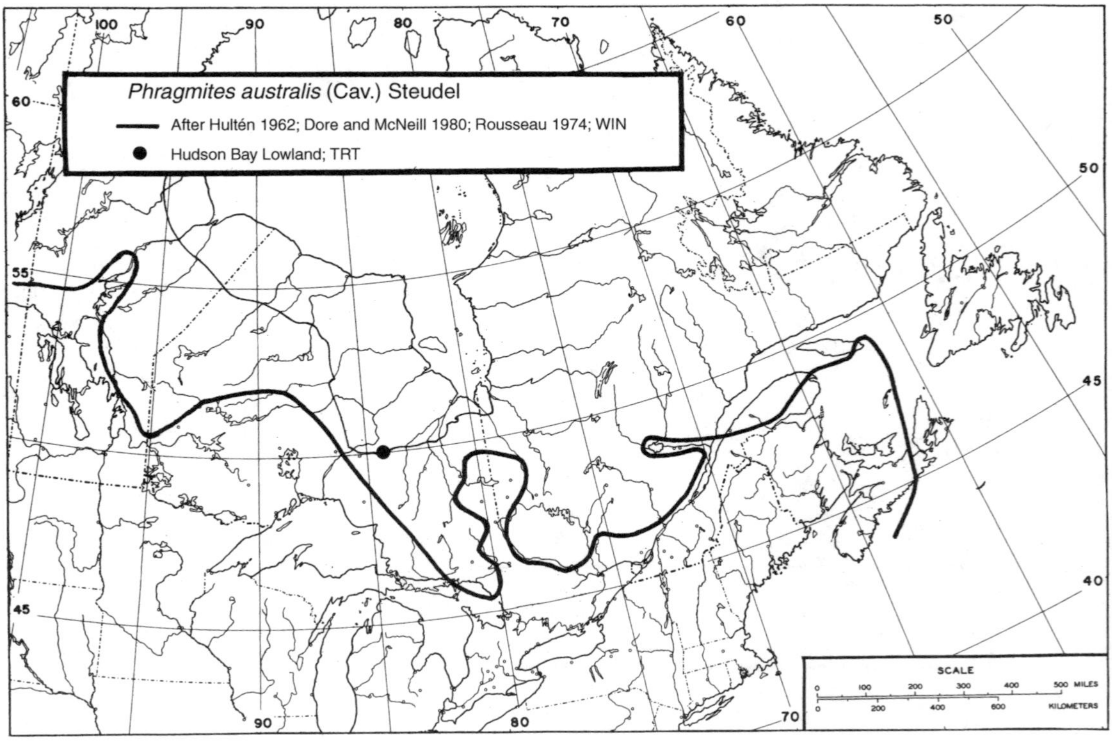

Fig. A111. Distribution of *Phyllodoce caerulea* in Canada.

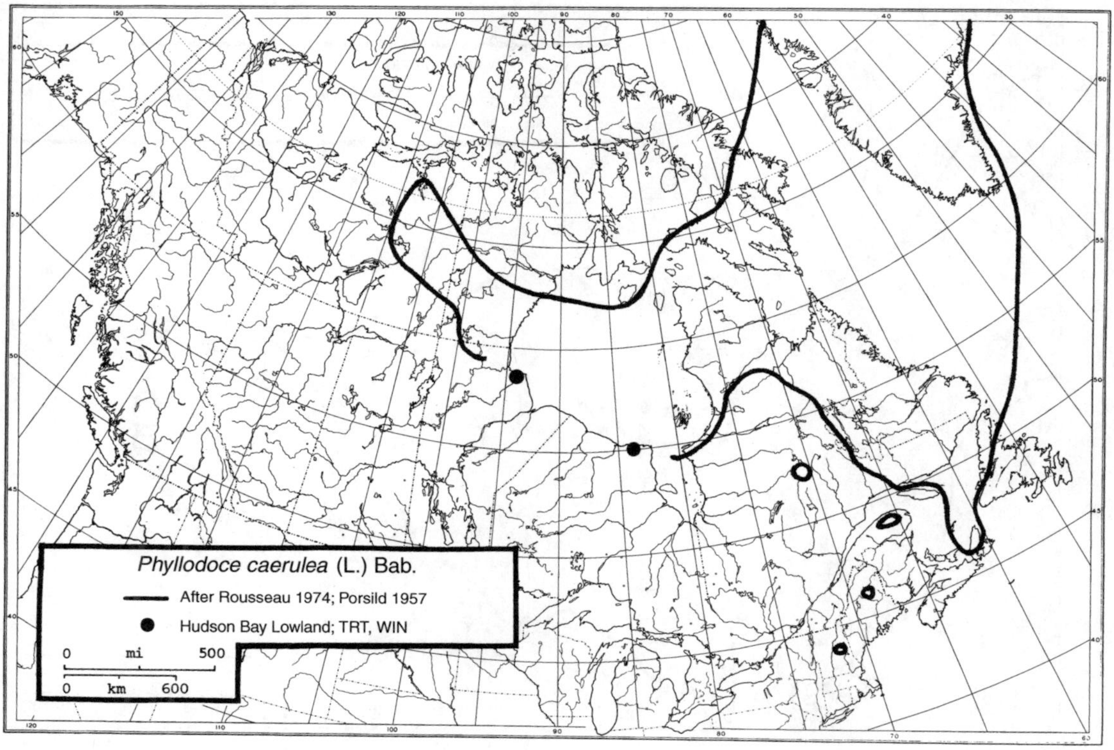

Fig. A112. Distribution of *Pinguicula villosa* in Canada.

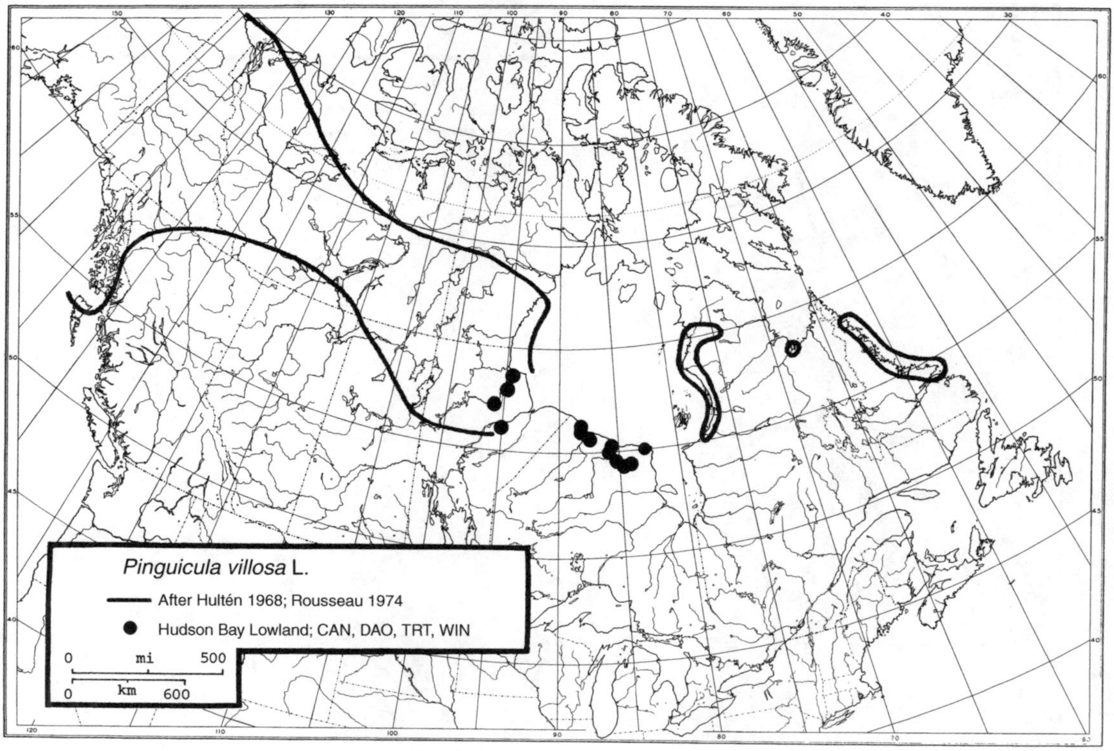

Fig. A113. Distribution of *Pinus banksiana* in eastern Canada.

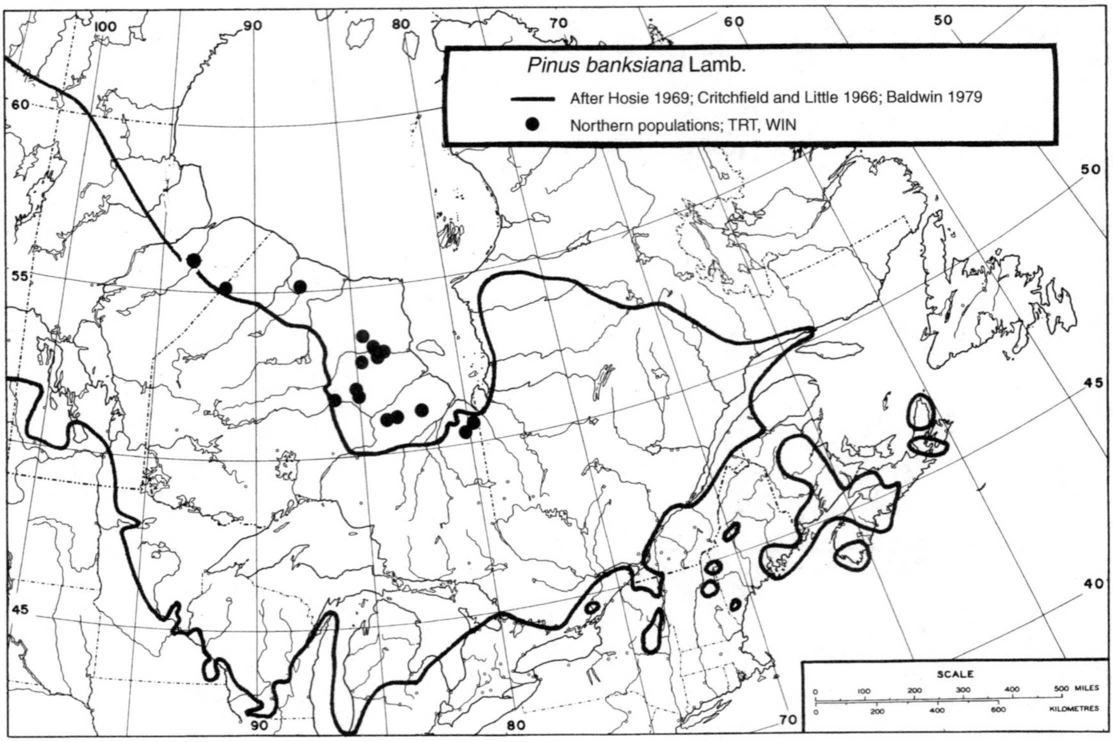

Fig. A114. Distribution of *Pinus* Subsect. *Contortae* in Canada.

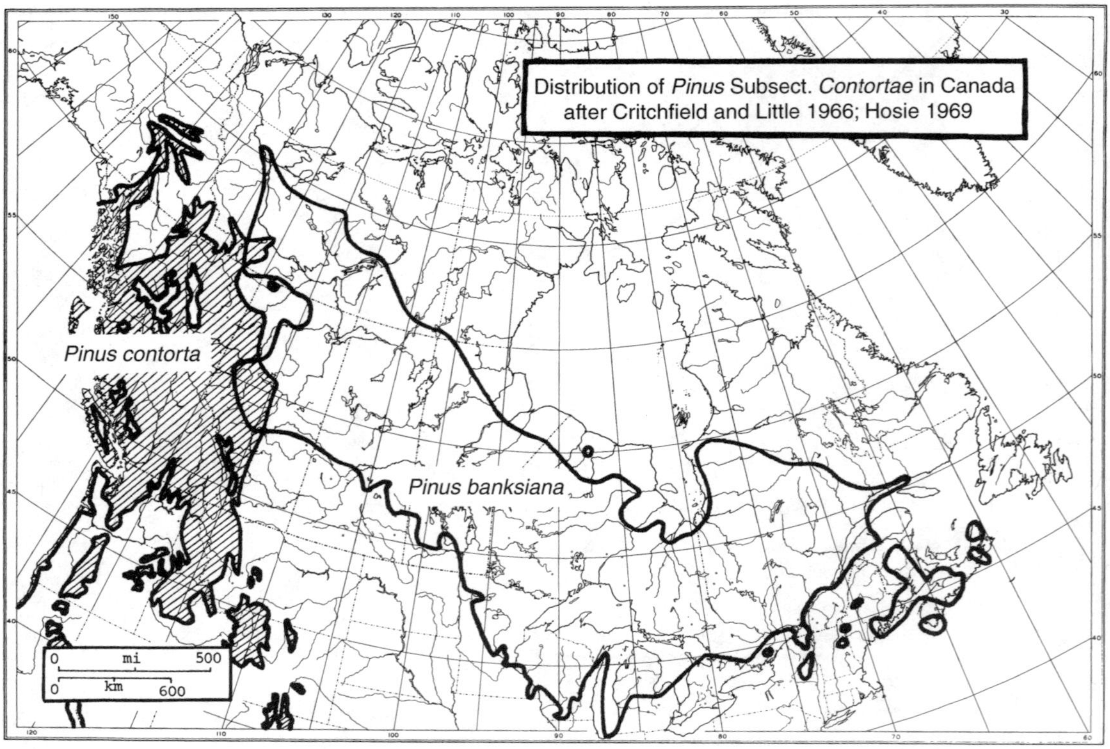

Fig. A115. Distribution of *Pinus resinosa* in eastern Canada.

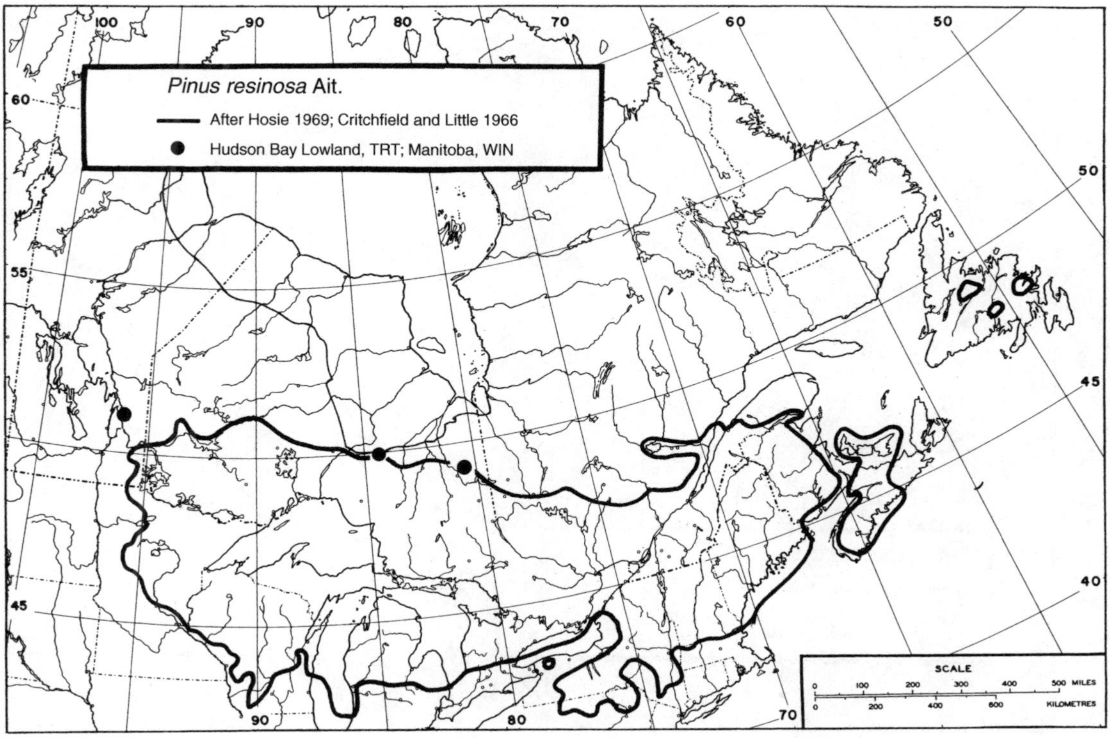

Fig. A116. Distribution of *Pinus strobus* in eastern Canada.

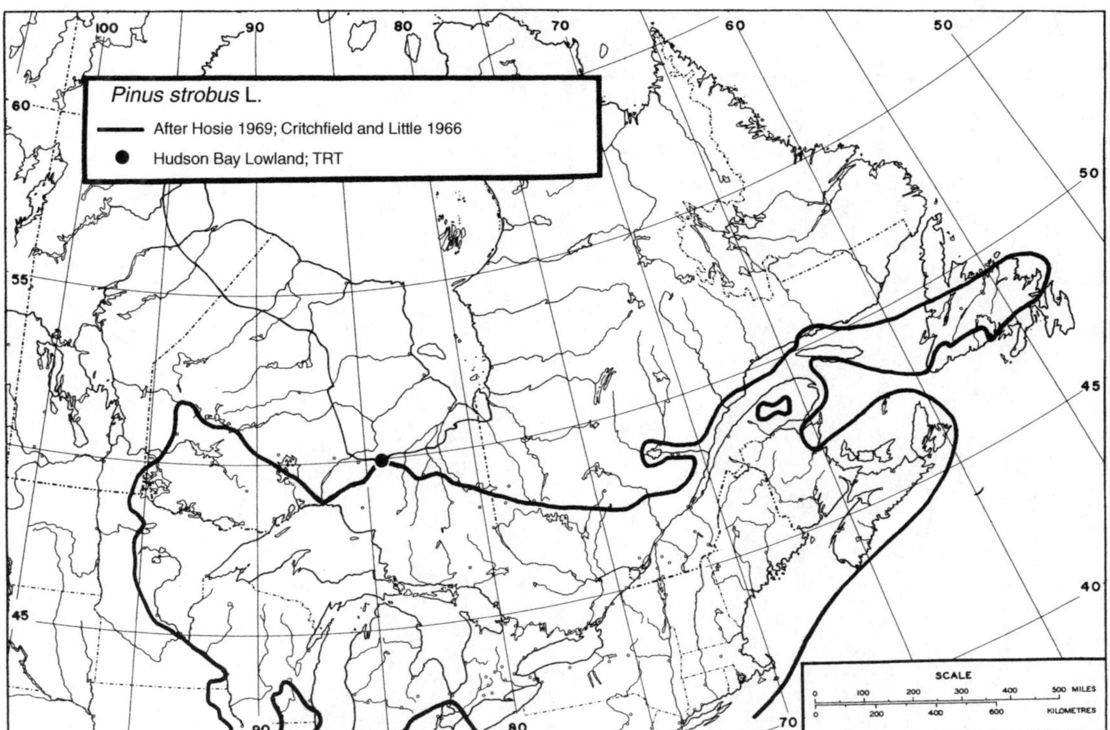

Fig. A117. Distribution of *Plantago maritima* in eastern Canada.

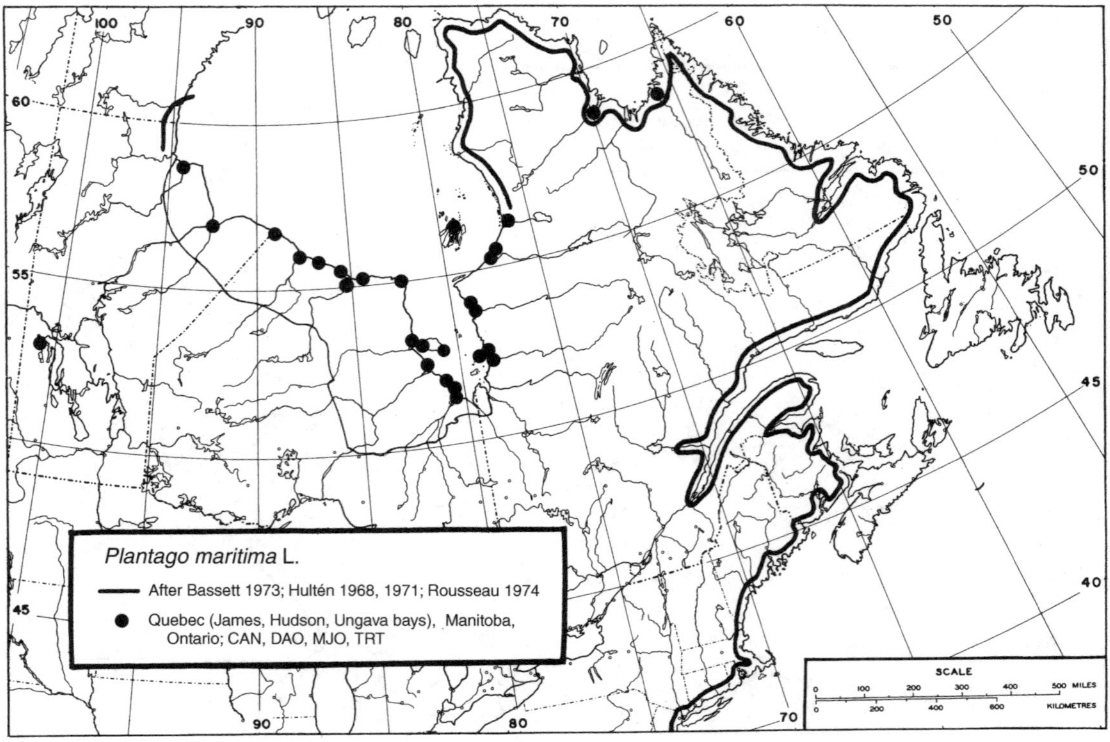

Fig. A118. Distribution of *Platanthera lacera* in eastern Canada.

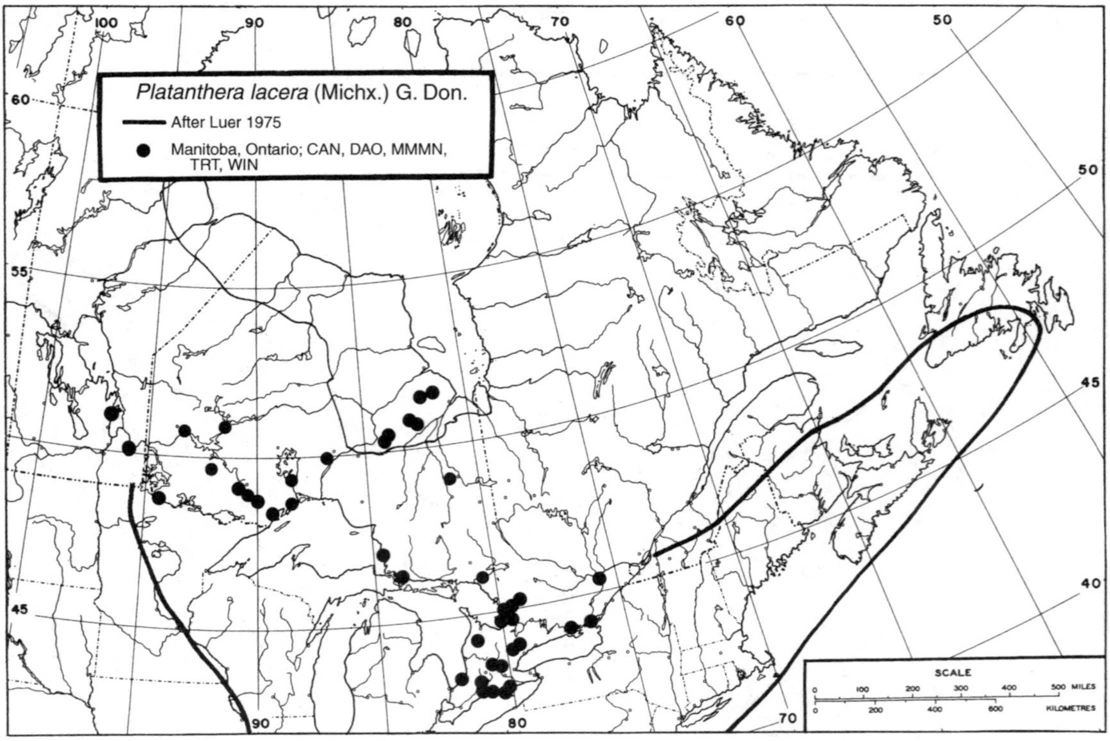

Fig. A119. Distribution of *Poa alpina* in eastern Canada.

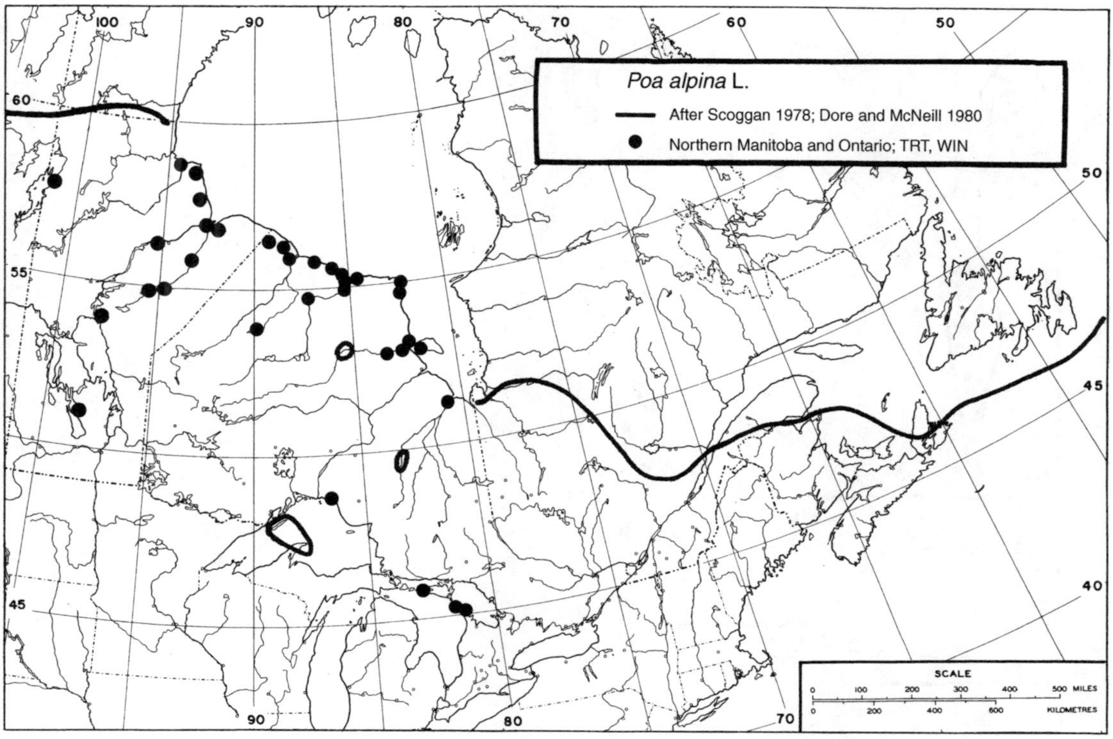

Fig. A120. Distribution of *Poa arctica* in eastern Canada.

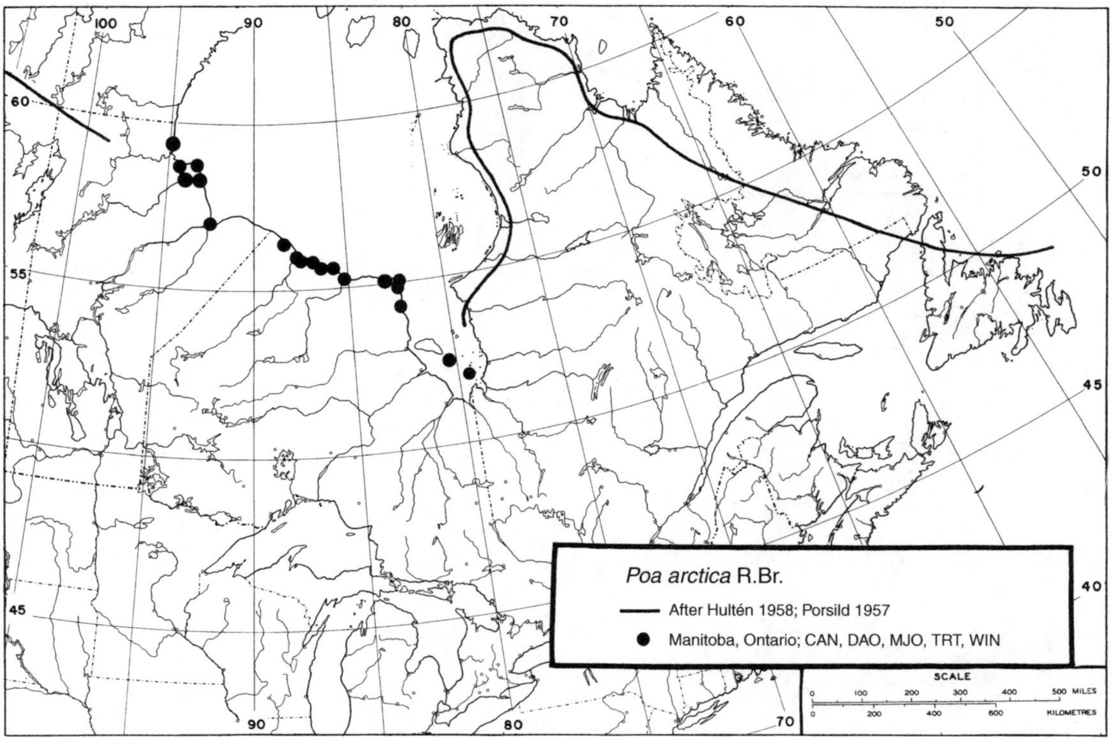

Fig. A121. Distribution of *Poa eminens* in Canada.

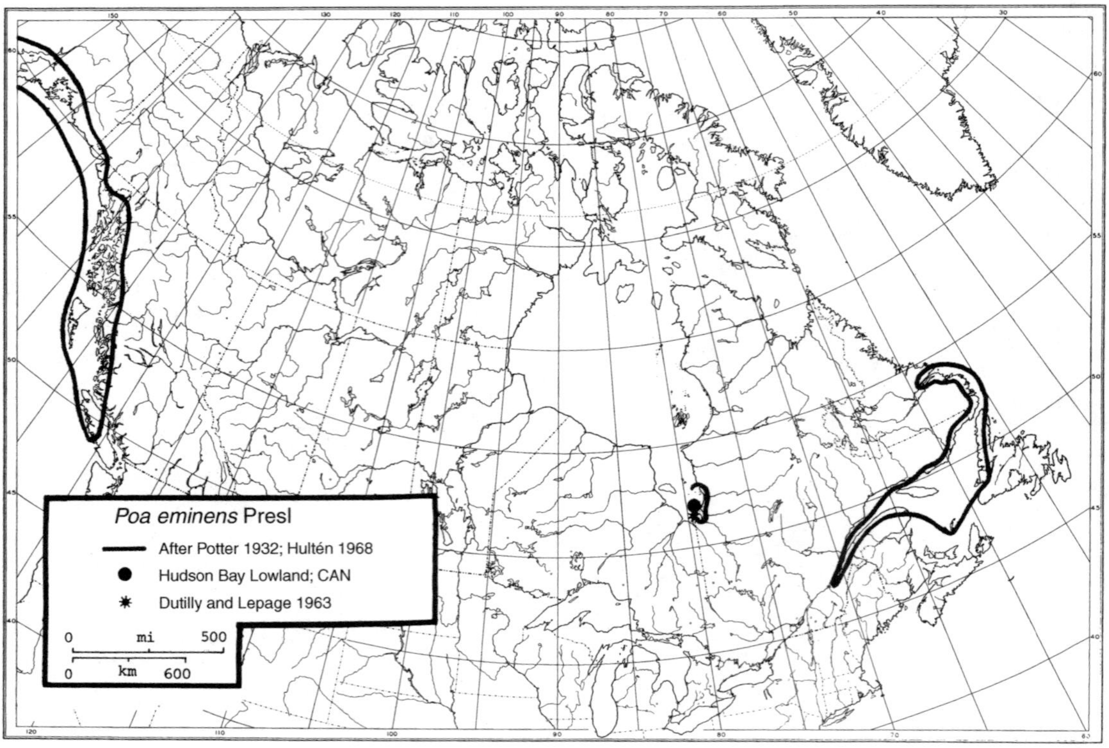

Fig. A122. Distribution of *Pogonia ophioglossoides* in eastern Canada.

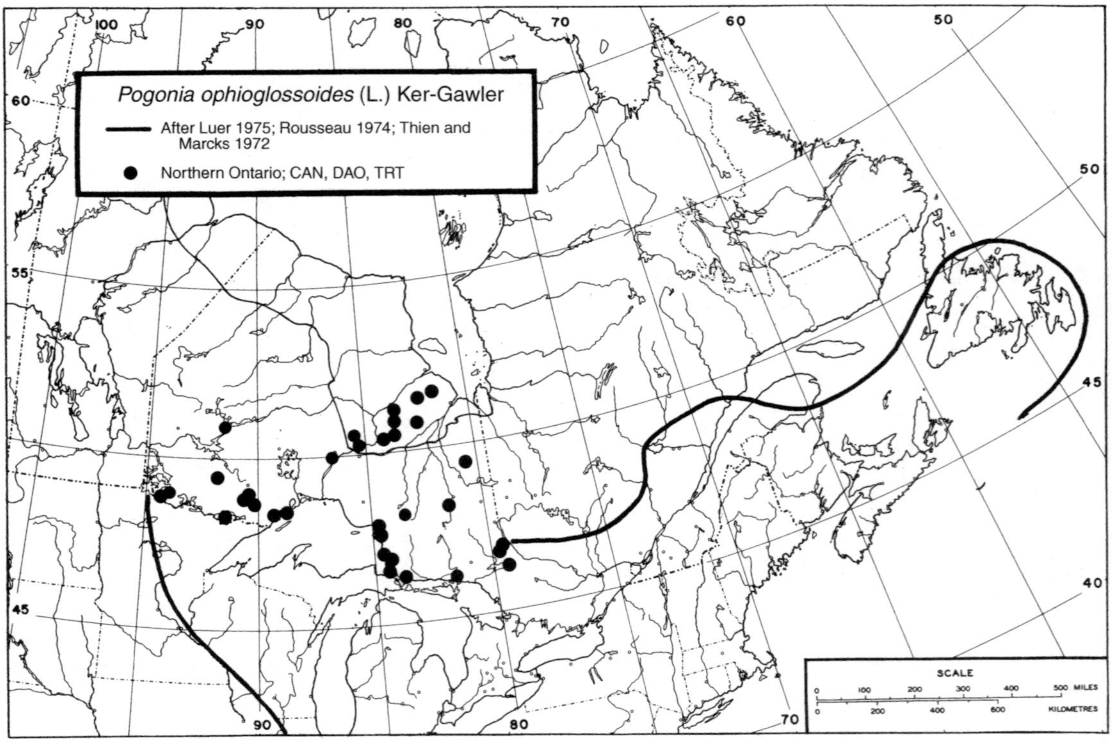

Fig. A123. Distribution of *Polygonum fowleri* in eastern Canada.

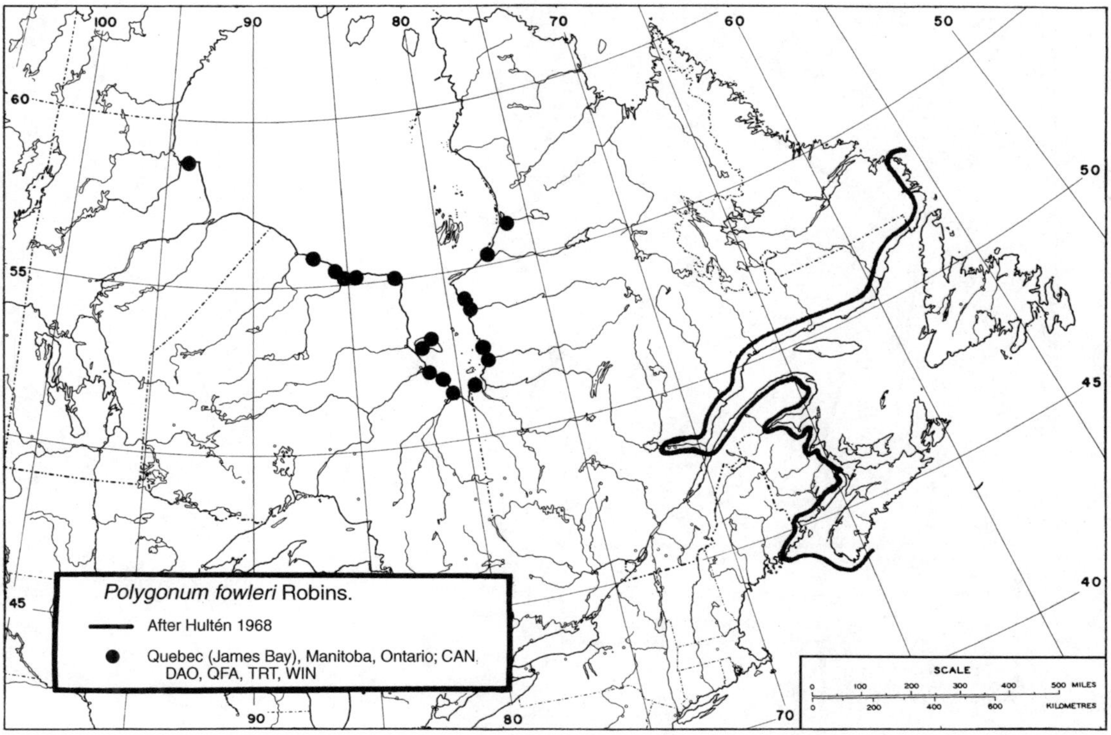

Fig. A124. Distribution of *Potentilla multifida* in Canada.

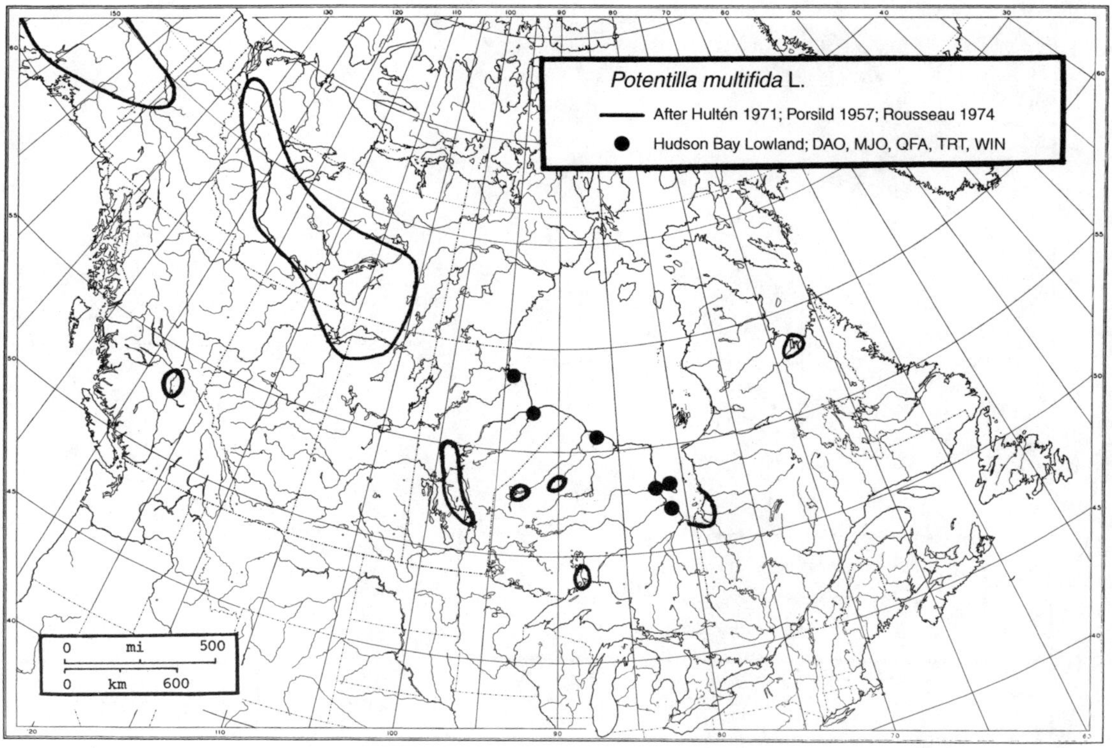

Fig. A125. Distribution of *Potentilla nivea* in Canada.

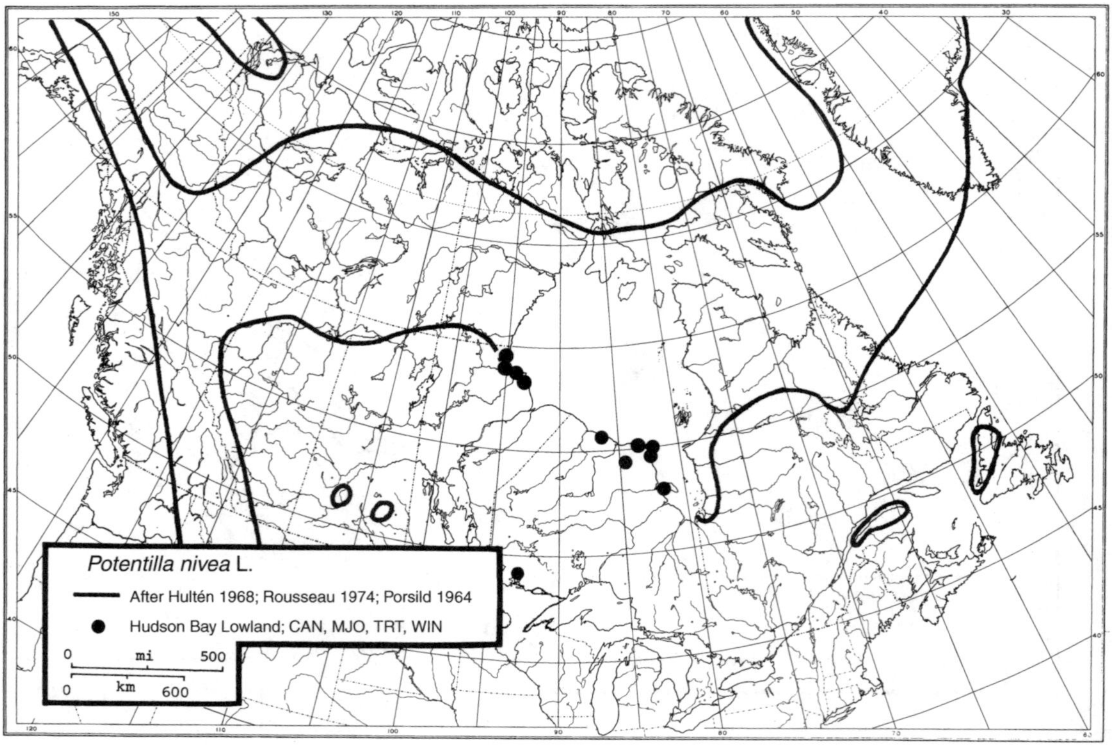

Fig. A126. Distribution of *Potentilla pulchella* in eastern Canada.

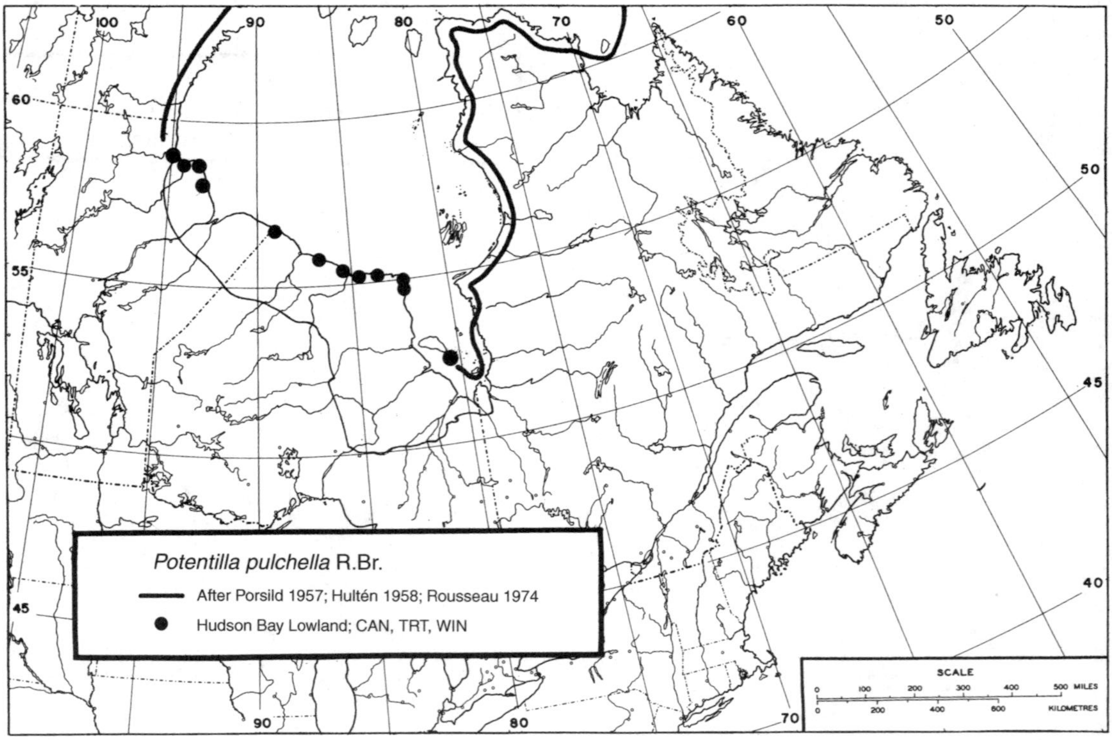

Fig. A127. Distribution of *Puccinellia lucida* in eastern Canada.

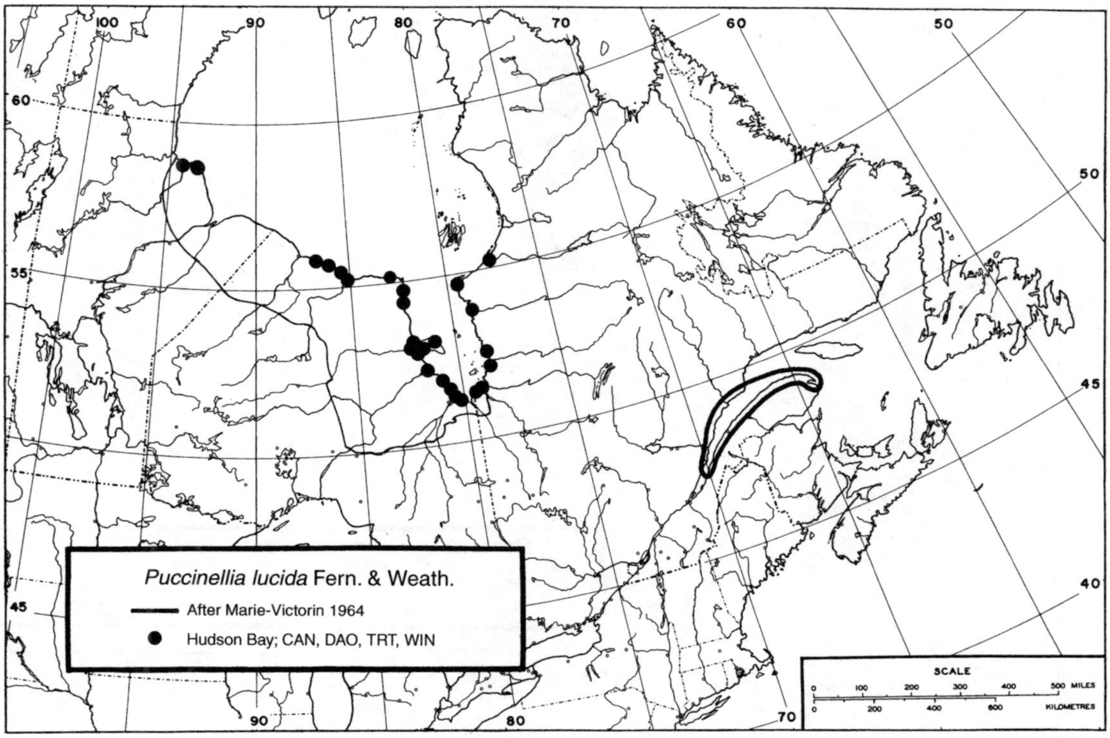

Fig. A128. Distribution of *Puccinellia phryganodes* in eastern Canada.

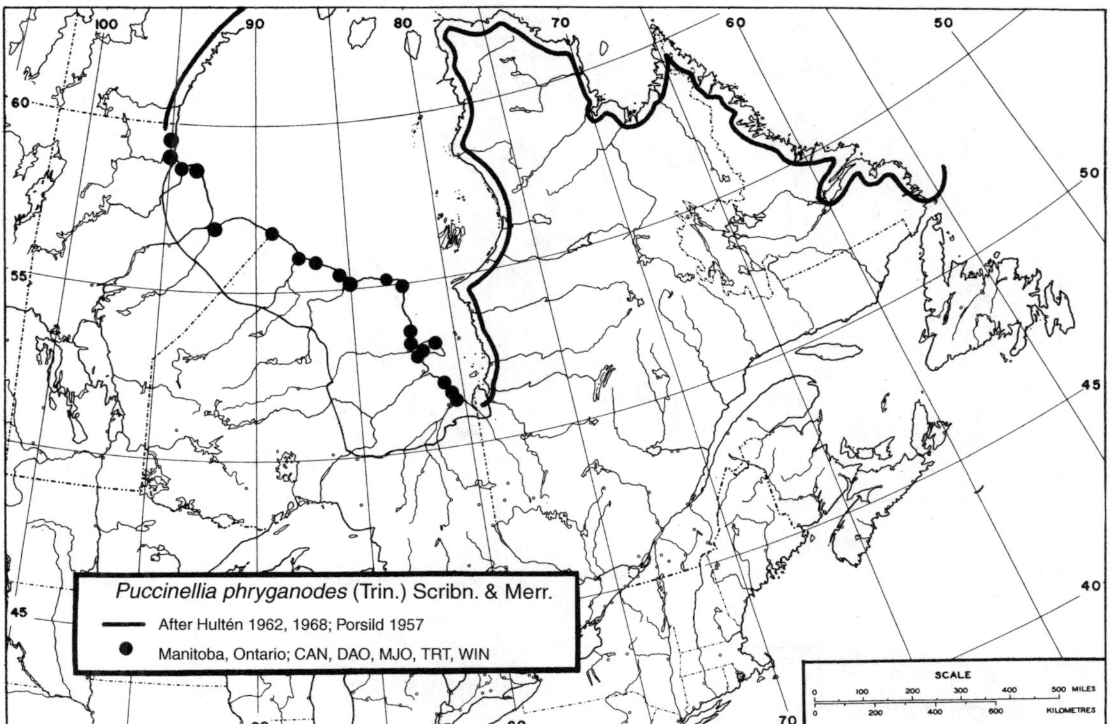

Fig. A129. Distribution of *Ranunculus hyperboreus* in eastern Canada.

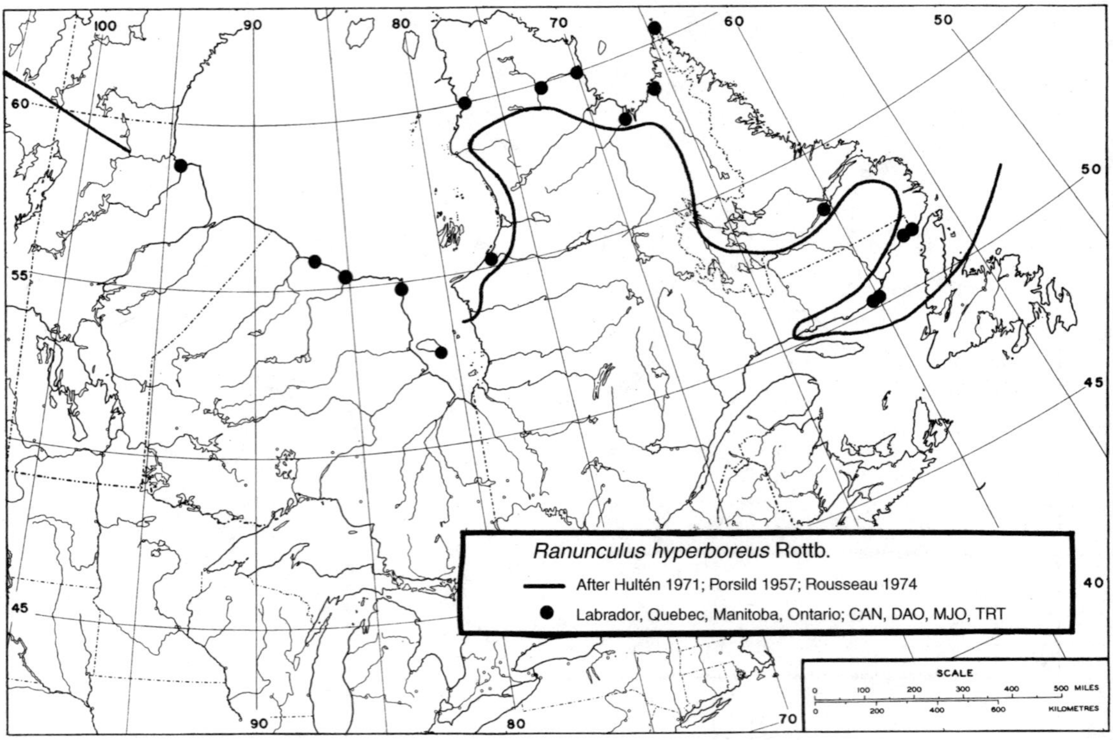

Fig. A130. Distribution of *Ranunculus pallasii* in Canada.

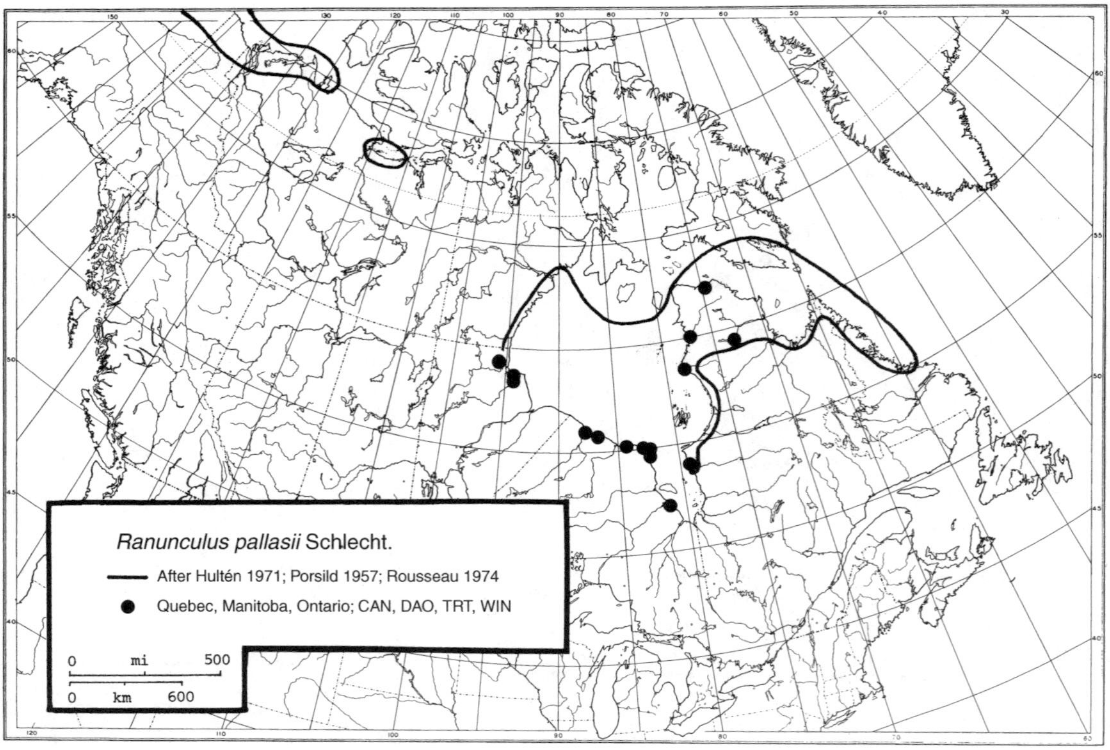

Fig. A131. Distribution of *Rhododendron lapponicum* in eastern Canada.

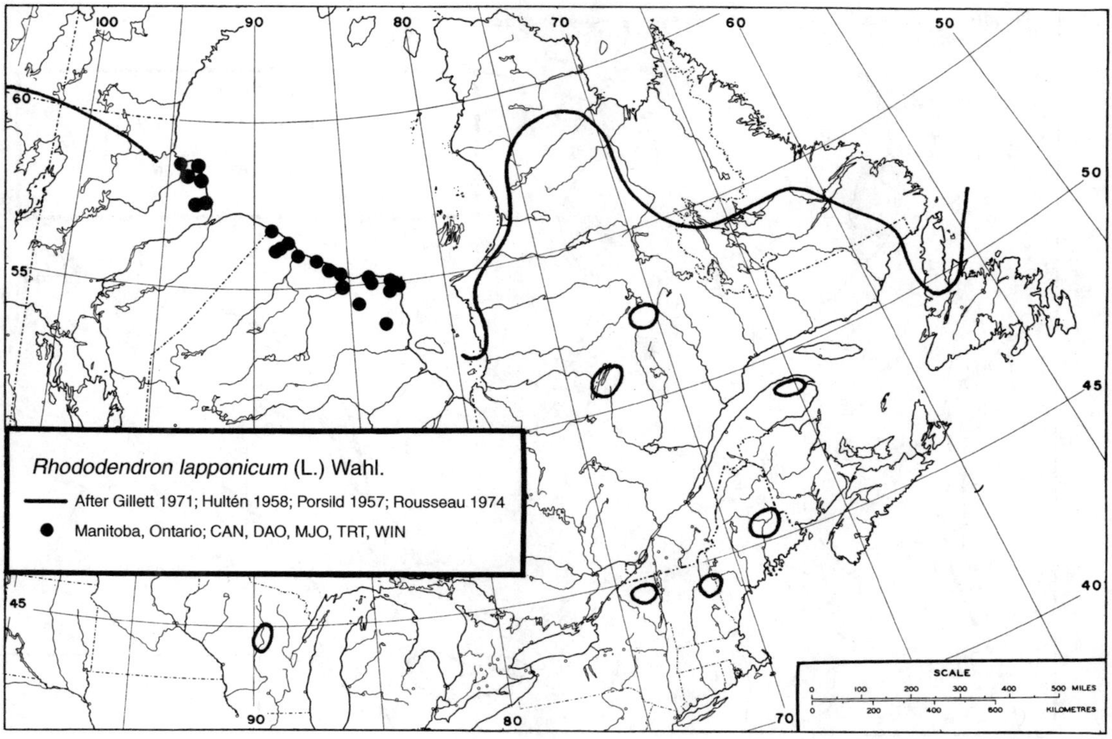

Fig. A132. Distribution of *Ruppia maritima* in eastern Canada.

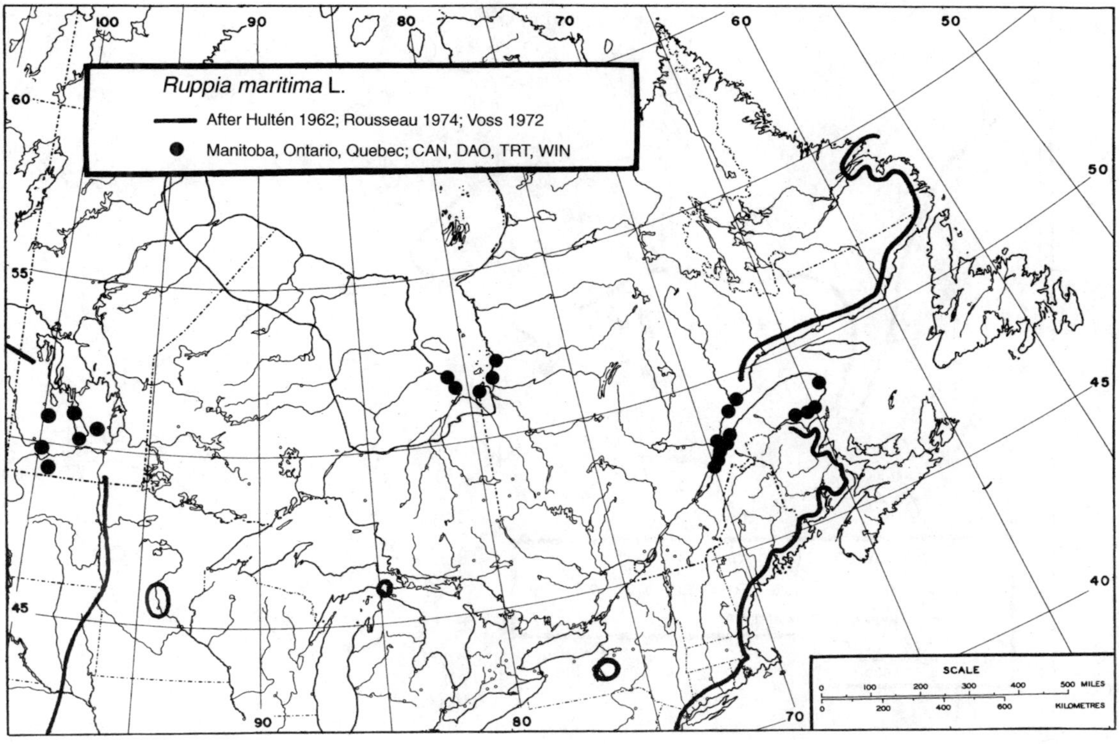

Fig. A133. Distribution of *Salicornia europaea* in eastern Canada.

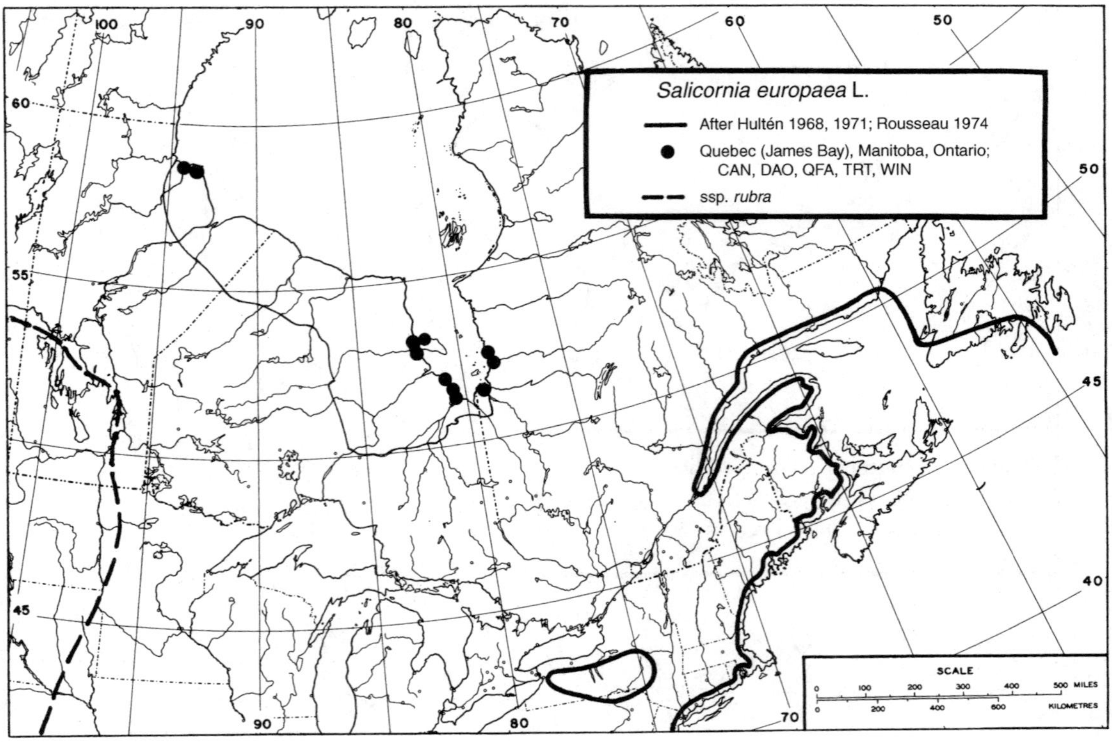

Fig. A134. Distribution of *Salix alaxensis* in Canada.

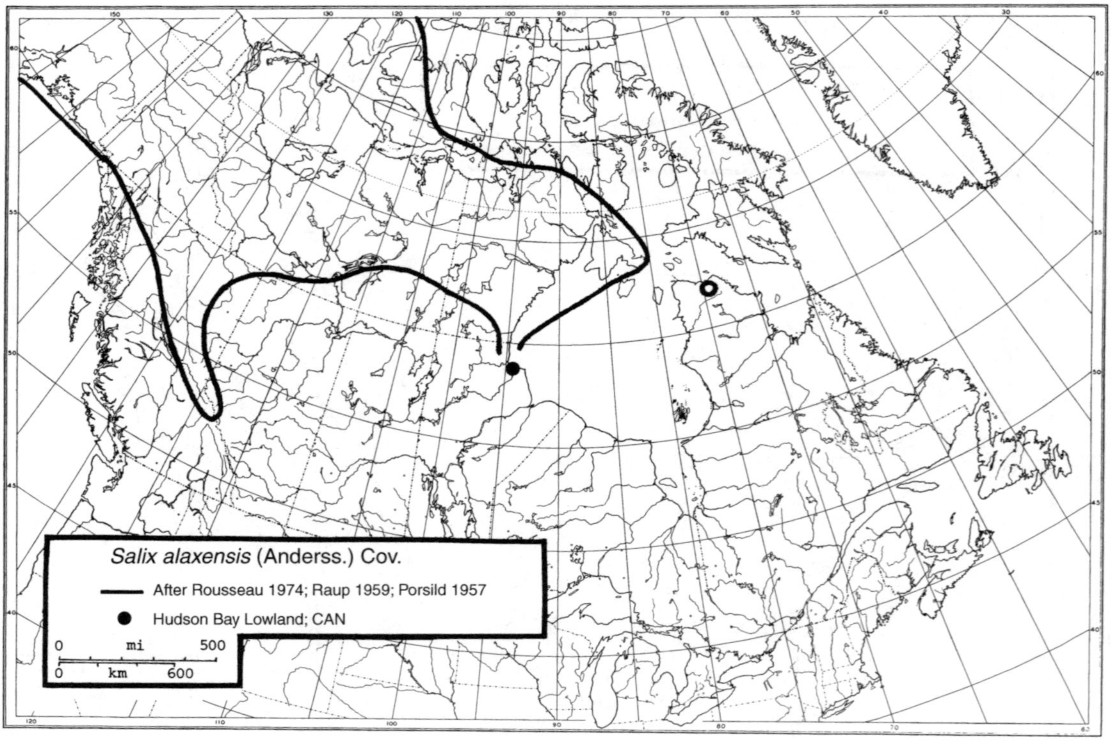

Fig. A135. Distribution of *Salix arbusculoides* in Canada.

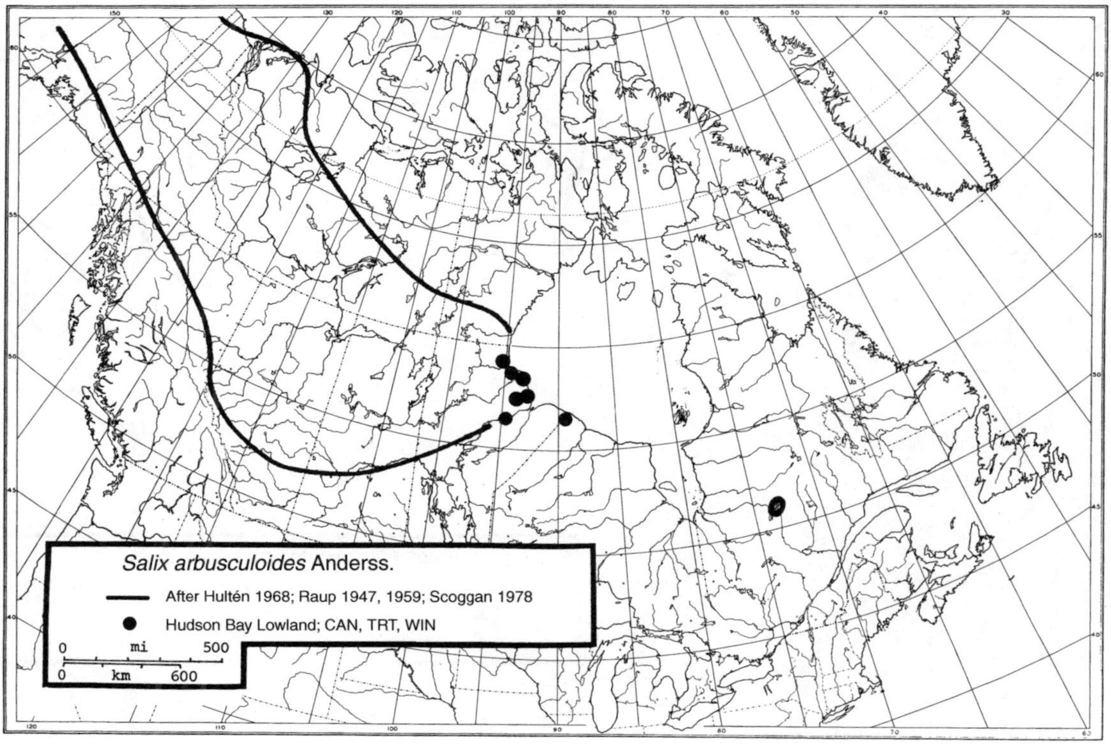

Fig. A136. Distribution of *Saxifraga paniculata* in eastern Canada.

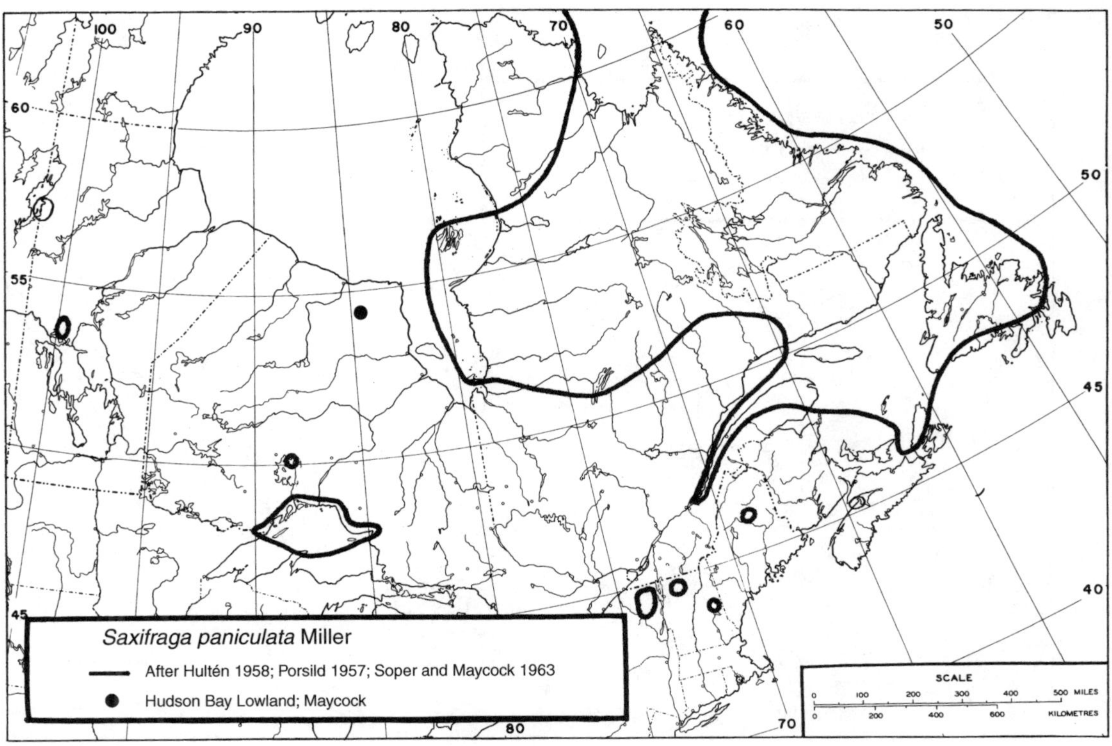

Fig. A137. Distribution of *Saxifraga tricuspidata* in eastern Canada.

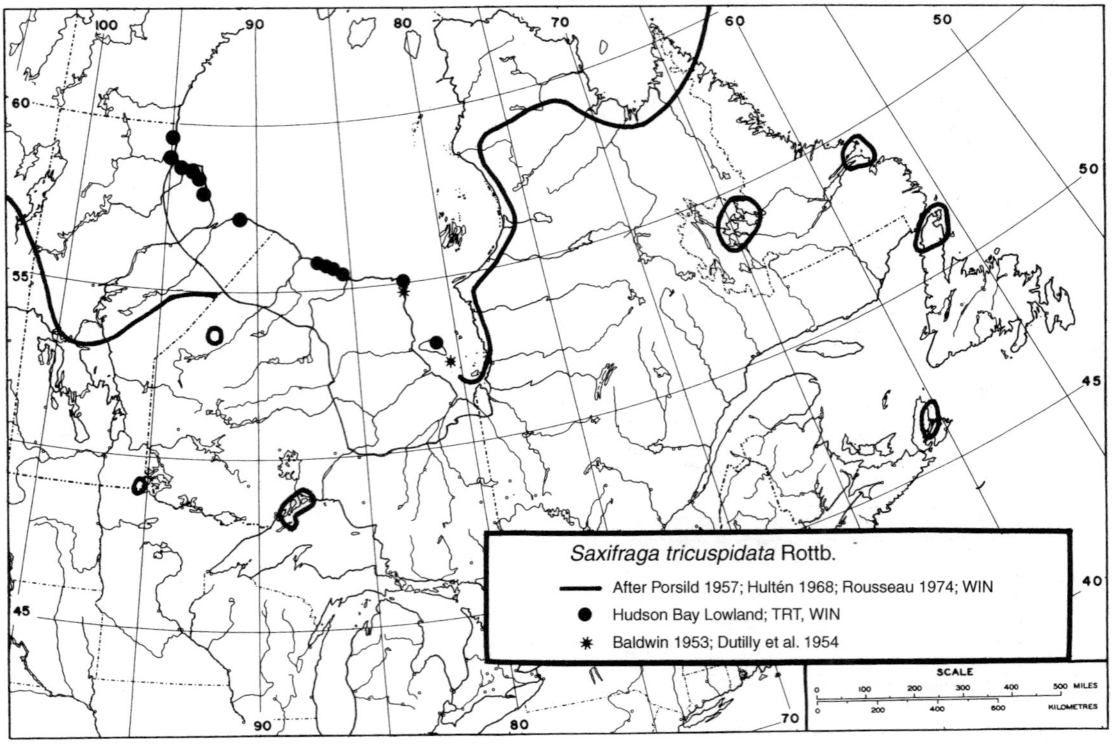

Fig. A138. Distribution of *Scirpus maritimus* in eastern Canada.

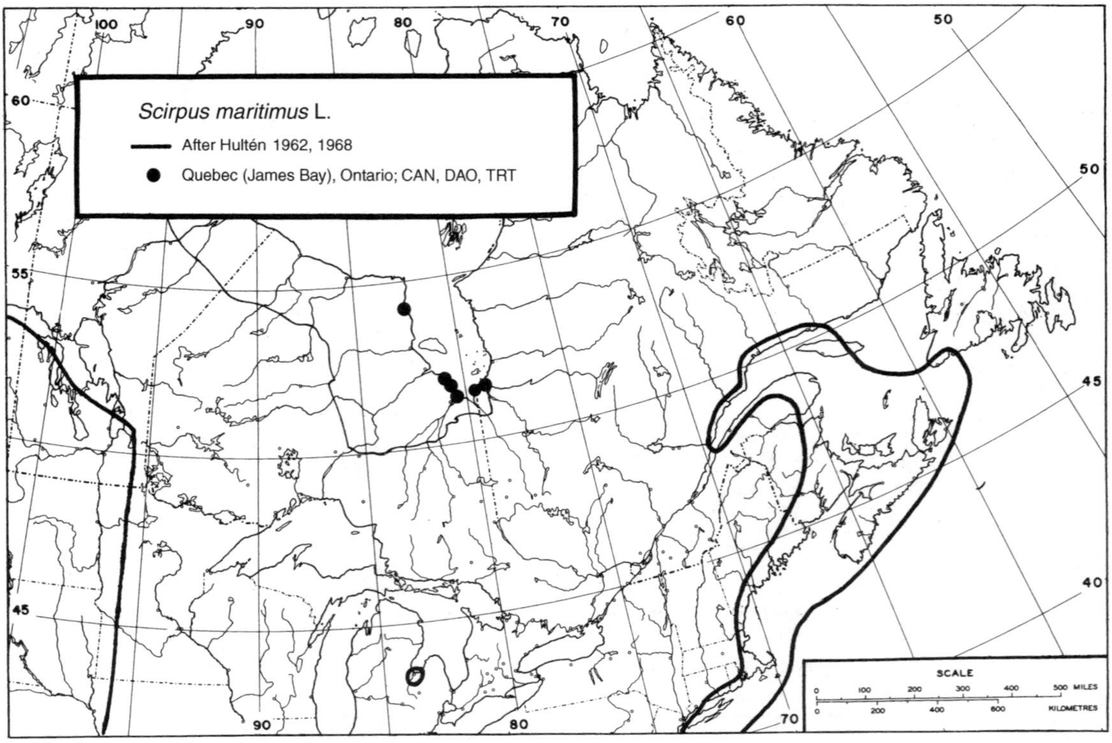

Fig. A139. Distribution of *Silene involucrata* in eastern Canada.

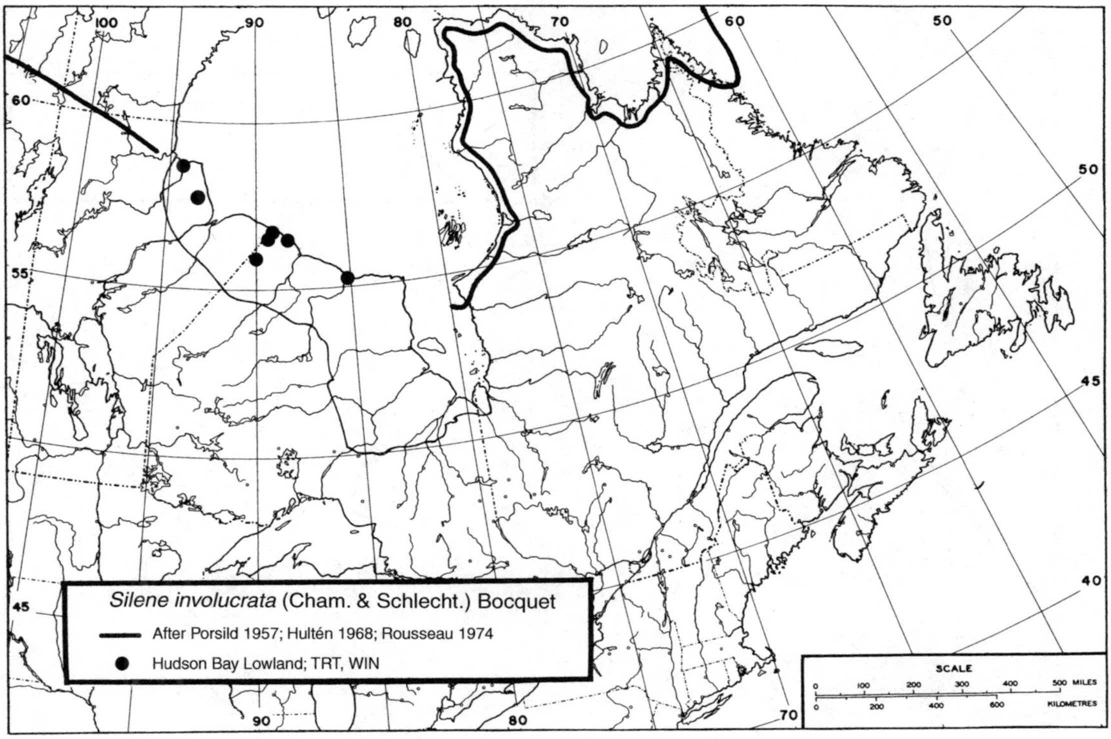

Fig. A140. Distribution of *Spartina pectinata* in eastern Canada.

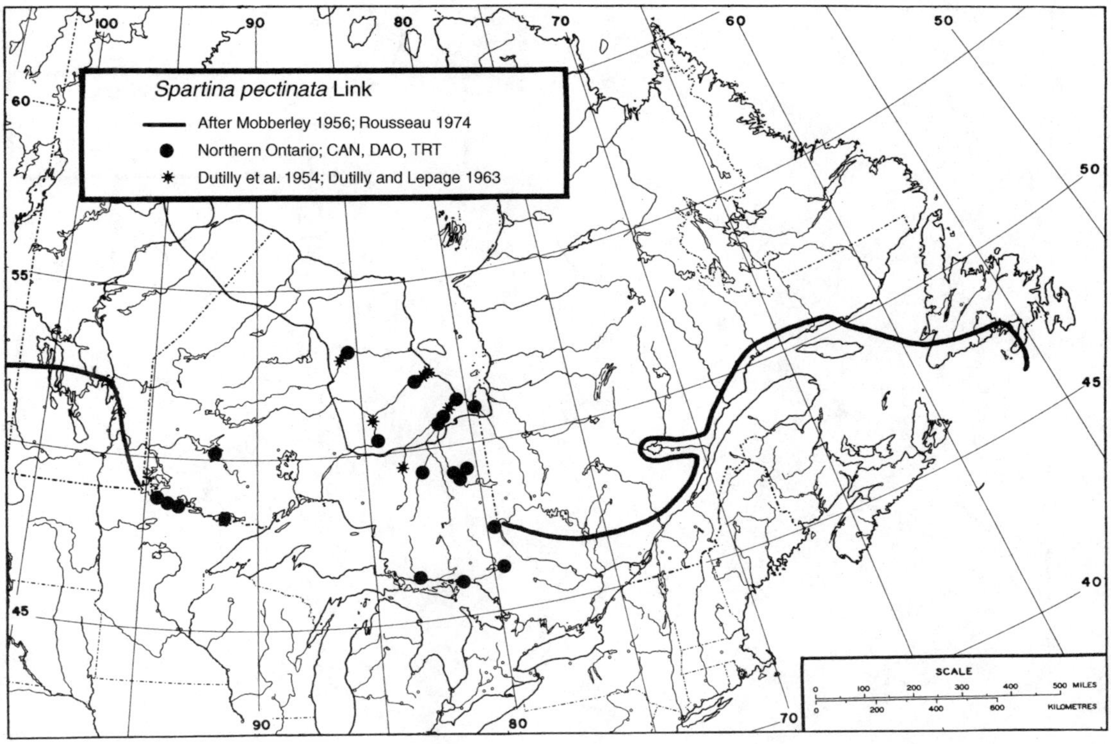

Fig. A141. Distribution of *Spergularia canadensis* in Canada.

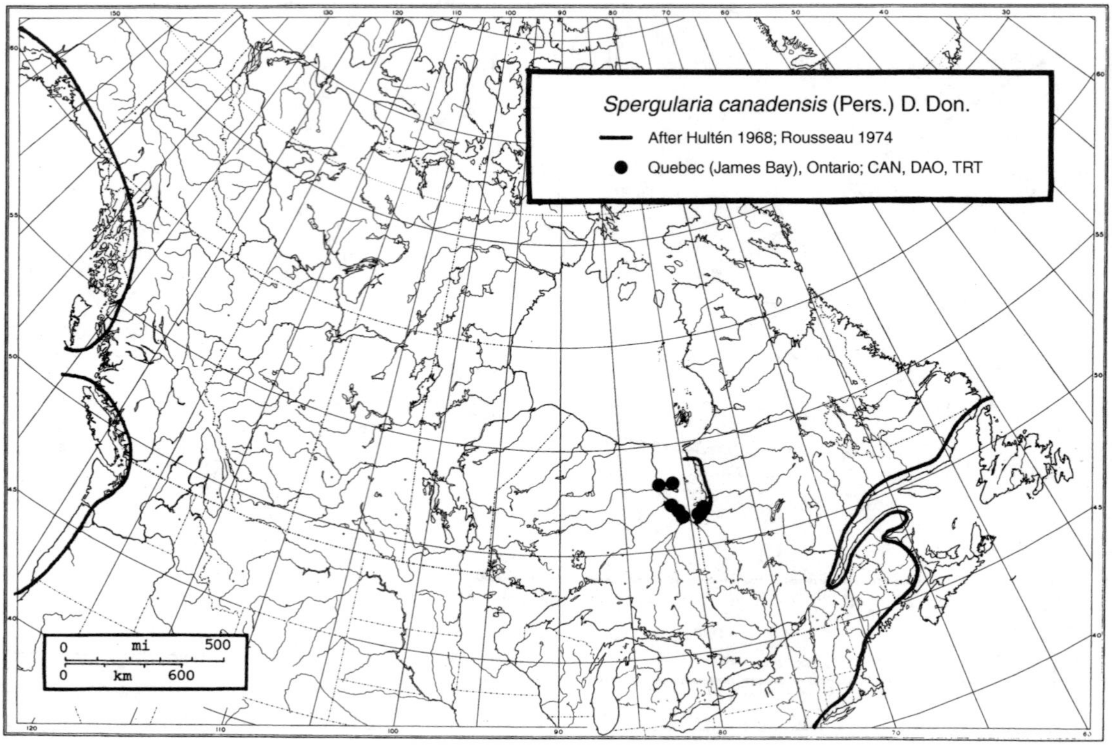

Fig. A142. Distribution of *Spergularia marina* in eastern Canada.

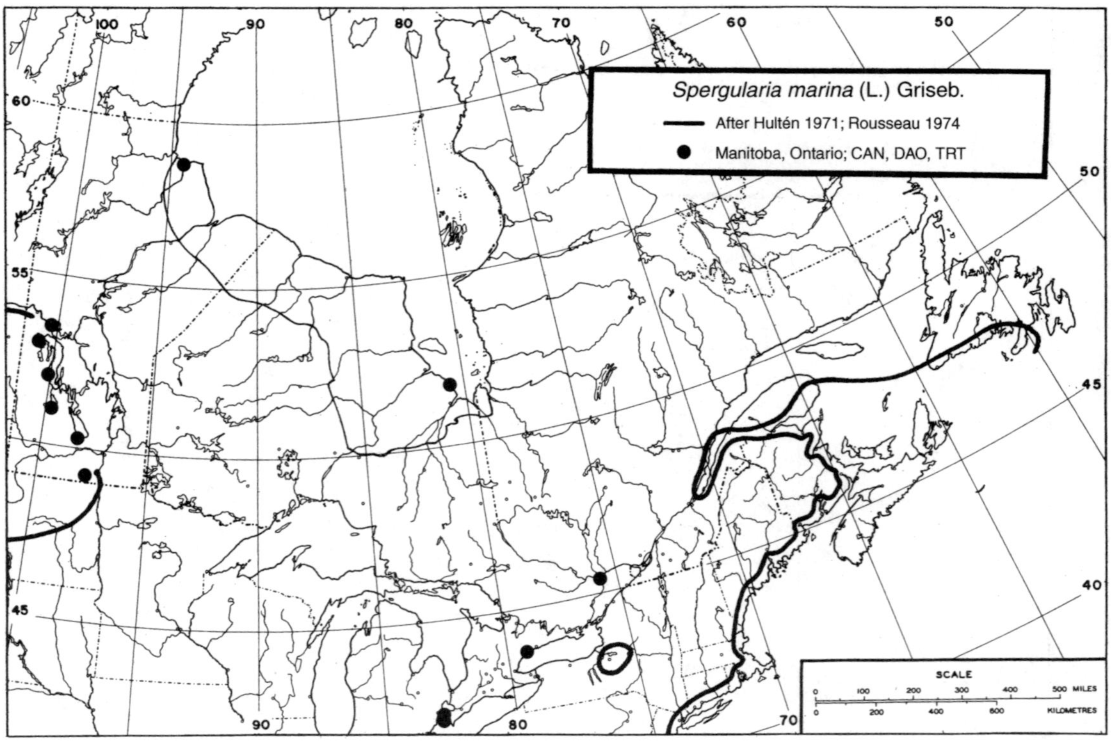

Fig. A143. Distribution of *Suaeda calceoliformis* in eastern Canada.

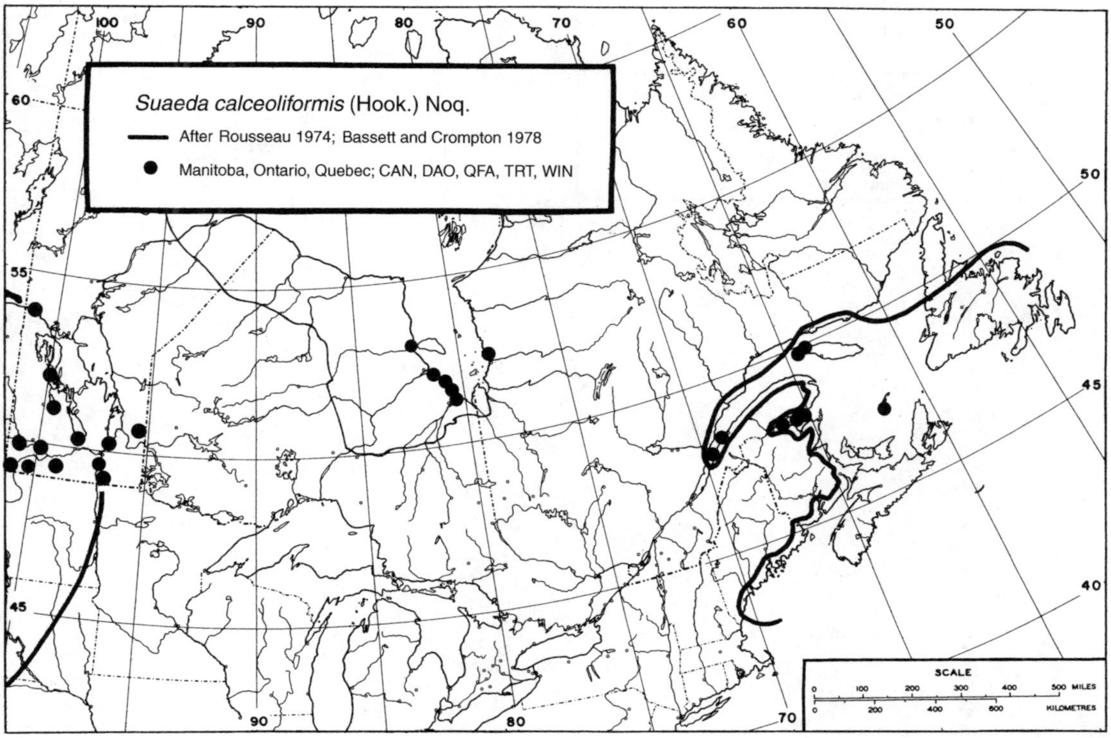

Fig. A144. Distribution of *Tripleurospermum phaeocephalum* in eastern Canada.

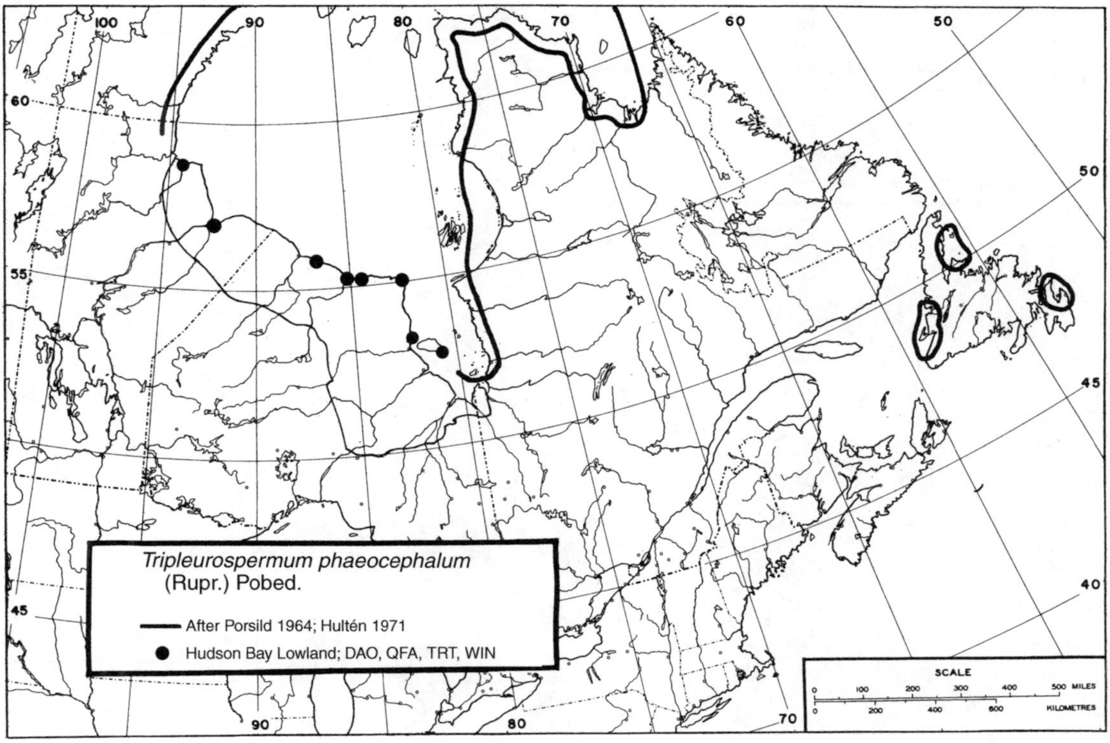

Fig. A145. Distribution of *Trisetum melicoides* in eastern Canada.

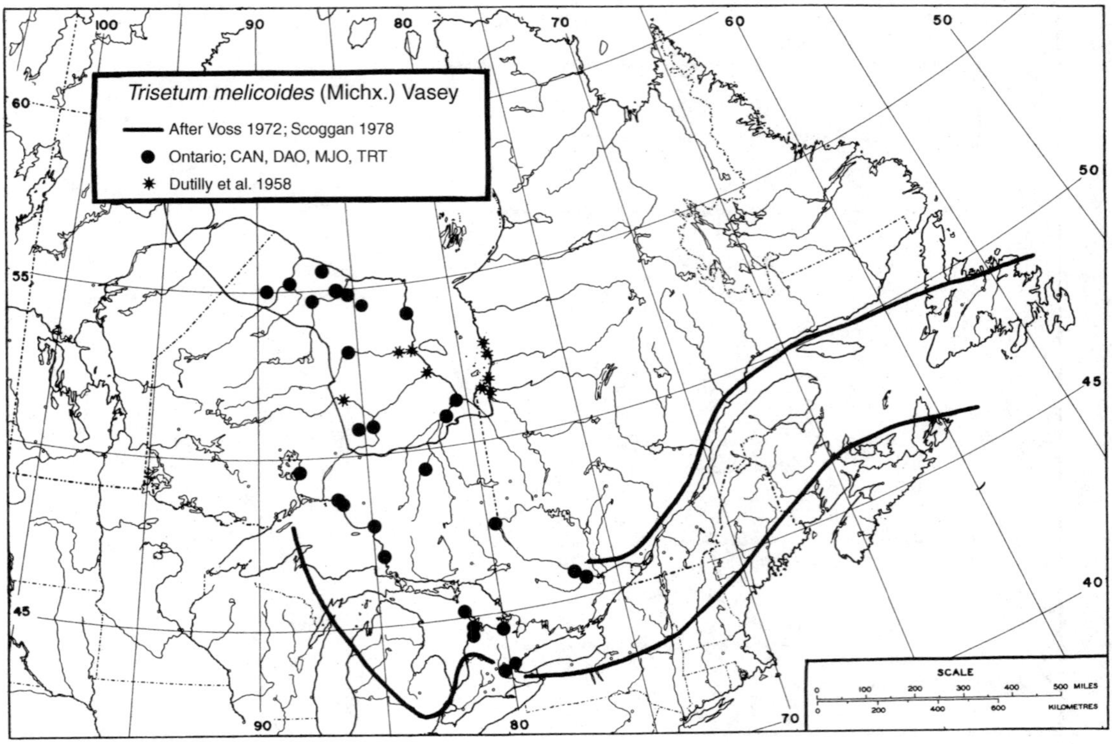

Fig. A146. Distribution of *Xyris montana* in eastern Canada.

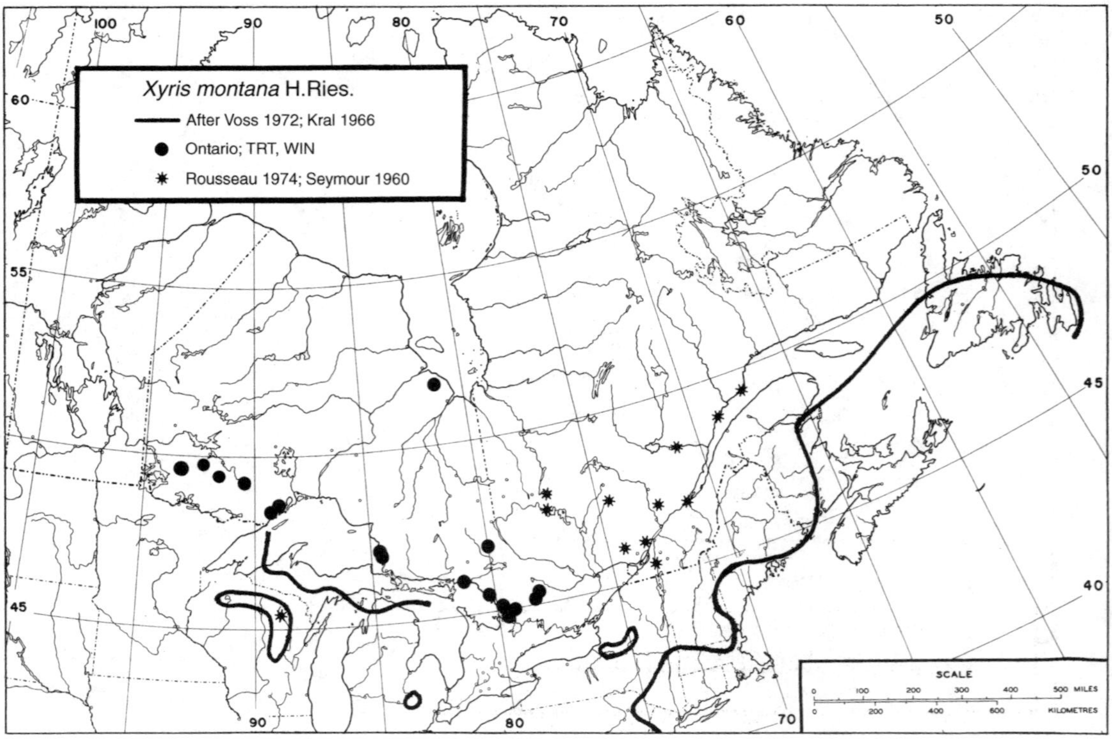

Fig. A147. Distribution of *Zannichellia palustris* in eastern Canada.

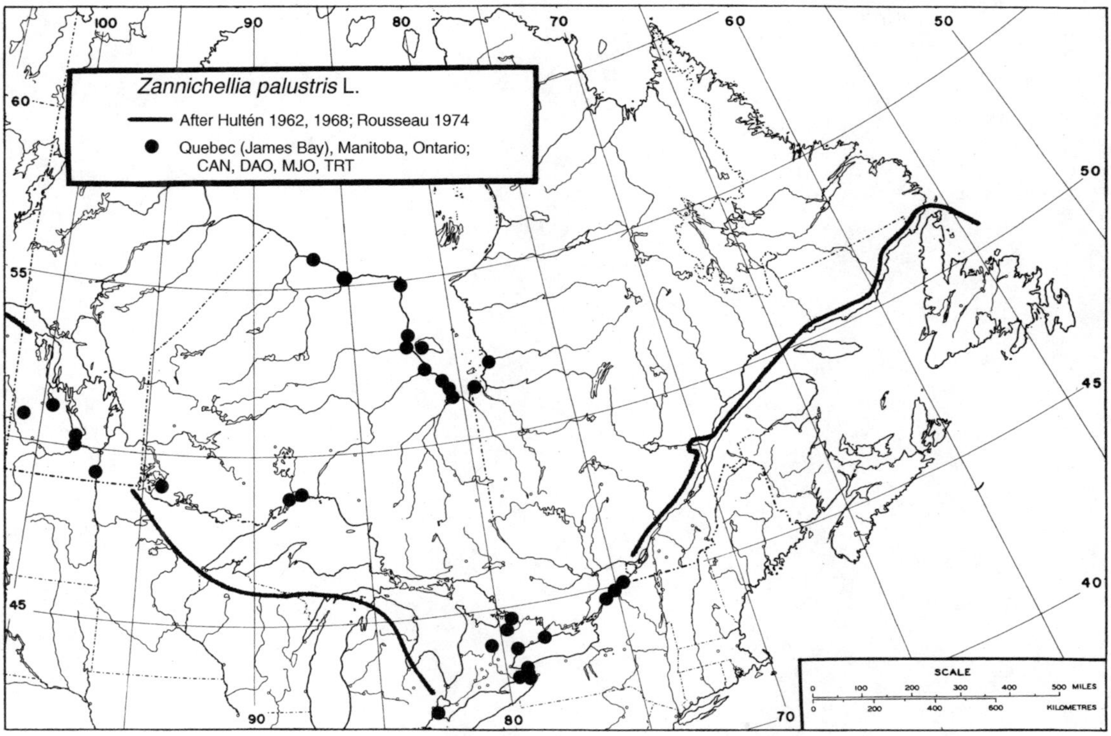

Fig. A148. Distribution of *Zostera marina* in Canada.

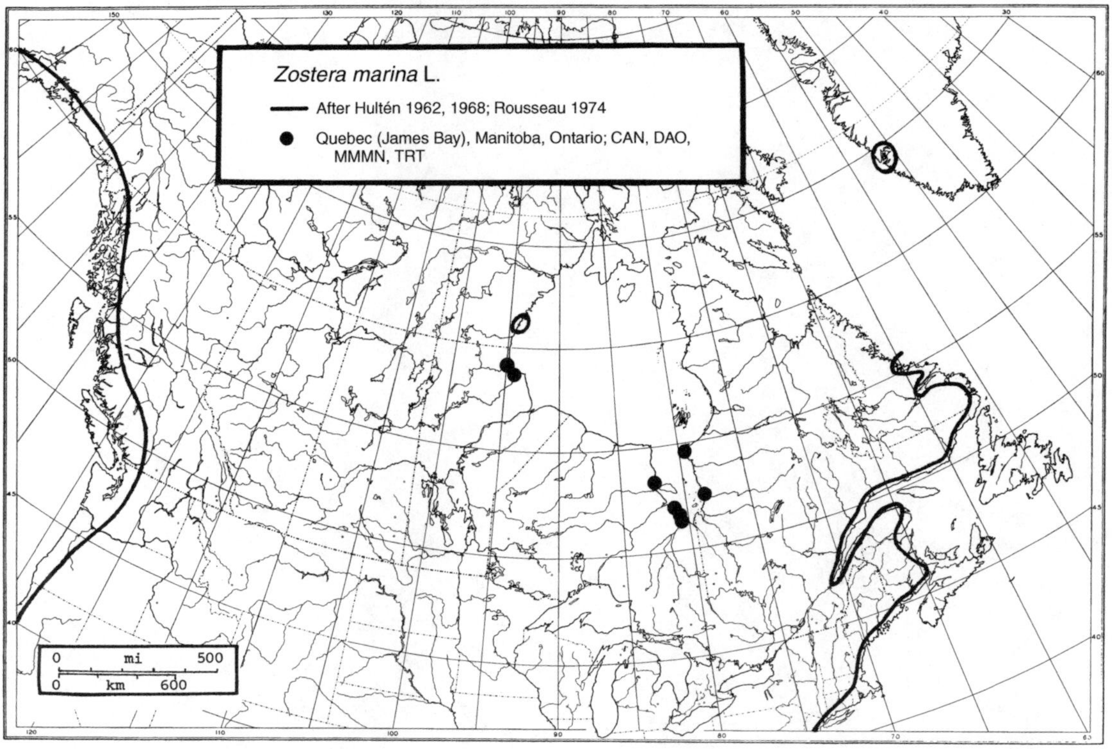

APPENDIX B

CATALOGUE OF THE VASCULAR PLANTS OF THE HUDSON BAY LOWLAND

This catalogue is a summary of species occurrence records for individual data collection areas in the Hudson Bay Lowland. The complete catalogue, containing full specimen collection data, literature records, and specimen data from herbaria, is maintained by the author. Stored with the complete catalogue are the collection locations and habitats visited in the course of the fieldwork undertaken for this study. Taxa included in the catalogue are annotated by the following:

1. whether the data collection record is based on collected specimens identified or verified by the author (X), or is based on accepted, site-specific sight or literature records (O);
2. the taxa's coded continental distribution (as indicated by Fig. 11 and Scoggan 1978);
3. the taxa's distribution across the Lowland; and
4. the taxa's frequency of occurrence in the Lowland.

The catalogue follows the order of families in Dalla Torre (1958) and Verdoorn (1938), with taxa organized alphabetically within families.

The nomenclature is referrable to Boivin (1966–1967), Hultén (1968), Scoggan (1978–1979), Porsild and Cody (1980), Voss (1972, 1985, 1996), Morton and Venn (1990), and Gleason and Cronquist (1991), where the species occur in the particular author's study area. Reference was also made to recent monographs (see References) and to the *Flora of North America* (volumes 2–3, 22–23, 26). The typical variety of each listed taxon is intended except as otherwise indicated or included.

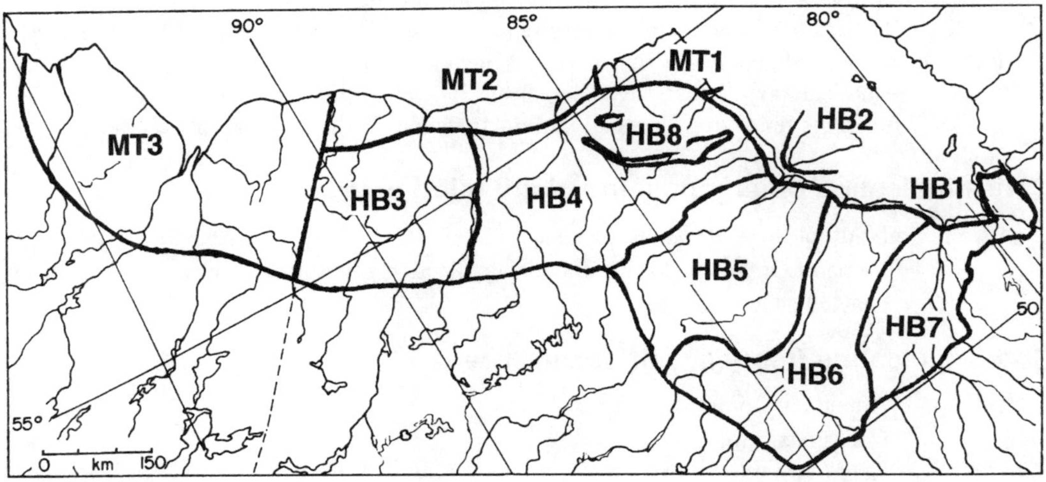

Codes indicating status of occurrence

In the data collection columns, the following symbols are used:

X	Hudson Bay Lowland taxa, based on voucher collections catalogued by author.
O	Accepted sight record or literature report, also catalogued by the author. For others, see Appendix C, Excluded Records.
P	Indicates an occurrence on the immediate boundary of the Lowland in Manitoba, specifically specimen or literature citations from Manitoba's Gillam area, from the Hayes River between the "Rock" and the Lowland, from the Gods River close to the Shield, or from the Hudson Bay coast just north of the Lowland. These 61 taxa are included in the flora of the adjacent data collection area on the basis of the high probability of them occurring there; 51 of them are already known elsewhere in the Lowland.
+	Indicates a specimen seen from Charlton Island but not found along the western James Bay coast or on Akimiski Island.

Codes indicating distribution patterns

W	Widespread (known from at least 3 of 5 coastal and 4 of 6 interior data collection areas).
(W)	Probably widespread (as yet under-sampled; maximum 2 data collection areas short of the above criteria).
R(S)	Restricted southern (HB6 and/or 7, and infrequent in 5 and/or 1 if there at all).
R(I)	Restricted interior (more than 15 km from coast or 25 km in estuarial sites).
R(SI)	Restricted southern interior.
R(N)	Restricted northern (north of a line between the lower Ekwan River, Akimiski Island, the Fawn River, and Gillam, Manitoba).
R(C)	Restricted coastal (up to 15 km from coast or 25 km in estuarial sites).
R(NC)	Restricted northern coastal (up to 20 km from the Hudson Bay coast, up to 40 km at Cape Henrietta Maria; south to Lake River).
R(NW)	Restricted northwestern (Manitoba's Lowland).
R(NWC)	Restricted northwestern coastal (Manitoba's Lowland coast).
R(NWI)	Restricted northwestern interior (interior Manitoba Lowland, south of the latitude of York Factory).
R(NE)	Restricted northeastern (Cape Henrietta Maria area).
R(LS)	Restricted Precambrian (Sutton Ridge outcropping and vicinity).
R(JB)	Restricted southwestern James Bay (coastal–estuarial sites south of Ekwan River).
(Int)	Introduced taxa in the Hudson Bay Lowland. Those introduced from native North American populations have annotations indicating their native continental distributions; the remaining Eurasian introductions have a dash (–) in that column.

Codes indicating other modifying distribution patterns

n	North of the Attawapiskat River.
t	South of Ontario's Hudson Bay data collection zones and not near the Hudson Bay coast in Manitoba.

Codes indicating frequency of occurrence

a	Abundant–Common
o	Occasional–Infrequent
r	Rare (currently known from less than 4 Lowland sites)

Taxa		James Bay Coast – HB1	Akimiski Island – HB2	Severn Basin – HB3	Winisk-Ekwan Basin – HB4	Attawapiskat Basin – HB5	Albany Basin – HB6	Moose Basin – HB7	Sutton Ridges – HB8	Cape Henrietta-Maria – MT1	Ont. Maritime Tundra – MT2	Manitoba Lowland – MT3	Continental Distribution	HBL Distribution and Frequency
LYCODPODIACEAE	Club-moss Family													
Lycopodium annotinum L.	Bristly Club-moss													
var. *annotinum*				X		X	X	X	X			X	aST/X	R(I)ot
var. *pungens* (LaPylaie) Desv. (f. *pungens* (Desv.) M.P.Pors.; incl. var. *alpestre* Hartm.)				X	X				X	X	X	X	aST/X	R(N)o
L. clavatum L.	Common Club-moss						X	X					aST/X	R(Sl)o
L. complanatum L. (var. *complanatum*; *Diphasiastrum c.* (L). Holub)	Flatbranch Club-moss	X	X	X	X	X	X	X	X		X	X	aST/X	(W)o
L. dendroideum Michx. (*L. obscurum* L. var *dendroideum* (Michx.) Eat.)	Round-branched Ground-pine					X		O	X				sT/X	R(Sl)o
L. sitchense Rupr. (*Diphasiastrum s.* (Rupr.) Holub; incl. *L. sabinifolium* Willd.)	Sitka Club-moss								X				ST/X	R(LS)r
L. selago L. (*Huperzia selago* (L.) Bernh. *s.l.*)	Mountain Club-moss													
var. *patens* (Beauv.) Desv. (f. *occidentale* (Clute) Boivin)		X					X		X	X			AST/X	R(N)a
var. *appresum* Desv. (f. *appressum* (Desv.) Gelert); incl. var. *miyoshianum* (Mak.) Mak. at HB8)					X		X		X	X	X	X	AST/X	R(N)a
SELAGINELLACEAE	Spike-moss Family													
Selaginella selaginoides (L.) Link	Spike-moss	X		X	X	X	X	X	X	X	X	X	aST/X	Wa
ISOETACEAE	Quillwort Family													
Isoetes echinospora Dur. (var. *braunii* (Dur.) Engelm., *I. muricata* Dur.)	Quillwort						X						aST/X	R(Sl)r
EQUISETACEAE	Horsetail Family													
Equisetum arvense L. (incl. var. *boreale* (Bong.) Ledeb. and *E.* Xlitorale Kuhl.)	Field Horsetail	X	X	X	X	X	X	X	X	X	X	X	AST/X	Wa
E. fluviatile L.	Water Horsetail	X	X	X	X	X	X	X	X	X	X	X	aST/X	Wa
E. hyemale L. (var. *pseudohyemale* (Farw.) Morton, var. *affine* (Engelm.) A.A.Eaton)	Scouring-rush						X						sT/X	R(Sl)r
E. palustre L.	Marsh Horsetail	X		O	X	X	X	X			X	X	aST/X	Wa
E. pratense Ehrh.	Meadow Horsetail	X		X	X	X	X	X		X		X	aST/X	(W)o

Taxa		James Bay Coast – HB1	Akimiski Island – HB2	Severn Basin – HB3	Winisk-Ekwan Basin – HB4	Attawapiskat Basin – HB5	Albany Basin – HB6	Moose Basin – HB7	Sutton Ridges – HB8	Cape Henrietta-Maria – MT1	Ont. Maritime Tundra – MT2	Manitoba Lowland – MT3	Continental Distribution	HBL Distribution and Frequency
E. scirpoides Michx.	Dwarf Scouring-rush			X	X	X	X	X	X	X	X	X	aST/X	Wo
E. sylvaticum L. (incl. var. *pauciramosum* Milde)	Wood Horsetail			X	X	X	X	X	X		X	X	aST/X	(W)o
E. variegatum Schleicher	Variegated Horsetail	X	X	X	X	X	X	X	X	X	X	X	aST/X	Wa
OPHIOGLOSSACEAE	Adder's-tongue Family													
Botrychium lunaria (L.) Sw.	Moonwort	X	X	O	X	O	O	O	X	X	X	X	aST/X	Wo
B. matricariaefolium A.Br.	Matricary Grape Fern	X											T/X	R(JB)r
B. minganense Vict.	Mingan Grape Fern	X	X	X		X	X			X		X	aST/X	(W)o
B. multifidum (Gmel.) Rupr.	Leathery Grape Fern							X					sT/X	R(SI)r
B. virginianum (L.) Sw.	Rattlesnake Fern	X		X	X	X	X	X				P	sT/X	ot
POLYPODIACEAE	Fern Family													
Athyrium filix-femina (L.) Roth (var. *michauxii* (Spreng.) Farw.)	Lady Fern						X	X					aST/X	R(SI)o
Cystopteris bulbifera (L.) Bernh.	Bulblet Fern							X					T/E	R(SI)r
C. fragilis (L.) Bernh. (incl. var. *dickieana* (Sim) Moore)	Fragile Fern				X	X		X	X			X	AST/X	(W)o
Dryopteris carthusiana (Vill.) H.P.Fuchs (*D. spinulosa* Watt)	Spinulose Wood Fern						X	X	X				aST/X	R(S)o
D. cristata (L.) A.Gray	Crested Wood Fern						X	O					T/X	R(SI)r
D. expansa (Presl) Fraser-Jenkins & Jermy (*D. assimilis* Walk., *D. dilatata* (Hoffm.) A.Gray)	Northern Wood Fern							X					aST/X	R(SI)r
D. fragrans (L.) Schott (var. *remotiuscula* Komarov)	Fragrant Cliff Fern								X				AST/X	R(LS)r
D. intermedia (Muhl.) A.Gray	Evergreen Wood Fern							X					aST/E	R(SI)r
Gymnocarpium dryopteris (L.) Newm.	Oak Fern					X	X	X	X			X	aST/X	R(I)ot
G. jessoense (Koidz.) Koidz. (ssp. *parvulum* Sarvela)	Nahanni Oak Fern								X			O	ST/X	R(LS)r
G. Xintermedium Sarvela (*G. dryopteris* X *jessoense* ssp. *parvulum*)								X						
Matteuccia struthiopteris (L.) Todaro (var. *pensylvanica* (Willd.) Mort.)	Ostrich Fern				O	X	X	X					sT/X	R(S)o
Onoclea sensibilis L.	Sensitive Fern						X						sT/E	R(SI)r

Taxa		James Bay Coast – HB1	Akimiski Island – HB2	Severn Basin – HB3	Winisk-Ekwan Basin – HB4	Attawapiskat Basin – HB5	Albany Basin – HB6	Moose Basin – HB7	Sutton Ridges – HB8	Cape Henrietta-Maria – MT1	Ont. Maritime Tundra – MT2	Manitoba Lowland – MT3	Continental Distribution	HBL Distribution and Frequency
Polypodium virginianum L. (*P. vulgare* L. var. *virginianum* (L.) Eaton; incl. *P. sibiricum* Sipl. at MT3)	Rock Polypody								X			X	sT/X	R(LS,NW)o
Pteridium aquilinum (L.) Kuhn. (var. *latiusculum* (Desv.) Underw.)	Bracken Fern						X						sT/X	R(SI)r
Thelypteris palustris (Salisb.) Schott (var. *pubescens* (Laws.) Fern.; *T. confluens* (Thunb.) Mort. var. *pubescens* (Laws.) Pringle)	Marsh Fern						X	X					T/E	R(SI)a
Woodsia glabella R.Br.	Smooth Woodsia								X				AST/X	R(LS)r
W. ilvensis (L.) R.Br.	Rusty Woodsia						X		X			P	aST/X	R(I)o
TAXACEAE	Yew Family													
Taxus canadensis Marsh.	Canada Yew						X						T/E	R(SI)r
CUPRESSACEAE	Cypress Family													
Juniperus communis L. (var. *depressa* Pursh)	Common Juniper	X	X	X	X	X	X	X	X	X	X	X	aST/X	Wa
J. horizontalis Moench	Creeping Juniper	X	X	X	X	X	X	X	X	O	X	X	sT/X	Wa
Thuja occidentalis L.	White Cedar					X	X	X					T/E	R(S)o
PINACEAE	Pine Family													
Abies balsamea (L.) Mill.	Balsam Fir				X	X	X	X					sT/X	R(I)ot
Larix laricina (DuRoi) K.Koch	Tamarack	X	X	X	X	X	X	X	X	X	X	X	ST/X	Wa
Picea glauca (Moench) Voss	White Spruce													
var. *glauca*		X	X	X	X	X	X	X	X	X	X	X	ST/X	Wa
var. *porsildii* Raup			X								X		Ss/W	R(NC)r
P. mariana (Mill.) BSP.	Black Spruce	O	O	X	X	X	X	X	X	X	X	X	ST/X	Wa
Pinus banksiana Lamb. (*P. divaricata* (Ait.) Dum.)	Jack Pine		X	X	X	X	O	O				P	ST/X	R(I)ot
P. resinosa Ait.	Red Pine						X						T/E	R(SI)r
P. strobus L.	White Pine						X						T/E	R(SI)r
TYPHACEAE	Cat-tail Family													
Typha latifolia L.	Common Cat-tail	X	X			X	X	X					ST/X	R(S)o
SPARGANIACEAE	Bur-reed Family													
Sparganium angustifolium Michx.		X	X		X	O	X	X	X			P	aST/X	ot
S. chlorocarpum Rydb.		X			X	X	X	X			O	X	ST/X	ot
S. eurycarpum Engelm.	Giant Bur-reed	X					X	X					sT/X	R(S)o

Taxa		James Bay Coast – HB1	Akimiski Island – HB2	Severn Basin – HB3	Winisk-Ekwan Basin – HB4	Attawapiskat Basin – HB5	Albany Basin – HB6	Moose Basin – HB7	Sutton Ridges – HB8	Cape Henrietta-Maria – MT1	Ont. Maritime Tundra – MT2	Manitoba Lowland – MT3	Continental Distribution	HBL Distribution and Frequency
S. fluctuans (Morong) Robins.									X				sT/X	R(LS)r
S. hyperboreum Laest.									X	X	X	X	aST/X	R(N)o
S. natans L. (*S. minimum* (Hartm.) Fries)	Small Bur-reed	X	X		X	X	X	X		X	X	X	ST/X	Wa
ZOSTERACEAE	Eel-grass Family													
Zostera marina L.	Eel-grass	X	X									X	aST/X	R(C)o
POTAMOGETONACEAE	Pondweed Family													
Potamogeton alpinus Balbis (incl. var. *tenuifolius* (Raf.) Ogden and var. *subellipticus* (Fern.) Ogden)	Alpine Pondweed	X	O		X	X	X	X	X	X	X	X	aST/X	Wa
P. epihydrus Raf.							X						ST/X	R(Sl)r
P. filiformis Pers. (incl. var. *borealis* (Raf.) St. John; *Stuckenia f.* (Pers.) Born.)		X	X		X	X	X	X	X	X	X	X	aST/X	Wa
P. friesii Rupr.		X			X	X	X			X	X	X	ST/X	(W)o
P. gramineus L.		X	X	X	X	X	X	X	X	X		X	aST/X	Wa
P. natans L.	Floating Pondweed	X				X	X	X	X			P	aST/X	ot
P. oakesianus Robbins							X						sT/E	R(Sl)r
P. obtusifolius Mert. and Koch												X	T/X	R(NWC)r
P. pectinatus L. (*Stuckenia p.* (L.) Born.)	Sago Pondweed	X			X	X	X	X	X	X	X	X	aST/X	Wa
P. praelongus Wulf.					X	X	X		X		X	X	ST/X	(W)o
P. pusillus L. (var. *tenuissimus* Mert. & Koch; *P. berchtoldii* Fieber; part of var. *pusillus sensu* Boivin)	Small Pondweed	X				X	X	X	X		X	X	ST/X	Wa
P. richardsonii (A. Benn.) Rydb. (*P. perfoliatus* L. var. *richardsonii* A. Benn.)		X	X	X	X	X	X	X	X		X	X	aST/X	Wa
P. strictifolius A. Benn. (*P. pusillus* L. var. *rutiloides* (Fern.) Boivin; Haynes 1974)												X	sT/X	R(NWC)r
P. subsibiricus Hagstr. (*P. porsildiorum* Fern.)										X			S/X	R(NE)r
P. vaginatus Turcz. (incl. *P. pectinatus* L. var. *interruptus* (Kit.) Aschers of reports; *Stuckenia vaginata* (Turcz.) Holub)		X			X	X	X	X	X		X	X	aST/X	Wa
P. zosteriformis Fern.						X	X	X	X		X		ST/X	(W)o

Taxa		James Bay Coast – HB1	Akimiski Island – HB2	Severn Basin – HB3	Winisk-Ekwan Basin – HB4	Attawapiskat Basin – HB5	Albany Basin – HB6	Moose Basin – HB7	Sutton Ridges – HB8	Cape Henrietta-Maria – MT1	Ont. Maritime Tundra – MT2	Manitoba Lowland – MT3	Continental Distribution	HBL Distribution and Frequency
RUPPIACEAE	Ditch-grass Family													
Ruppia maritima L.	Ditch-grass	X											sT/X	R(JB)r
ZANNICHELLIACEAE	Horned Pondweed Family													
Zannichellia palustris L.	Horned Pondweed	X	X							X	X		sT/X	R(C)o
JUNCAGINACEAE	Arrow-grass Family													
Scheuchzeria palustris L. (var. *americana* Fern.)	Bog-bean		X	X	X	X	X	X	X				ST/X	R(I)a
Triglochin maritima L.	Maritime Arrow-grass	X	X	O	X	X	X	X	X	X	X	X	aST/X	Wa
T. palustris L.	Narrow Arrow-grass	X	X	X	X	X	X	X	X	X	X	X	aST/X	Wa
ALISMATACEAE	Water-plaintain Family													
Alisma triviale Pursh (*A. plantago-aquatica* L. var. *americanum* Schult.& Schult.f.)	Water-plantain	X					X	X					sT/X	R(S)o
Sagittaria cuneata Sheldon		X			X	X	X	X			X	X	ST/X	Wo
S. latifolia Willd.	Common Arrowhead	X					X	X					T/X	R(S)o
HYDROCHARITACEAE	Frog's-bit Family													
Elodea canadensis Michx.	Canada Waterweed	O		O	X	X	X	X					T/X	R(I)ot
E. nuttallii (Planch.) St. John		O											T/X	R(JB)r
POACEAE	Grass Family													
X*Agroelymus hirtiflorus* (Hitchc.) Bowden (*Agropyron trachycaulum* X *Elymus innovatus*.; *Elyleymus hirtiflorus* (Hitchc.) Barkw. & Dew., X*Agroelymus ontariensis* Lepage)		X		X		X					X		ST/W	o
X*Agrohordeum macounii* (Vasey) Lepage (*Agropyron trachycaulum* X *Hordeum jubatum*; X*Elyhordeum macounii* (Vasey) Barkw. & Dew.)							X						aST/W	r
Agropyron repens (L.) Beauv. (*Elymus repens* (L.) Gould, *Elytrigia repens* (L.) Nevski)	Quack Grass	X					X	X				X	–	(Int)o
A. trachycaulum (Link) Malte (*Elymus trachycaulus* (Link) Gould)	Wheat Grass													
var. *trachycaulum* (*A. pauciflorum* (Schwein.) Hitchc.; incl. *A. t.* var. *majus* (Vasey) Fern.)		X	X	X	X	X	X	X	X	O	X	X	aST/X	Wa

Taxa	James Bay Coast – HB1	Akimiski Island – HB2	Severn Basin – HB3	Winisk-Ekwan Basin – HB4	Attawapiskat Basin – HB5	Albany Basin – HB6	Moose Basin – HB7	Sutton Ridges – HB8	Cape Henrietta-Maria – MT1	Ont. Maritime Tundra – MT2	Manitoba Lowland – MT3	Continental Distribution	HBL Distribution and Frequency
var. *glaucum* (Pease and Moore) Malte (*A. subsecundum* (Link) Hitchc.)	X				X	X	X		X		X	aST/X	(W)o
var. *novae-angliae* (Scribn.) Fern. (*A. pauciflorum* ssp. *novae-angliae* (Scribn.) Melderis)	X	X	X	X	X	X	X	X	O	X		aST/X	Wa
A. violaceum (Hornem.) Lange (*A. latiglume* (Scribn. and Sm.) Rydb, *Elymus trachycaulus* (Link) Gould ssp. *violaceus* (Hornem.) A. & D.Löve)			X	X				X		X	X	ASs/X	R(N)o
Agrostis gigantea Roth — Redtop (*A. stolonifera* L. var. *major* (Grand.) Farw.; *A. alba* L., in part)	X					X	X					–	(Int)at
A. mertensii Trin. — Red Bent Grass (ssp. *borealis* (Hartm.) Tzvel.; *A. borealis* Hartm.)								X			X	aST/X	R(LS,NWC)o
A. scabra Willd. — Rough Hair Grass (var. *septentrionalis* Fern.; *A. hyemalis* (Walt.) BSP. var. *tenuis* (Tuck.) Gl.; incl. a viviparous form)	X		X	X	X	X	X	X	O	X	X	ST/X	Wa
A. stolonifera L. — Creeping Bent					X	X	X				X	–	(Int)o
Alopecurus aequalis Sobol — Short-awned Foxtail				X	X	X	X		X		X	aST/X	(W)o
A. alpinus Sm. — Alpine Foxtail (ssp. *alpinus sensu* Hultén)									X	X	X	AST/X	R(NC)o
Andropogon gerardii Vitman — Big Bluestem						X						T/X	R(SI)r
Arctagrostis latifolia (R.Br.) Griseb. — Polar Grass									X	X	X	AS/X	R(NC)o
Arctophila fulva (Trin.) Anderss. (*Colpodium fulvum* (Trin.) Griseb.)									X	X	X	ASs/X	R(NC)o
Avena fatua L. — Wild Oats											X	–	(Int)r
A. sativa L. — Oats							X					–	(Int)r
Beckmannia syzigachne (Steud.) Fern. — Slough Grass (var. *uniflora* (Scribn.) Boivin; ssp. *baicalensis* (Kuzn.) Hult.)	X		O	X	X	O	X		X	X	X	ST/X	Wo
Bromus ciliatus L. — Fringed Brome (*B. canadensis* Michx. *sensu* Voss 1972, McNeill 1976, not Scoggan 1978; incl. *B. dudleyi* Fern. of regional reports)	X		O	X	X	X	X		X		P	T/E	Wa

Taxa		James Bay Coast – HB1	Akimiski Island – HB2	Severn Basin – HB3	Winisk-Ekwan Basin – HB4	Attawapiskat Basin – HB5	Albany Basin – HB6	Moose Basin – HB7	Sutton Ridges – HB8	Cape Henrietta-Maria – MT1	Ont. Maritime Tundra – MT2	Manitoba Lowland – MT3	Continental Distribution	HBL Distribution and Frequency
B. inermis Leyss.	Smooth Brome	X			X	X	X	X				X	–	(Int)o
B. japonicus Thunb.	Japanese Brome							X					–	(Int)r
B. kalmii A.Gray	Kalm's Brome						X						T/E	R(SI)r
B. pumpellianus Scribn. (*B. inermis* Leyss. ssp. p. (Scribn.) Wagnon)	Pumpelly Brome											P	aST/W	R(NW)r
Calamagrostis canadensis (Michx.) P.Beauv. (incl. var. *scabra* (Kunth) Hitchc. (ssp. *langsdorfii* (Link) Hult.))	Canada Bluejoint	X	X	X	X	X	X	X	X	X	X	X	aST/X	Wa
C. deschampsioides Trin.											X	X	aST/X	R(NC)r
C. inexpansa A.Gray (*C. stricta* ssp. *inexpansa* (A.Gray) C.W.Greene)	Northern Reed Grass	X	X	X	X	X	X	X	X	X	X	X	aST/X	Wa
C. lapponica (Wahl.) Hartm. (var. *nearctica* Porsild)						X			X	O		X	aST/X	on
C. purpurascens R.Br.	Purple Reed Grass				X						X		AST/X	R(N)r
C. stricta (Timm) Koeler (*C. neglecta* (Ehrh.) Gaertn., Mey. & Schreb.; incl. var. *brevior* (Laest.) Kearney (*C. holmii* Lange))	Reed Grass	X	X	X	X	X	X	X	X	X	X	X	aST/X	Wa
Catabrosa aquatica (L.) P.Beauv. (incl. var. *uniflora* A.Gray and var. *laurentiana* Fern.)	Brook-grass	X		X		X	X	X	X	X	X	X	aST/X	Wo
Cinna latifolia (Trevir.) Griseb.	Wood Reedgrass	X				X	X	X					sT/X	R(S)o
Dactylis glomerata L.	Orchard Grass							X					–	(Int)r
Danthonia intermedia Vasey	Timber Oat Grass			X			X		X				sT/X	R(I)ot
D. spicata (L.) P.Beauv.	Poverty Oat Grass					X	X		X				aST/X	R(S)o
Deschampsia cespitosa (L.) P.Beauv.														
var. *cespitosa* (incl. var. *glauca* (Hartm.) Lindm.f. and var. *abbei* Boivin)	Tufted Hair Grass	X	X	X	X	X	X	X	X	X	O	X	AST/X	Wa
var. *maritima* Vasey (var. *littoralis* (Reut.) Richter)		X								X	X	X	AST/X	R(C)a
Dupontia fisheri R.Br. (var. *psilosantha* (Rupr.) Trautv.; incl. f. *aristata* (Malte) Lepage)		X	X							X	X	X	AS/X	R(C)an

Taxa		James Bay Coast – HB1	Akimiski Island – HB2	Severn Basin – HB3	Winisk-Ekwan Basin – HB4	Attawapiskat Basin – HB5	Albany Basin – HB6	Moose Basin – HB7	Sutton Ridges – HB8	Cape Henrietta-Maria – MT1	Ont. Maritime Tundra – MT2	Manitoba Lowland – MT3	Continental Distribution	HBL Distribution and Frequency
Elymus canadensis L. (incl. var. *albanensis* Lepage)	Canada Wild Rye	X					X	X					sT/X	R(S)o
E. innovatus Beal				X	X	X		X	X		X	X	ST/W	Wo
E. mollis Trin. (incl. var. *villosissimus* (Scribn.) Hult.; *E. arenarius* L. var. *villosus* E.Mey., *E. arenarius* ssp. *mollis* (Trin.) Hult., *Leymus mollis* (Trin.) Pilg.)	Dune Grass	X	X					X		X	X	X	aST/X	R(C)a
E. virginicus L.	Virginia Wild Rye	X					X	X					T/X	R(S)o
E. wiegandii Fern.	Wiegand Wild Rye	X					X	X					sT/X	R(S)o
Festuca elatior L. (*F. arundinacea* Schreb.)	Tall Fescue						X						–	(Int)r
F. baffinensis Polunin	Baffin Fescue									X			AS/X	R(NE)r
F. brachyphylla Schult. (var. *brachyphylla* (Schult.) Piper; *F. ovina* L. var. *brevifolia* (R.Br.) S.Wats.)		X	X	X					X	X	X	X	AST/X	(W)o
F. pratensis Huds. (*F. elatior* L., in part)	Meadow Fescue	X						X					–	(Int)o
F. rubra L.	Red Fescue													
var. *rubra* (incl. var. *arenaria* Osbeck, ssp. *arctica* (Hackel) Gov., and var. *mutica* Hartm.)		X	X	X	X	X	X	X	X	X	X	X	aST/X	Wa
var. *prolifera* (Piper) Piper. (f. *lasiolepis* (Fern.) A.Löve, *F. prolifera* (Piper) Fern.)				X	X	X			X	X	X	X	aST/X	Won
F. saximontana Rydb. (*F. ovina* L. var. *saximontana* (Rydb.) Gl., *F. brachyphylla* Schult. var. *rydbergii* (St.-Yves.) Cronq.; Frederiksen 1982)	Rocky Mountain Fescue		X							X	X	O	aST/X	R(NC)o
F. trachyphylla (Hackel) Krajina (*F. ovina* L. var. *duriuscula* auct.; *F. longifolia* Thuill.)	Hard Fescue							X					–	(Int)r
Glyceria borealis (Nash) Batch.	Northern Manna Grass	X			X	X	X	X		X			sT/X	(W)o
G. grandis S.Wats. (*G. maxima* (Hartm.) Holmb. var. *americana* (Torr.) Boivin; ssp. *grandis* (S.Wats.) Hult.)	Tall Manna Grass	X			X	X	X	X				X	ST/X	ot

Taxa		James Bay Coast – HB1	Akimiski Island – HB2	Severn Basin – HB3	Winisk-Ekwan Basin – HB4	Attawapiskat Basin – HB5	Albany Basin – HB6	Moose Basin – HB7	Sutton Ridges – HB8	Cape Henrietta-Maria – MT1	Ont. Maritime Tundra – MT2	Manitoba Lowland – MT3	Continental Distribution	HBL Distribution and Frequency
G. striata (Lam.) A.Hitchc. (var. *stricta* (Scribn.) Fern.)	Fowl Manna Grass	X	X	X	X	X	X	X	X			X	ST/X	(W)o
Hierochloe alpina (Sw.) Roem. & Schult. (var. *orthantha* (Sor.) Hult.)	Alpine Holy Grass								X				AST/X	R(LS)r
H. odorata (L.) P.Beauv. (*H. hirta* (Schrank) Borbas ssp. *arctica* (Presl) Weim.)	Sweet Grass	X	X	X	X	X	X	X	X	X	X	X	aST/X	Wa
H. pauciflora R.Br.	Arctic Holy Grass									X	X	X	ASs/X	R(NC)o
Hordeum jubatum L.	Foxtail	X	X	X	X		X	X		X	X	X	aST/X	Wa
Koeleria macrantha (Ledeb.) Schult. (*K. cristata* auct.)	June Grass											X	sT/X	R(NWI)r
Lolium persicum Boiss. & Hohen.	Persian Darnell											X	–	(Int)r
Muhlenbergia glomerata (Willd.) Trin. (var. *cinnoides* (Link) Hern.; part of *M. racemosa* (Michx.) BSP. var. *cinnoides* (Link) Boivin)	Muhly Grass				X	X	X	X	X				sT/X	R(I)a
M. mexicana (L.) Trin.	Satin Grass						X	X					T/X	R(SI)o
M. richardsonis (Trin.) Rydb. (*M. squarrosa* (Trin.) Rydb.)	Richardson Muhly				X	X	X	X					sT/X	R(SI)o
Oryzopsis asperifolia Michx.	Rough-leaf Mountain-rice			X	X	X	X	O	X			P	T/X	R(I)ot
O. pungens (Torr.) A.Hitchc.	Sharp-leaf Oryzopsis			X	X		X		X	X	X	P	sT/X	Wo
Panicum capillare L.	Common Witch Grass						X						T/X	R(SI)r
P. lanuginosum Ell. (var. *implicatum* (Scribn.) Fern.)	Hairy Panic Grass						X	X					T/X	R(SI)r
P. lindheimeri Nash							X	X					T/X	R(SI)r
P. virgatum L.	Switch Grass						X						T/E	R(SI)r
Phalaris arundinacea L.	Reed Canary Grass	X		X	X	X	X	X					ST/X	ot
Phleum alpinum L. (*P. commutatum* Gaud. var. *americanum* (P.Fourn.) Hult.)	Alpine Timothy	X								X	X		aST/X	R(C)r
P. pratense L.	Timothy	X			O		X	X			X	X	–	(Int)r
Phragmites australis (Cav.) Trin. (*P. communis* Trin.)	Reed Grass						X						sT/X	R(SI)r
Poa alpina L. (*P. commutata* Gandoger)	Alpine Blue Grass	X	X	X	X	X		X		X	X	X	aST/X	Wo

Taxa		James Bay Coast – HB1	Akimiski Island – HB2	Severn Basin – HB3	Winisk-Ekwan Basin – HB4	Attawapiskat Basin – HB5	Albany Basin – HB6	Moose Basin – HB7	Sutton Ridges – HB8	Cape Henrietta-Maria – MT1	Ont. Maritime Tundra – MT2	Manitoba Lowland – MT3	Continental Distribution	HBL Distribution and Frequency
P. annua L.	Annual Blue Grass					X		X	X				–	(Int)o
P. arctica R.Br. (var. *arctica*; incl. ssp. *cespitans* (Simmons) Nannf.)	Arctic Blue Grass	+	X		X					X	X	X	AST/X	R(N)a
P. compressa L.	Wire Grass				O		X	X			X		–	(Int)o
P. eminens Presl		X											aST/E,W	R(JB)r
P. flexuosa Sm. (*P. laxa* Haenke ssp. *flexuosa* (Sm.) Hyl.)												X	aST/E	R(NWC)r
P. glauca Vahl (incl. ssp. *glaucantha* (Gaud.) Lindm.)	Glaucous Poa	X	X	X	X	X	X	X	X	X	X	X	AST/X	Wo
P. interior Rydb. (allied with *P. palustris*; incl. specimens identified as *P. nemoralis* L., such as by Lepage (1966), which were corrected by pers. comm. 1977 although most were actually *P. glauca* (QFA))	Interior Poa			X			X						sT/X	ot
P. palustris L.	Fowl Meadow Grass	X	O	X	X	X	X	X	X	X	X	X	sT/X	Wa
P. pratensis L.	June Grass													
var. *alpigena* Fries		X	X	X	X	X		X	X	X	X	X	AST/X	Wa
var. *pratensis*		X			X	X	X	X	X	X	X	X	ST/X	Wo(+Int)
P. saltuensis Fern. & Wieg.	Two-rayed Poa						X						T/E	R(Sl)r
P. trivialis L.	Rough Blue Grass						X						–	(Int)r
Puccinellia ambigua Sor. (*P. paupercula* (Holm.) Fern. & Weath. var. *alaskana* (Scribn. & Mer.) Fern. & Weath.)		X								X	X	X	ST/E	R(C)o
P. langeana (Berl.) Sor. (*P. paupercula* (Holm.) Fern. & Weath.; *P. pumila* (Vasey) Hitchc. var. *fernaldii* Hult.)											X	X	AST/X	R(NC)r
P. lucida Fern. and Weath. (incl. *P. macra* Fern. & Weath., *P. laurentiana* Fern. & Weath., *P. distans* (L.). Parl., *P. deschampsioides* Sor., *P. coarctata* Fern. & Weath., and *P. nuttalliana* (Sch.) Hitchc. of regional reports)	Shining Saltmarsh Grass	X	X			X		X		X	X	X	ST/E	R(C)a
P. phryganodes (Trin.) Scribn. & Merr.	Goose Grass	X	X							X	X	X	ASs/X	R(C)a

Taxa	James Bay Coast – HB1	Akimiski Island – HB2	Severn Basin – HB3	Winisk-Ekwan Basin – HB4	Attawapiskat Basin – HB5	Albany Basin – HB6	Moose Basin – HB7	Sutton Ridges – HB8	Cape Henrietta-Maria – MT1	Ont. Maritime Tundra – MT2	Manitoba Lowland – MT3	Continental Distribution	HBL Distribution and Frequency
P. vaginata (Lange) Fern. & Weath. (incl. *P. angustata*, (Krecz.) Kit. of regional reports and *P. hauptiana* (Krecz.) Kit. of Lepage 1966 (CAN))										X	X	AS/X	R(NC)r
Schizachne purpurascens (Torr.) Swallen — Purple Melic Grass	X		X	X		X	X	X	X			sT/X	ot
Setaria viridis (L.) P.Beauv. — Green Foxtail											X	–	(Int)r
Spartina pectinata Link — Tall Cord Grass	X				X	X	X					sT/X	R(S)o
Sphenopholis intermedia (Rydb.) Rydb. — Slender Wedge Grass (*S. obtusata* (Michx.) Scribn. var. *major* (Torr.) Erdm.)	X		X	X	X	X	X				X	sT/X	ot
Trisetum melicoides (Michx.) Scribn. — Melic-like Trisetum	X		X	X	X	X	X	X	X	X		T/E	Wo
T. spicatum (L.) K.Richter — Spike Trisetum													
var. *spicatum* (*T. triflorum* (Bigel.) A. & D.Löve; incl. var. *maidenii* (Grand.) Fern.)		X						X	X	X		ASs/X	R(NC)o
var. *molle* (Michx.) Beal (*T. molle* (Michx.) Kunth)	X		X	X	X	X	X	X	X	X	X	AST/X	Wo
var. *pilosiglume* Fern. (*T. molle* ssp. *pilosiglume* (Fern.) Pavl.)				X	X	X	X		X	X		AST/X	(W)o
Triticum aestivum L. — Wheat							X				X	–	(Int)r
Vahlodea atropurpurea (Wahl.) E.Fries (*Deschampsia atropurpurea* (Wahl.) Scheele)									X			aST/X	R(NE)r
CYPERACEAE — Sedge Family													
Carex adelostoma Krecz. (*C. buxbaumii* var. *alpicola* Hartm., *C. morrisseyi* Porsild)											X	S/X	R(NWC)r
C. aquatilis Wahl. — Water Sedge	X	X	X	X	X	X	X	X	X	X	X	AST/X	Wa
C. arcta F.Boott — Clustered Sedge			X									sT/X	R(I)r
C. arctata P.Boott — Compressed Sedge							X					ST/E	R(SI)r
C. atherodes Spreng. — Slough Sedge					X	X					P	ST/X	R(I)rt
C. atratiformis Britt. (ssp. *raymondii* (Calder) Scoggan; f. *wolfii* (Kneucker) Kük.; incl. *C. atrata* of regional reports)			X	X	X			X			P	ST/X	R(I)ot

Taxa		James Bay Coast – HB1	Akimiski Island – HB2	Severn Basin – HB3	Winisk-Ekwan Basin – HB4	Attawapiskat Basin – HB5	Albany Basin – HB6	Moose Basin – HB7	Sutton Ridges – HB8	Cape Henrietta-Maria – MT1	Ont. Maritime Tundra – MT2	Manitoba Lowland – MT3	Continental Distribution	HBL Distribution and Frequency
C. atrofusca Schk.										X	X	X	ASs/X	R(NC)o
C. aurea Nutt.	Golden-fruit Sedge	X	X	X	X	X	X	X	X	X	X	X	ST/X	Wa
C. backii F.Boott												P	T/X	R(NWI)
C. bebbii (L.Bailey) Fern.	Bebb Sedge	X				X	X	X				P	sT/X	R(S)o
C. bicolor All.	Bicolour Sedge	X			X					X	X	X	aST/X	R(N)o
C. bigelowii Torr.			X						X			X	AST/X	R(N)o
C. brunnescens (Pers.) Poir. (incl. var. *brunnescens* and var. *sphaerostachya* (Tuck.) Kük.)	Brownish Sedge	X	X			X	X	X	X		X	P	aST/X	Wo
C. buxbaumii Wahl.	Brown Bog Sedge				X	X	X	X	X	X		X	aST/X	(W)o
C. canescens L. (incl. var. *subloliacea* Laest. (*C. lapponica* O.Lang), e.g., at HB8)	Hoary Sedge			X	X	X	X	X	X	X	X	X	aST/X	Wa
C. capillaris L.	Hair-like Sedge													
var. *capillaris* (incl. var. *elongata* Olney)		X	X		X	X	X	X	X	X	X	X	AST/X	Wa
var. *krausei* (Boeck.) Krantz (var. *porsildiana* Pol. *sensu* Boivin)										X			AS/X	R(NE)r
C. capitata L. (incl var. *arctogena* (Sm.) Hult.)			X						X	X	X	X	aST/X	R(N)a
C. chordorrhiza L.f.	Creeping Sedge	X	X	O	X	X	X	X	X	X	X	X	aST/X	Wa
C. concinna R.Br.		X	X		X	X	X	X	X	X	X	X	ST/X	Wa
C. crawei Dewey	Crawe Sedge					X	X	X					T/X	R(SI)o
C. crawfordii Fern.	Crawford Sedge							X					ST/X	R(S)r
C. deflexa Hornem.							X		X	X		X	aST/X	R(I)ot
C. deweyana Schwein.	Short-scale Sedge						X	X					sT/X	R(S)
C. diandra Schrank	Panicled Sedge	X	X	X	X	X	X	X	X	X	X	X	ST/X	Wa
C. disperma Dewey	Soft-leaf Sedge			X	X	X	X	X	X	X		X	ST/X	a(W)
C. eburnea F.Boott	Bristle-leaf Sedge			X	X	X	X	X	X	O	X	X	ST/X	Wa
C. exilis Dewey	Coast Sedge				X	X	X	X	X				sT/E	R(SI)o
C. flava L.	Yellow Sedge	X	X			X	X	X	X				ST/X	ot
C. foenea Willd. (*C. aenea* Fern.)	Bronze Sedge			X	X		X	X	X		X	P	ST/X	(W)o
C. garberi Fern. (incl. var. *bifaria* Fern.)	Elk Sedge	X	X	O	X	X	X	X	X	X	X	X	ST/X	Wa

Taxa		James Bay Coast – HB1	Akimiski Island – HB2	Severn Basin – HB3	Winisk-Ekwan Basin – HB4	Attawapiskat Basin – HB5	Albany Basin – HB6	Moose Basin – HB7	Sutton Ridges – HB8	Cape Henrietta-Maria – MT1	Ont. Maritime Tundra – MT2	Manitoba Lowland – MT3	Continental Distribution	HBL Distribution and Frequency
C. glacialis Mack.				O		X				X	X	X	AST/X	R(N)o
C. glareosa Wahl.		X	X							X	X	X	AST/X	R(C)a
(incl. var. *glareosa* and var. *amphigena* Fern.)														
C. granularis Muhl.	Meadow Sedge						X	X					T/E	R(Sl)o
(var. *haleana* (Olney) Porter)														
C. gynocrates Wormsk.	Bog Sedge	X	X	X	X	X	X	X	X	X	X	X	aST/X	Wa
(*C. dioica* L. var. *gynocrates* (Wormsk.) Ostenf.)														
C. heleonastes L.f.				X	X	X	X			X	X	X	ST/X	Wo
C. houghtoniana Torr.	Houghton Sedge						X	X				P	T/X	R(I)ot
C. interior Bailey	Inland Sedge	X	X	X	X	X	X	X	X		X		sT/X	Wa
C. intumescens Rudge	Bladder Sedge						X						T/E	R(Sl)r
(incl. var. *fernaldii* Bailey of regional reports)														
C. lasiocarpa Ehrh.	Hairy-fruit Sedge	O			X	X	X	X					ST/X	at
(var. *americana* Fern.)														
C. lenticularis Michx.					X	X	X						ST/X	R(I)ot
C. leptalea Wahl.	Bristle-stalk Sedge	X	X	X	X	X	X	X	X	X	X	X	ST/X	Wa
C. leptonervia Fern.	Fine-nerve Sedge							X					T/E	R(Sl)r
C. limosa L.		X	X	X	X	X	X	X	X	X	X	X	ST/X	Wa
C. livida (Wahl.) Willd.	Livid Sedge		X		X	X	X	X	X	O	X	X	ST/X	Wa
(incl. var. *radicaulis* Paine)														
C. loliacea L.						X		X					ST/X	R(Sl)r
C. mackenziei Krecz.		X	X							X	X	X	aST/X	R(C)a
C. magellanica Lam.	Stunted Sedge	X	X	X	X	X	X	X	X	X	X	X	aST/X	Wa
(ssp. *irrigua* (Wahl.) Hult.; *C. paupercula* Michx.)														
C. marina Dewey		O								X	X	X	aSs/X	R(NC)o
(*C. amblyorhyncha* Krecz.)														
C. maritima Gunn.										X	X	X	AST/X	R(NC)o
(incl. *C. dutillyi* O'Neill & Duman from Churchill)														
C. microglochin Wahl.		O	X		X	X			X	X	X	X	aST/X	R(N)a
C. misandra R.Br.										X			AST/X	R(NC)r
(*C. fuliginosa* Schk.)														
C. nardina Fries										X		X	AST/X	R(NC)r
C. norvegica Retz.														
var. *norvegica*			X	X					X	X	X	X	aST/E	R(N)o
var. *inferalpina* (Wahl.) Boivin (C. *media* R.Br.)						X		X	X		X	X	ST/X	(W)o

Taxa		James Bay Coast – HB1	Akimiski Island – HB2	Severn Basin – HB3	Winisk-Ekwan Basin – HB4	Attawapiskat Basin – HB5	Albany Basin – HB6	Moose Basin – HB7	Sutton Ridges – HB8	Cape Henrietta-Maria – MT1	Ont. Maritime Tundra – MT2	Manitoba Lowland – MT3	Continental Distribution	HBL Distribution and Frequency
C. oligosperma Michx.	Few-seed Sedge		X		X	X	X	X	X				sT/X	R(I)at
C. ormostachya Wieg. (*C. laxiflora* Lam. var. *ormostachya* (Wieg.) Gl.)								X					T/E	R(Sl)r
C. paleacea Wahl.		X	X									X	ST/E	R(C)a
C. pauciflora Lightf.	Few-flower Sedge	O			O	X	X	X	X				ST/X	at
C. peckii Howe (*C. nigromarginata* Schw. var. *elliptica* (F.Boott) Gl.)								X					ST/X	R(S)r
C. pedunculata Muhl.					X			X					T/X	R(S)r
C. pellita Muhl. (*C. lanuginosa* Michx.; *C. lasiocarpa* Ehrh. var. *latifolia* Boeck.)	Woolly Sedge	X		X	X	X	X	X	X			X	T/X	a(W)
C. prairea Dewey	Prairie Sedge	X				X	X	X					T/X	R(S)o
C. praticola Rydb.		X	X					X	X				aST/X	ot
C. projecta Mack.	Spreading Sedge	X					X	X					sT/E	R(S)o
C. rariflora (Wahl.) Smith		X	X	X	X	X	X		X	X	X	X	aST/X	Wa
C. recta F.Boott (*C. salina* Wahl. var. *kattegatensis* (Fries) Almq.)		X									X	X	aST/E	R(C)o
C. retrorsa Schwein.	Retrorse Sedge	X				X	X	X					sT/X	R(S)o
C. richardsonii R.Br.	Richardson Sedge							X					sT/X	R(Sl)r
C. rossii F.Boott											X		ST/W	R(N)r
C. rostrata Stokes (*s. str.*)			X		X	X	X		X		X	X	aST/X	R(I)o
C. rotundata Wahl.												X	aST/W	R(NWC)r
C. rupestris All.										X		X	AST/X	R(NWC)r
C. salina Wahl. (*s. str.*)		X								X	X		aST/E	R(C)o
C. sartwellii Dewey		X						X					sT/X	R(S)r
C. saxatilis L. (incl. var. *miliaris* (Michx.) Bailey and var. *major* Olney (*C. physocarpa* Presl); and *C. rotundata* of Ontario reports)		X	X	X	X	X	O		X	X	X	X	AST/X	Wa
C. scirpoidea Michx. (incl. var. *scirpiformis* (Mack.) O'Neill & Duman)		X		X	X	X	X	X	X	X	X	X	aST/X	Wa
C. scoparia Schk.								X					T/X	R(S)r
C. sterilis Willd. (*C. muricata* L. var. *sterilis* (Carey) Gl.)		X		X	X	X	X						sT/X	ot
C. stipata Muhl.	Awl-fruit Sedge						X	X					sT/X	R(S)o

Taxa		James Bay Coast – HB1	Akimiski Island – HB2	Severn Basin – HB3	Winisk-Ekwan Basin – HB4	Attawapiskat Basin – HB5	Albany Basin – HB6	Moose Basin – HB7	Sutton Ridges – HB8	Cape Henrietta-Maria – MT1	Ont. Maritime Tundra – MT2	Manitoba Lowland – MT3	Continental Distribution	HBL Distribution and Frequency
C. stricta Lam.	Tussock Sedge					X	X	X					T/E	R(S)o
C. subspathacea Wormsk. (*C. salina* Wahl. var. *subspathacea* (Wormsk.) Tuckerm.)		X	X							X	X	X	aST/X	R(C)a
C. tenuiflora Wahl.	Sparse-flower Sedge	X		X	X	X	X	X	X	O	X	X	aST/X	Wa
C. trisperma Dewey	Three-fruit Sedge	O	X			X	X	X					ST/X	R(S)o
C. umbellata Schk. (var. *brevirostra* F.Boott; *C. abdita* Bickn.)	Umbellate Sedge						X						sT/X	R(Sl)r
C. ursina Dewey												X	ASs/X	R(NWC)r
C. utriculata F.Boott (*C. rostrata* Stokes var. *utriculata* (F.Boott) Bailey)	Beaked Sedge	X	X	X	X	X	X	X	X	X	X	X	aST/X	Wa
C. vaginata Tausch (*C. saltuensis* Bailey)		X	X	X	X	X	X	X	X	X	X	X	aST/X	Wa
C. viridula Michx.	Greenish Sedge		X		X	X	X	X					aST/X	R(Sl)o
C. williamsii Britt.	Williams Sedge									X	X	X	aST/X	R(NC)o
Cladium mariscoides (Muhl.) Torr.	Twig-rush						X						T/E	R(Sl)r
Eleocharis acicularis (L.) Roem. & Schult.	Slender Spike-rush	X	X		X	X	X	X	X			X	aST/X	Wa
E. elliptica Kunth (*E. tenuis* (Willd.) Schult. var. *borealis* (Sven.) Gl.; incl. *E. compressa* Sull. of regional reports, misapplied)	Elliptic Spike-rush					X	X	X					T/X	R(S)o
E. erythropoda Steud. (*E. calva* Torr.)	Red-base Spike-rush	X			X	X	X	X					sT/X	R(S)o
E. kamtschatica (C.A.Meyer) Kom.		X									X		ST/W	R(JB,NC)o
E. pauciflora (Lightf.) Link (var. *fernaldii* Sven.; *E. quinqueflora* (Hartm.) Schwarz)		X			X	X	X			X	X	X	aST/X	Wo
E. smallii Britt. (incl. *E. macrostachya* of regional reports; both variants of *E. palustris* (L.) Roem. & Schult. *s.l.*)	Small Spike-rush	X	X	O	X	X	X	X	X	X	X	X	ST/X	Wa
E. uniglumis (Link) Schult. (*E. halophila* Fern & Brack.)		X			X						X	X	ST/X	R(C)o
Eriophorum angustifolium Honck.	Thin-scale Cotton-grass	X	X	X	X	X	X	X	X	X	X	X	AST/X	Wa

Taxa	James Bay Coast – HB1	Akimiski Island – HB2	Severn Basin – HB3	Winisk-Ekwan Basin – HB4	Attawapiskat Basin – HB5	Albany Basin – HB6	Moose Basin – HB7	Sutton Ridges – HB8	Cape Henrietta-Maria – MT1	Ont. Maritime Tundra – MT2	Manitoba Lowland – MT3	Continental Distribution	HBL Distribution and Frequency
E. brachyantherum Trautv. & C.A.Meyer (*E. opacum* (Bjornstr.) Fern.)			X	X	X	X	X	X	X	X	X	aST/X	Wa
E. callitrix Cham.			O						X	X	X	aSs/X	R(NC)o
E. chamissonis C.A.Meyer Rusty Cotton-grass													
var. *chamissonis* (*E. russeolum* Fries ssp. *rufescens* (Anders.) Hyl.)	X	X		X	X	O	X	X	X		O	aST/X	Wa
var. albidum (Nyl.) Fern. (f. *subalbidum* (Lindb.f.) Blomgr.)	X			X	X	X	X		X	X	X	aST/X	Wo
E. gracile Koch Slender Cotton-grass	X			X	X	X	X			X	X	ST/X	Wa
E. scheuchzeri Hoppe			X						X	X	X	AST/X	R(N)o
E. tenellum Nutt. Conifer Cotton-grass		X			X	X	X					T/E	R(Sl)o
E. vaginatum L. Tussock Cotton-grass (var. *spissum* (Fern.) Boivin)		X	X	X	X	X	X	X	X	X	X	aST/X	Wa
E. virginicum L. Tawny Cotton-grass					X	X	X					sT/X	R(Sl)o
E. viridicarinatum (Engelm.) Fern. Dark-scale Cotton-grass		X	X	X	X	X	X	X	X	X	X	ST/X	Wa
Kobresia myosuroides (Vill.) Fiori & Paol. (*K. bellardii* (All.) Degl.)									X	X	X	ASs/X	R(NC)o
K. simpliuscula (Wahl.) Mack. (var. *americana* Duman)		X		X				X	X	X	X	AST/X	R(N)a
Rhynchospora alba (L.) M.Vahl White Beak-rush				X	X	X	X	X				sT/X	R(I)a
R. fusca (L.) Ait.f. Brown Beak-rush						X						T/E	R(Sl)r
Scirpus acutus Muhl. Hard-stem Bulrush (*Schoenoplectus acutus* (Muhl.) Löve & Löve; *S. lacustris* L. ssp. *glaucus* (Sm.) Hartm.)	X				X	X	X					ST/X	R(S)o
S. atrovirens Willd. Black Bulrush						X	X					T/X	R(Sl)o
S. cespitosus L. Deer-grass (var. *callosus* Bigel.; *Trichophorum* *cespitosum* (L.) Schur)	O	X	X	X	X	X	X	X	X	X	X	aST/X	Wa
S. cyperinus (L.) Kunth Wool-grass (var. *brachypodus* (Fern.) Gilly; *S.atrotinctus* Fern.)						X	X					T/E	R(S)o
S. hudsonianus (Michx.) Fern. (*Trichophorum alpinum* (L.) Pers.)	O	X	X	X	X	X	X	X	O	X	X	ST/X	Wa
S. maritimus L. Maritime Bulrush (var. *paludosus* (Nels.) Kük.; *Bolboschoenus* *maritimus* (L.) Palla)	X								X			ST/X	R(JB)o

Taxa		James Bay Coast – HB1	Akimiski Island – HB2	Severn Basin – HB3	Winisk-Ekwan Basin – HB4	Attawapiskat Basin – HB5	Albany Basin – HB6	Moose Basin – HB7	Sutton Ridges – HB8	Cape Henrietta-Maria – MT1	Ont. Maritime Tundra – MT2	Manitoba Lowland – MT3	Continental Distribution	HBL Distribution and Frequency
S. microcarpus C.Presl (var. *rubrotinctus* (Fern.) Jones)	Red-sheath Bulrush	X		X	X	X	X	X				X	sT/X	ot
S. pungens Vahl (*S. americanus* of regional reports)	Threesquare	X					X	X					sT/X	R(S)o
S. rufus (Huds.) Schrad. (var. *neogaeus* Fern.; *Blysmus rufus* (Huds.) Link)		X	X								X	X	ST/X	R(JB,NWC)o
S. validus Vahl (var. *creber* Fern.; *Schoenoplectus tabernaemontani* (C.Gmel.) Palla; *S. lacustris* L. ssp. *v.* (Vahl) Koyama)	Soft-stem Bulrush	X				X	X	X					ST/X	R(S)o
ARACEAE	Arum Family													
Acorus americanus (Raf.) Raf.	Sweet-flag							X					sT/X	R(SI)r
Calla palustris L.	Wild Calla	X			X	X	X	X				P	ST/X	ot
LEMNACEAE	Duck-weed Family													
Lemna minor L.	Common Duck-weed	X						X			X	X	ST/X	o
L. trisulca L.	Star Duck-weed	X				X		X	X	X	X	X	ST/X	Wo
XYRIDACEAE	Yellow-eyed-grass Family													
Xyris montana Ries	Yellow-eyed-grass						X						T/E	R(SI)r
ERIOCAULONACEAE	Pipewort Family													
Eriocaulon aquaticum (Hill) Druce (*E. septangulare* With.)	Pipewort						X						sT/E	R(SI)r
JUNCACEAE	Rush Family													
Juncus alpinoarticulatus Chaix (*J. alpinus* Vill. var. *rariflorus* Hartm. (*J. nodulosus* Wahl.) and var. *alpinus*)	Alpine Rush	X	X	X	X	X	X	X	X	X	X	X	aST/X	Wa
J. arcticus Willd.	Arctic Rush				X					X	X	X	AST/X	R(NC)o
J. articulatus L.	Jointed Rush							X					T/X	R(S)r
J. balticus Willd. (var. *littoralis* Engelm.; J. *arcticus* var. *littoralis* (Engelm.) Boivin; incl. var. *stenocarpus* Buch. & Fern.)	Baltic Rush	X	X	X	X	X	X	X	X	X	X	X	aST/X	Wa
J. biglumis L.												X	AST/X	(NWC)r
J. brevicaudatus (Engelm.) Fern.								X					ST/X	R(S)r
J. bufonius L. (incl. var. *halophilus* Buch. and Fern.; *J. ranarius* Song. & Perr.)	Toad Rush	X	X	O	O	X	X	X		X	X	X	aST/X	Wa

Taxa		James Bay Coast – HB1	Akimiski Island – HB2	Severn Basin – HB3	Winisk-Ekwan Basin – HB4	Attawapiskat Basin – HB5	Albany Basin – HB6	Moose Basin – HB7	Sutton Ridges – HB8	Cape Henrietta-Maria – MT1	Ont. Maritime Tundra – MT2	Manitoba Lowland – MT3	Continental Distribution	HBL Distribution and Frequency
J. castaneus Sm. (incl. f. *pallidus* (Hook.) Boivin)			X		X					X	X	X	AST/X	R(N)o
J. dudleyi Wieg.	Dudley Rush	X			X	X	X	X					sT/X	ot
J. ensifolius Wikstr.								X					sT/W	R(JB)r
J. filiformis L.				O		X	X	X					aST/X	R(I)ot
J. longistylis Torr.	Long-style Rush	X			X		X	X					sT/X	R(S)o
J. nodosus L.	Knot Rush	X			X	X	X	X				X	ST/X	Wo
J. pelocarpus E.Meyer	Brown-fruit Rush	X				X							sT/E	R(S)r
J. pylaei Laharpe (*J. effusus* var. *p.* (Laharpe) Fern. & Wieg.)	Soft Rush							X					T/X	R(S)r
J. stygius L. (var. *americanus* Buch.)					X	X	X	X	X				ST/X	R(I)ot
J. subtilis E.Meyer					X								aST/E	R(I)rt
J. tenuis Willd.	Path Rush							X					sT/X	R(SI)r
J. triglumis L. (var. *albescens* Lange; *J. albescens* (Lange) Fern.)		X	X	X	X	O	X		X	X	X	X	AST/X	Wo
J. vaseyi Engelm.	Vasey Rush											P	sT/X	R(NWI)
Luzula acuminata Raf. (*L. pilosa* (L.) Willd. var. *americana* Roem. & Schult.)	Pointed Wood-rush							X					sT/X	R(SI)r
L. confusa Lindeb.									X			X	AST/X	R(LS,NWC)r
L. multiflora (Retz.) Lej. (var. frigida (Buch.) Sam.; incl. *L. groenlandica* Böcher; *L. campestris* (L.) DC. *sensu* Boivin)	Common Wood-rush									X	X	X	aST/X	R(NC)a
L. parviflora (Ehrh.) Desv. (incl. var. *parviflora* and var. *melanocarpa* (Michx.) Buch.)	Small Wood-rush	X	X	X				X	X	X	X	X	aST/X	Wo
L. wahlenbergii Rupr. (*L. spadicea* (All.) DC.)												X	aST/X	R(NW)r
LILIACEAE	Lily Family													
Allium schoenoprasum L. (var. *sibiricum* (L.) Hartm.; incl. var. *laurentianum* Fern.)	Wild Chives	X		X			X	X			X	X	ST/X	(W)o
Clintonia borealis (Ait.) Raf.	Bluebead-lily						X	X					sT/E	R(SI)o

Taxa		James Bay Coast – HB1	Akimiski Island – HB2	Severn Basin – HB3	Winisk-Ekwan Basin – HB4	Attawapiskat Basin – HB5	Albany Basin – HB6	Moose Basin – HB7	Sutton Ridges – HB8	Cape Henrietta-Maria – MT1	Ont. Maritime Tundra – MT2	Manitoba Lowland – MT3	Continental Distribution	HBL Distribution and Frequency
Lilium philadelphicum L.	Wood Lily	X					X	X				P	ST/X	R(S)a
Maianthemum canadense Desf.	Canada Mayflower													
var. *canadense*								X					ST/X	R(Sl)o
var. *interius* Fern.						X	X	O					ST/X	R(Sl)o
M. stellatum (L.) Link (*Smilacina stellata* (L.) Desf.)	Starry Solomon-seal	X		X	X	X	X	X			O	X	ST/X	Wa
M. trifolium (L.) Slob. (*Smilacina trifolia* (L.) Desf.)	Three-leaf Solomon-seal	O	X	X	X	X	X	X	X	X	X	X	ST/X	Wa
Prosartes trachycarpa S.Wats. (*Disporum trachycarpum* (Wats.) Benth. & Hook.)	Mandarin						X						ST/X	R(Sl)r
Streptopus amplexifolius (L.) DC. (var. *americanus* Schult.)	Twisted-stalk							O					aST/X	R(Sl)r
S. lanceolatus (Ait.) Rev. (incl. var. *longipes* (Vail) Rev.)	Rose Twisted-stalk						X	X					sT/X	R(Sl)o
Tofieldia glutinosa (Michx.) Pers. (*Triantha g.* Baker)	Sticky False Asphodel	X	X	X	X	X	X	X	X	X	X	X	sT/X	Wa
T. pusilla (Michx.) Pers.	Small False Asphodel		X	X	X	X			X	X	X	X	aST/X	Wa
Trillium cernuum L.	Nodding Trillium						X	X					T/E	R(Sl)o
IRIDACEAE	Iris Family													
Iris veriscolor L.	Blue Flag	X	X		X	X	X	X					sT/E	(W)o
Sisyrinchium montanum Greene	Blue-eyed-grass	X		X	X	X	X	X		O	X	X	aST/X	Wo
ORCHIDACEAE	Orchid Family													
Amerorchis rotundifolia (Banks) Hult.	Round-leaf Orchis	O	X	X	X	X	X	X	X	X	X	X	aST/X	Wo
Arethusa bulbosa L.	Arethusa					X	X	X					sT/E	R(S)o
Calypso bulbosa (L.) Oakes (var. *americana* (R.Br.) Luer)	Calypso	X	X				X	X					ST/X	R(S)o
Coeloglossum viride (L.) Hartm.	Bract Orchid													
var. *viride* (*Habenaria viridis* (L.) R.Br. var. *interjecta* Fern.)											X		aSs/X	R(N)r
var. *virescens* (Muhl.) Luer (ssp. *bracteatum* (Muhl.) Hult.; *H. viridis* var. *bracteata* (Willd.) A.Gray)		X		X	X	X	X	X		X		X	aST/X	Wo
Corallorhiza striata Lindl.	Striped Coral-root						X	O					T/X	R(Sl)r
C. trifida Chat.	Early Coral-root	X	O	X	X	X	X	X	X	X	X	X	aST/X	Wa

Taxa		James Bay Coast – HB1	Akimiski Island – HB2	Severn Basin – HB3	Winisk-Ekwan Basin – HB4	Attawapiskat Basin – HB5	Albany Basin – HB6	Moose Basin – HB7	Sutton Ridges – HB8	Cape Henrietta-Maria – MT1	Ont. Maritime Tundra – MT2	Manitoba Lowland – MT3	Continental Distribution	HBL Distribution and Frequency
Cypripedium parviflorum Salisb.	Yellow Lady-slipper													
var. *makisin* (Farw.) Shev. (*C. parviflorum* of reports)		X			X	X	X	X	X			P	ST/X	ot
var. *planipetalum* Fern.						X							ST/E	R(I)r
var. *pubescens* (Willd.) O.W.Knight						X	X	X	X				ST/X	R(I)ot
C. passerinum Rich.	Franklin Lady-slipper	X	X	X	X	X	X	X		X	X	X	ST/X	Wo
C. reginae Walt.	Showy Lady-slipper						X						T/E	R(Sl)r
Goodyera repens (L.) R.Br. (var. *ophioides* Fern.)	Dwarf Rattlesnake-plantain	X	X			X	X	X					ST/E	R(S)
Listera auriculata Wieg.	Auricled Twayblade							X					T/E	R(S)r
L. borealis Morong	Northern Twayblade	X	X			X		O			X	X	ST/X	(W)o
L. cordata (L.) R.Br.	Heartleaf Twayblade	X	X		X	X		X			X	X	aST/X	Wo
Malaxis monophyllos (L.) Sw. (var. *brachypoda* (Gray) Morris & Eames)	White Adder's-mouth	X											sT/X	R(JB)r
M. paludosa (L.) Sw.	Bog Adder's-mouth						X						sT/W	R(Sl)r
M. unifolia Michx.	Green Adder's-mouth						X	X					T/E	R(Sl)o
Platanthera aquilonis Shev. (incl. *P. huronensis* (Nutt.) Lindl., and *P. hyperborea* of reports)	Tall Northern Green Orchid	X	X	X	X	X	X	X	X	X	X	X	aST/X	Wa
P. dilatata (Pursh) Lindl.	Tall White Bog Orchid	X		X	X	X	X	X	X	X	X	X	sT/X	W
P. lacera (Michx.) G.Don	Ragged Fringe Orchid						X	X					T/E	R(Sl)o
P. obtusata (Banks) Lindl.	Blunt-leaf Orchid	X	X	X	X	X	X	X		X	X	X	aST/X	Wa
P. orbiculata (Pursh) Lindl.	Round-leaf Orchid							X					T/X	R(Sl)r
P. psycodes (L.) Lindl.	Purple Fringe Orchid							X					T/E	R(S)r
Pogonia ophioglossoides (L.) Ker-Gawler	Rose Pogonia						X	X					T/E	R(Sl)o
Spiranthes romanzoffiana Cham.	Hooded Ladies'-tresses	X	X	X	X	X	X	X	X	X	X	X	ST/X	Wa
SALICACEAE	Willow Family													
Populus balsamifera L.	Balsam Poplar	X	X	X	X	X	X	X	X	X	X	X	ST/X	Wa
P. tremuloides Michx.	Trembling Aspen		X	X	X	X	X	X	X		X	X	ST/X	Wa
P. Xjackii Sarg. (*P. balsamifera* X *P. deltoides* Marsh.)							X						ST/X	R(Sl)r

Taxa		James Bay Coast – HB1	Akimiski Island – HB2	Severn Basin – HB3	Winisk-Ekwan Basin – HB4	Attawapiskat Basin – HB5	Albany Basin – HB6	Moose Basin – HB7	Sutton Ridges – HB8	Cape Henrietta-Maria – MT1	Ont. Maritime Tundra – MT2	Manitoba Lowland – MT3	Continental Distribution	HBL Distribution and Frequency
Salix alaxensis (Anderss.) Cov. (incl. var. *longistylis* (Rydb.) Sch.)												X	aST/W	R(NWC)r
S. arbusculoides Anderss.											X	X	ST/W	R(N)r
S. arctica Pall. (incl. *S. anglorum* Cham. of regional reports)	Arctic Willow								X	X	X		AST/X	R(N)a
S. arctophila Cock.	Labrador Willow	X	X		X					X	X	X	aST/X	R(N)a
S. athabascensis Raup (*S. fallax* Raup; *S. pedicellaris* var. *athabascensis* (Raup) Boivin)	Athabasca Willow											X	aST/W	R(NWC)r
S. ballii Dorn (*S. myrtillifolia* var. *brachypoda* Fern.)											X		ST/E	R(N)r
S. bebbiana Sarg.	Beaked Willow	X	O	O	X	X	X	X		X	X	X	ST/X	Wa
S. brachycarpa Nutt. (incl. *S. fullertonensis* Schneid. *sensu* Porsild)		X	X			X				X	X	X	aST/X	R(C)a
S. calcicola Fern. & Wieg. (*S. lanata* L. ssp. *calcicola* (Fern. & Wieg.) Hult.)										X	X	X	aST/X	R(NC)a
S. candida Fluegge	Sage Willow	X	X	X	X	X	X	X	X	X	X	X	ST/X	Wa
S. cordata Michx.	Heart-leaf Willow	X		X	X	X	X	X		X	X		ST/E	Wa
S. discolor Muhl.	Pussy Willow	O						X				P	sT/X	R(S)r
S. eriocephala Michx. (*S. rigida* Muhl.; *S. cordata* var. *rigida* (Muhl.) Carey; incl. *S. rigida* var. *angustata* (Pursh) Fern.)		X				X	X	X					ST/E	R(S)o
S. exigua Nutt. (*S. interior* Rowlee)	Sandbar Willow	X			X	X	X	X			X	X	ST/X	Wo
S. glauca L. (var. *callicarpaea* (Trautv.) Boch.; *S. cordifolia* Pursh var. *callicarpaea* (Trautv.) Fern., and var. *macounii* (Rydb.) Schneider)		X	X	X	X	X		X	X	X	X	X	aST/X	Wa
S. humilis Marsh.	Upland Willow					X	X	X	X				T/E	R(I)ot
S. lucida Muhl.	Shining Willow	X			X	X	X	X			X	X	ST/E	Wo
S. lutea Nutt.		X					X	X				X	ST/W	R(S,NWC)o
S. maccalliana Rowlee		X				X	X	X	X		X	X	ST/X	Wo
S. myricoides Muhl. (S. glaucophylloides Fern.; incl. *S.* X*laurentiana* Fern. of regional reports, and *S. g.* var. *albovestita* Ball)	Blue-leaf Willow	X	X			X		X					ST/E	R(S)o

Taxa		James Bay Coast – HB1	Akimiski Island – HB2	Severn Basin – HB3	Winisk-Ekwan Basin – HB4	Attawapiskat Basin – HB5	Albany Basin – HB6	Moose Basin – HB7	Sutton Ridges – HB8	Cape Henrietta-Maria – MT1	Ont. Maritime Tundra – MT2	Manitoba Lowland – MT3	Continental Distribution	HBL Distribution and Frequency
S. myrtillifolia Anderss.														
var. *myrtillifolia*		X	X	X	X	X	X	X	X	X	X	X	ST/X	Wa
var. *cordata* (Anderss.) Dorn (*S. novae-angliae* Anderss.)				X	X	X	X		X	X	X	X	ST/X	Wa
S. pedicellaris Pursh (incl. var. *tenuescens* Fern. and var. *hypoglauca* Fern.)	Bog Willow	O	X	O	X	X	X	X	X	X	X	X	ST/X	Wa
S. pellita Anderss.		X		X	X	X		X		X	X	X	ST/E	Wa
S. pentandra L.	Bay-leaf Willow							X					–	(Int)o
S. petiolaris Sm. (*S. gracilis* Anderss.)	Slender Willow						X	X					ST/X	R(Sl)r
S. planifolia Pursh (S. *phylicifolia* L. ssp. *planifolia* (Pursh) Hiit.)		X	X	X	X	X	X	X	X	X	X	X	aST/X	Wa
S. pseudomonticola C.Ball (*S. monticola* Bebb. of regional reports; *S. padophylla* Rydb.)					X	X	X	X				X	ST/X	(W)o
S. pyrifolia Anderss.	Balsam Willow					X	X	X					ST/X	R(Sl)o
S. reticulata L.	Net-veined Willow	X	X							X	X	X	aST/X	R(NC)a
S. serissima (Bailey) Fern.	Autumn Willow	X	X		X	X	X	X			X	X	ST/X	Wa
S. vestita Pursh	Hairy Willow			X	X				X	X	X	X	aST/X	R(N)o
MYRICACEAE	Bayberry Family													
Myrica gale L.	Sweet Gale	X	X	X	X	X	X	X	X	X	X	X	ST/X	Wa
BETULACEAE	Birch Family													
Alnus crispa (Ait.) Pursh (*A. viridis* (Chaix) DC. ssp. *crispa* (Ait.) Turrill)	Green Alder			X	X	X	X	X	O		X	X	aST/X	a(W)
Alnus rugosa (DuRoi) Spreng. (var. *americana* (Regel) Fern.; *A. incana* (L.) Moench ssp. *rugosa* (DuRoi) Claus.)	Speckled Alder	X		X	X	X	X	X		X	X	X	ST/X	Wa
Betula glandulosa Michx. (*s. str.*)	Dwarf Birch				X				X	X	X	X	aST/X	R(N)o
B. neoalaskana Sarg. (*B. resinifera* (Regel.) Britt.)	Alaska Birch											X	ST/W	R(NWI)r
B. occidentalis Hook. (*B. fontinalis* Sarg.; *B. minor* (Tuckerm.) Fern.)	River Birch					X	X		X			X	ST/X	o
B. papyrifera Marsh. (incl. var. *papyrifera*, var. *cordifolia* (Regel) Fern., var. *macrostachya* Fern., var. *pensilis* Fern., var. *commutata* (Regel) Fern.)	White Birch			X	X	X	X	X	X				ST/X	R(I)o

Taxa		James Bay Coast – HB1	Akimiski Island – HB2	Severn Basin – HB3	Winisk-Ekwan Basin – HB4	Attawapiskat Basin – HB5	Albany Basin – HB6	Moose Basin – HB7	Sutton Ridges – HB8	Cape Henrietta-Maria – MT1	Ont. Maritime Tundra – MT2	Manitoba Lowland – MT3	Continental Distribution	HBL Distribution and Frequency
B. pumila L. (var. *glandulifera* Regel; *B. glandulosa* var. *glandulifera* (Regel) Gl.)	Bog Birch	X	X	X	X	X	X	X	X	X	X	X	ST/X	Wa
B. X*sandbergii* Britt. (*B. papyrifera* X *B. pumila*)						X	X	X					ST/X	R(Sl)o
Corylus cornuta Marsh.	Beaked Hazel						X	X					T/X	R(Sl)r
ULMACEAE	Elm Family													
Ulmus americana L.	White Elm						X	X					T/E	R(Sl)r
URTICACEAE	Nettle Family													
Urtica dioica L. (var. *procera* (Muhl.) Wedd.; ssp. *gracilis* (Ait.) Sel.)	Stinging Nettle	X		O	X	X	X	X	X	X	X	X	ST/X	Wo
U. urens L.	Dog Nettle											O	–	(Int)r
VISCACEAE	Mistletoe Family													
Arceuthobium pusillum Peck	Dwarf Mistletoe					O	O	X	O				T/E	R(Sl)o
SANTALACEAE	Sandalwood Family													
Geocaulon lividum (Rich.) Fern. (*Comandra livida* Rich.)	Toadflax	X	X	O	X	X	O	X	X	O	X	X	ST/X	Wa
ARISTOLOCHIACEAE	Birthwort Family													
Asarum canadense L.	Wild Ginger						X	X					T/E	R(Sl)r
POLYGONACEAE	Buckwheat Family													
Koenigia islandica L.										X		X	aST/X	R(NC)r
Polygonum achoreum Blake	Striate Knotweed						X	X				O	ST/X	R(S)o
P. amphibium L. (var. *stipulaceum* Coleman; *P. natans* (Michx.) Eat.)	Water Smartweed				X	X	X	X	X			X	ST/X	o
P. caurianum Robins. (ssp. *hudsonianum* Wolf & McNeill; incl. *P. aviculare* L., *P. arenastrum* Bor. and *P. heterophyllum* Lindm. of regional reports)		X		O		X	X	X		X		O	ST/E	Wo
P. convolvulus L.	Wild Buckwheat						X	X				X	–	(Int)o
P. fowleri Robins. (incl. *P. boreale* (Lange) Sm., *P. buxiforme* Small *sensu* Lepage)		X	X							X	X	X	aST/X	R(C)a
P. lapathifolium L.	Willow Weed													
var. *lapathifolium*												X	ST/X	R(NW)r

Taxa		James Bay Coast – HB1	Akimiski Island – HB2	Severn Basin – HB3	Winisk-Ekwan Basin – HB4	Attawapiskat Basin – HB5	Albany Basin – HB6	Moose Basin – HB7	Sutton Ridges – HB8	Cape Henrietta-Maria – MT1	Ont. Maritime Tundra – MT2	Manitoba Lowland – MT3	Continental Distribution	HBL Distribution and Frequency
var. *salicifolium* Sibth. (incl. *P. tomentosum* Schrank and *P. scabrum* Moench of regional reports)				0	X	X	0	X				X	ST/X	(W)o
P. persicaria L.	Lady's-thumb				X								–	(Int)r
P. ramosissimum Michx. (incl. *P. exsertum* Sm. and *P. prolificum* (Sm.) Robins. of regional reports)	Bushy Knotweed	X											T/X	R(JB)o
P. viviparum L. (*Bistorta vivipara* (L.) S.F.Gray)	Alpine Bistort	X	X	X	X	X	X	X	X	X	X	X	AST/X	Wa
Rumex crispus L.	Curled Dock											P	–	(Int)
R. maritimus L. (var. *fueginus* (Phil.) Dusen)	Golden Dock	X				X	O	X			X	X	ST/X	(W)o
R. occidentalis S.Wats. (incl. var. *occidentalis* (*R. fenestratus* Greene) and var. *labradoricus* (Rech.f.) Lepage)	Western Dock	X	X	X	X	X	X	O		X	X	X	ST/X	Wa
Rumex orbiculatus Gray (*R. brittanica* L.)	Great Water Dock	X						X					T/X	R(JB)r
R. salicifolius Weinm.	Willow-leaf Dock													
var. *triangulivalvis* (Danser) Lepage (ssp. *t.* Danser *sensu* Scoggan; *R. mexicanus* Meisn. var. *angustifolius* (Meisn.) Boivin)		X		X		X	X	X		X	X	X	ST/X	Wo
var. *subarcticus* (Lepage) Boivin (part of ssp. *salicifolius sensu* Scoggan; *R. subarcticus* Lepage)		X											ST/X	R(JB)r
CHENOPODIACEAE	Goosefoot Family													
Atriplex glabriuscula Edmonst. (*A. patula* L. var. *oblanceolata* (Vict. & Rouss.) Boivin)		X								X	X	X	ST/E	R(C)o
A. patula L. (var. *subspicata* (Nutt.) Rydb.)	Spearscale	X	X		X							X	ST/X	R(C)o
Axyris amaranthoides L.	Russian Pigweed											X	–	(Int)r
Chenopodium album L. (incl. *C. berlandieri* Moq.)	Lamb's Quarters	X					X	X				X	–	(Int)o
C. capitatum (L.) Asch.	Strawberry Blite		X	X			X	X			X	X	ST/X	(W)o
C. gigantospermum Aellen (*C. hybridum* L. of reports)	Maple-leaf Goosefoot											X	ST/X	R(NWC)r

Taxa		James Bay Coast – HB1	Akimiski Island – HB2	Severn Basin – HB3	Winisk-Ekwan Basin – HB4	Attawapiskat Basin – HB5	Albany Basin – HB6	Moose Basin – HB7	Sutton Ridges – HB8	Cape Henrietta-Maria – MT1	Ont. Maritime Tundra – MT2	Manitoba Lowland – MT3	Continental Distribution	HBL Distribution and Frequency
C. glaucum L. (var. *salinum* (Standl.) Boivin; var. *pulchrum* Aellen)	Oak-leaf Goosefoot	X	X			X	X				X	X	ST/W	R(JB,NW)o
C. pratericola Rydb. (*C. leptophyllum* (Nutt.) S.Wats. of regional reports)												X	sT/W	(Int)r
C. rubrum L.	Alkali Blite											X	sT/X	(Int)r
Monolepis nuttalliana (Roem. & Schult.) Greene	Poverty Weed											X	ST/W	(Int)r
Salicornia europaea L. (incl. *S. maritima* Wolff & Jeff. on James Bay, and *S. borealis* Wolff & Jeff. at Churchill)	Glasswort	X	X									X	sT/X	R(JB,NWC)o
Salsola kali L. (var. *tenuifolia* Tausch)	Saltwort											X	–	(Int)r
Suaeda calceoliformis (Hook.) Moq. (*S. maritima* var. *americana* (Pers.) Boivin; *S. depressa* (Pursh) S.Wats.)	Plains Sea-blite	X	X										aST/X	R(JB)o
S. maritima (L.) Dum.	White Sea-blite											X	aST/X	R(NWC)r
AMARANTHACEAE	Amaranth Family													
Amaranthus retroflexus L.	Rough Pigweed											X	–	(Int)r
PORTULACACEAE	Purslane Family													
Montia fontana L. (var. *fontana*; *M. lamprosperma* Cham.)	Blinks	X								X	X	X	aST/X	R(C)o
CARYOPHYLLACEAE	Pink Family													
Arenaria humifusa Wahl.				X	X					X	O	X	ASs/X	R(N)o
A. lateriflora L. (*Moehringia l.* (L.) Fenzl)	Grove Sandwort	X	X	O	X	X		X	X	X	X	X	ST/X	Wo
A. serpyllifolia L.	Thyme-leaf Sandwort							X					–	(Int)r
Cerastium alpinum L. (ssp. *lanatum* (Lam.) Asch. & Graeb.; *C. a.* var. *a.* *sensu* Boivin; *C. arcticum* of reports)									X	X	X	X	AST/X	R(N)o
C. fontanum Baumg. (*C. vulgatum* L.)	Mouse-ear Chickweed				X			X	X				–	(Int)r
C. nutans Raf.	Nodding Chickweed											P	sT/X	R(NWI)
Honkenya peploides (L.) Ehrh. (*Arenaria peploides* L. var. *diffusa* Horn. *sensu* Boivin; var. *peploides* *sensu* Scoggan)	Seaside Sandwort	X	X							X	X	X	aST/X	R(C)a

Taxa	James Bay Coast – HB1	Akimiski Island – HB2	Severn Basin – HB3	Winisk-Ekwan Basin – HB4	Attawapiskat Basin – HB5	Albany Basin – HB6	Moose Basin – HB7	Sutton Ridges – HB8	Cape Henrietta-Maria – MT1	Ont. Maritime Tundra – MT2	Manitoba Lowland – MT3	Continental Distribution	HBL Distribution and Frequency
Minuartia dawsonensis (Britt.) Mattf. (*Arenaria stricta* Michx. var. *d.* (Britt.) Scoggan, var. *litorea* (Fern.) Boivin)	X	X	X	X	X	X	X	X	X	X	X	aST/X	Wo
M. groenlandica (Retz.) Ostenf. Greenland Sandwort								X				aST/E	R(LS)r
M. rubella (Wahl.) Hiern Arctic Sandwort	X								X	X	X	AST/X	R(C)o
M. stricta (Sw.) Hiern (*Arenaria s.* Michx. var. *uliginosa* (Schleich.) Boivin)			X						X	X	X	aST/X	R(N)o
Sagina nodosa (L.) Fenzl Pearlwort (ssp. *borealis* Crow; var. *nodosa* *sensu* Boivin and Scoggan)	X	X	X	X	X				X	X	X	aST/X	Wo
Silene acaulis L. Moss-campion	+											AST/X	
S. cserei Baumg. Catchfly							X					–	(Int)r
S. involucrata (Cham. & Schlecht.) Bocq. (*Lychnis furcata* Fern.; *Melandrium affine* J.Vahl)									X	X	X	AST/X	R(N)o
S. noctiflora L. Night Catchfly											X	–	(Int)r
S. pratensis (Raf.) Godr. & Gren. White Campion (*S. alba* (Mill.) E.H.Krause; *S. latifolia* Poir.; *Lychnis alba* L.)											O	–	(Int)r
S. uralensis (Rupr.) Bocq. (var. *mollis* (Cham. & Schlecht.) Bocq.; *Lychnis apetala* L.; *Melandrium a.* (L.) Fenzl)									X	X	X	AST/X	R(N)o
S. vulgaris (Moench) Garcke Bladder Campion (*S. cucubalus* Wibel)						X						–	(Int)r
Spergularia canadensis (Pers.) D.Don Northern Sand-spurrey	X	X										sT/X	R(JB)o
S. marina (L.) Griseb. Saltmarsh Sand-spurrey	X										X	sT/X	R(C)r
Stellaria borealis Bigel. Northern Stitchwort (*S. calycantha* (Ledeb.) Bong.; incl. var. *bongardiana* Fern. and var. *floribunda* Fern.)		X	O	X	X	X	X	X	X	X	X	aST/X	Wo
S. crassifolia Ehrh. Fleshy Stitchwort	X		X	X	X	X	X	X	X	X	X	aST/X	Wo
S. edwardsii R.Br. Edwards Stitchwort (incl. var. *crassipes* (Hult.) Boiv.; part of *S. longipes s.l.*)									X		X	ASs/X	R(NC)o
S. graminea L. Common Stitchwort							X					–	(Int)r
S. humifusa Rottb. Salt-marsh Stitchwort	X	X							X	X	X	AST/X	R(C)a

Taxa		James Bay Coast – HB1	Akimiski Island – HB2	Severn Basin – HB3	Winisk-Ekwan Basin – HB4	Attawapiskat Basin – HB5	Albany Basin – HB6	Moose Basin – HB7	Sutton Ridges – HB8	Cape Henrietta-Maria – MT1	Ont. Maritime Tundra – MT2	Manitoba Lowland – MT3	Continental Distribution	HBL Distribution and Frequency
S. laeta Rich. (var. *laeta* and var. *altocaulis* (Hult.) Boivin (*S. monantha* Hult.); part of *S. longipes s.l.*)	Bright Stitchwort			X						X	X	X	AST/X	R(N)o
S. longifolia Muhl. (incl. var. *eciliata* Boivin and f. *atrata* Boivin)	Long-leaf Stitchwort	X	X	X	X	X	X	X	X	X	X	X	ST/X	Wa
S. longipes Goldie (incl. *S. stricta* Rich. and *S. subvestita* Greene)	Long-stalk Stitchwort	X	X	X	X	X		X	X	X	X	X	AST/X	Wa
S. media (L.) Vill.	Common Chickweed							O			X	X	–	(Int)o
Vaccaria hispanica (Mill.) Rausch. (*Saponaria vaccaria* L.)	Cow Cockle											X	–	(Int)r
CERATOPHYLLACEAE	Hornwort Family													
Ceratophyllum demersum L.	Common Coontail	X											ST/X	R(JB)r
NYMPHACEAE	Water-lily Family													
Nuphar microphylla (Pers.) Fern. (*N. luteum* (L.) Sibth. ssp. *pumila* (Timm) Beal, *N. pumilum* (Timm) DC.)	Small Pond Lily						X						ST/E	R(SI)r
N. X*rubrodisca* Morong (*N. microphylla* X *N. variegata*)							X						ST/E	R(SI)r
N. variegata Durand (*N. lutea* ssp. *variegata* (Engelm.) Beal)	Bullhead Lily			X	X	X	X	X	X			P	ST/X	R(I)o
Nymphaea leibergii Morong (*N. tetragona* Georgi var. *l.* (Morong) Boivin)	Dwarf Water-lily						X						ST/X	R(SI)r
N. odorata Ait.	Fragrant Water-lily						X						T/E	R(SI)r
RANUNCULACEAE	Buttercup Family													
Actaea rubra (Ait.) Willd. (incl. f. *neglecta* (Gillm.) Robins.)	Red Baneberry	X	X	X	X	X	X	X	X	X	X	X	ST/X	Wo
Anemone canadensis L.	Canada Anemone	X	X	X	X	X	X	X			X	X	sT/X	Wa
A. multifida Poir. (var. *hudsoniana* DC.; incl. f. *sanguinea* (Pursh) Fern.)	Cut-leaf Anemone	X	X	X	O	X		O	X	X	X	X	ST/X	Wa
A. parviflora Michx.		X	X	X	X	X	X	O	X	X	X	X	aST/X	Wo
A. quinquefolia L. (var. *bifolia* Farw.; var. *interior* of regional reports)	Wood Anemone					X	X	X					T/X	R(SI)o
A. richardsonii Hook.										X		X	aST/X	R(NC)r
A. virginiana L. (var. *riparia* (Fern.) Boivin; var. *alba* (Oakes) Wood)	Tall Anemone					X	X	X					sT/X	R(SI)o

Taxa		James Bay Coast – HB1	Akimiski Island – HB2	Severn Basin – HB3	Winisk-Ekwan Basin – HB4	Attawapiskat Basin – HB5	Albany Basin – HB6	Moose Basin – HB7	Sutton Ridges – HB8	Cape Henrietta-Maria – MT1	Ont. Maritime Tundra – MT2	Manitoba Lowland – MT3	Continental Distribution	HBL Distribution and Frequency
Aquilegia brevistyla Hook.	Small-flower Columbine			X	X	X	X		X			X	ST/W	R(I)o
Caltha natans Pall.					X								ST/W	R(I)rn
C. palustris L.	Marsh Marigold	X	X	O	O	X	X	X	X	X	X	X	aST/X	Wa
Clematis occidentalis (Horn.) DC. (*C. verticillaris* DC.)	Purple Clematis							X					T/X	R(SI)r
Coptis trifolia (L.) Salisb. (*C. groenlandica* (Oeder) Fern.)	Goldthread				X	X	X	X	X			P	aST/X	R(I)ot
Ranunculus abortivus L. (var. *acrolasius* Fern.)	Small-flower Crowfoot		X					X		X	X	P	ST/X	o
R. acris L.	Common Buttercup	X				X		X	X		X	X	–	(Int)o
R. aquatilis L. (*sensu Flora of North America*, vol. 3)	Water Crowfoot	X	X	X	X	X	X	X	X	X	X	X	aST/X	Wa
R. cymbalaria Pursh	Seaside Crowfoot	X	X			O		X		X	X	X	aST/X	R(C)a
R. flammula L.	Creeping Spearwort													
var. *reptans* (L.) Meyer (*R. reptans* L. var. *filiformis* (Michx.) DC.)		X	X	X	X	X	X	X	X	X		X	aST/X	Wa
var. *ovalis* (Bigel.) Benson (*R. reptans* var. *ovalis* (Bigel.) Torr. & Gray)						X	X						aST/X	R(S)r
R. gmelinii DC. (var. *hookeri* (D.Don) Bens.; ssp. *purshii* (Rich.) Hult.)	Yellow Water Crowfoot	X	X		X	X	X	X	X	X	X	X	aST/X	Wa
R. hispidus Michx. (var. *caricetorum* (Greene) T. Duncan; *R. septentrionalis* Poir.)	Hispid Buttercup	X					X	X					sT/E	R(S)o
R. hyperboreus Rottb.	Arctic Buttercup		X							X	X	X	AST/X	R(NC)o
R. lapponicus L.	Lapland Buttercup				X	X	X	X	X	X	X	X	aST/X	Wa
R. macounii Britt.	Marsh Crowfoot	X	X	X		X	X	X			X	X	ST/X	Wo
R. nivalis L.	Snow Buttercup											X	AS/X	R(NWC)r
R. pallasii Schlecht.		X								X	X	X	AST/X	R(C)o
R. pedatifidus Sm. (var. *affinis* (R.Br.) Benson; var. *leiocarpus* of regional reports)			X							X	X	X	AST/X	R(C)o
R. pensylvanicus L.f.	Bristly Crowfoot	X						X				X	sT/X	o
R. scleratus L. (var. *multifidus* Nutt.)	Celery-leaf Buttercup	X				X					X	X	ST/X	R(C)o

Taxa	James Bay Coast – HB1	Akimiski Island – HB2	Severn Basin – HB3	Winisk-Ekwan Basin – HB4	Attawapiskat Basin – HB5	Albany Basin – HB6	Moose Basin – HB7	Sutton Ridges – HB8	Cape Henrietta-Maria – MT1	Ont. Maritime Tundra – MT2	Manitoba Lowland – MT3	Continental Distribution	HBL Distribution and Frequency
Thalictrum dasycarpum Fisch. & Avé-Lall. Purple Meadow Rue	X				X	X	X					sT/X	R(S)o
T. sparsiflorum Turcz.						X					X	ST/W	R(JB,NWC)r
(var. *richardsonii* (Gray) Boivin; incl. var. *viridis* Boivin)													
T. venulosum Trel. Border Meadow Rue													
var. *venulosum*	X	X	X	X	X		X		X	X	X	ST/X	Wa
var. *confine* (Fern.) Boivin	X		X	X	X	X	X					ST/E	ot
(*T. confine* Fern.)													
FUMARIACEAE Fumitory Family													
Corydalis aurea Willd. Golden Corydalis						X	X					ST/X	R(SI)r
C. sempervirens (L.) Pers. Pink Corydalis			X	X	X	X	X	X			X	ST/X	R(I)ot
BRASSICACEAE Mustard Family													
Arabis alpina L. Alpine Rock Cress			X					X		X		aST/E	R(N)o
A. arenicola (Rich.) Gelert													
var. *arenicola*			X						X	X		ASs/X	R(N)r
var. *pubescens* (S.Wats.) Gelert		X	X						X	X	X	ASs/X	R(N)o
A. divaricarpa Nels.											P	ST/X	R(NW)
A. glabra (L.) Bernh. Tower Mustard											P	–	(Int)
A. hirsuta (L.) Scop. Hairy Rock Cress	X		X	X	X	X	X	X		X	P	ST/X	Wo
(var. *pycnocarpa* (Hopk.) Roll.)													
A. holboellii Hornem.											X	aST/X	R(NWI)r
(var. *retrofracta* (Graham) Rydb.)													
Armoracia lacustris (A.Gray) Al-Sheh. & Bates						X						T/E	R(SI)r
(*A. aquatica* (Eat.) Wieg.) Lake Cress													
Barbarea orthoceros Ledeb. Northern Winter Cress										O	X	aST/X	R(NC)r
(part of *B. vulgaris s.l.*)													
B. vulgaris R.Br. Yellow Rocket							O				X	–	(Int)r
Brassica juncea (L.) Czern. Brown Mustard							X				X	–	(Int)r
B. kaber (DC.) Wheel. Charlock						O					X	–	(Int)r
(var. *pinnatifida* (Stokes) Wheel.; *Sinapis arvensis* L. aggr. sp.)													
B. rapa L. Field Mustard						O					X	–	(Int)r
(*B. campestris* L.)													
Braya humilis (C.A.Meyer) Robins.	X		X	X	X	X	X		X	X	X	AST/X	Wo
(var. *interior* (Boch.) Boivin)													

Taxa	James Bay Coast – HB1	Akimiski Island – HB2	Severn Basin – HB3	Winisk-Ekwan Basin – HB4	Attawapiskat Basin – HB5	Albany Basin – HB6	Moose Basin – HB7	Sutton Ridges – HB8	Cape Henrietta-Maria – MT1	Ont. Maritime Tundra – MT2	Manitoba Lowland – MT3	Continental Distribution	HBL Distribution and Frequency
Camelina microcarpa Andrz. Small-seed False Flax							X					–	(Int)r
Capsella bursa-pastoris (L.) Medic. Shepherd's-purse	X				X	X	X	X	X	X	X	–	(Int)a
Cardamine pensylvanica Muhl.			X	O	X	X	X					sT/X	R(I)ot
C. pratensis L. Cuckooflower													
var. *angustifolia* Hook.	X	X							X	X	X	AST/X	R(C)o
var. *palustris* Wimm. & Graeb.	X			X	X		X	X	X	X	X	AST/X	Wo
Cochlearia officinalis L. (*s.l.*) Scurvy-grass									X		X	AST/X	R(NC)r
Conringia orientalis (L.) Dum. Hare's-ear Mustard (*Erysimum orientale* (L.) R.Br.)											X	–	(Int)r
Descurainia richardsonii (Sw.) O.Schulz Tansy Mustard							X				X	ST/X	(Int)r
D. sophia (L.) Webb Flixweed							X				X	–	(Int)r
D. sophioides (Fisch.) O.Schulz											X	aS/W	(Int)r
Draba alpina L. (var. *nana* Hook. *sensu* Scoggan)									X	X	X	AST/X	R(NC)o
D. aurea M.Vahl (incl. *D. luteola* Greene var. *minganense* Vict.)	X	X							X	X	X	aST/X	R(C)a
D. cinerea Adams									X			ASs/X	R(NE)r
D. glabella Pursh (D. *hirta* auct.)	X	X					X		X	X	X	AST/X	R(C)a
D. incana L. (incl. var. *confusa* (Ehrh.) Lilj.)	X	X		X				X	X	X	X	aST/E	R(N)a
D. lactea Adams (*D. fladnizensis* Wulf. var. *heterotricha* (Lindbl.) Ball)											X	AST/X	R(NWC)r
D. nemorosa L. Wood Draba (incl. var. *leiocarpa* Lindbl.)										X	X	ST/X	R(NC)r
D. nivalis Lilj.									X		X	ST/X	R(NC)r
D. norvegica Gunn.									X			AST/E	R(NE)r
Eruscastrum gallicum (Willd.) O.Schulz Dog Mustard											X	–	(Int)r
Erysimum cheiranthoides L. Wormseed Mustard (incl. ssp. *altum* Ahti)	X	X	X	X	X	X	X		X	X	X	aST/X	Wa
Eutrema edwardii R.Br.											X	AS/X	R(NWC)r

Taxa		James Bay Coast – HB1	Akimiski Island – HB2	Severn Basin – HB3	Winisk-Ekwan Basin – HB4	Attawapiskat Basin – HB5	Albany Basin – HB6	Moose Basin – HB7	Sutton Ridges – HB8	Cape Henrietta-Maria – MT1	Ont. Maritime Tundra – MT2	Manitoba Lowland – MT3	Continental Distribution	HBL Distribution and Frequency
Hutchinsia procumbens (L.) Desv.												X	ST/X	R(NWC)r
Lepidium bourgeauanum Thell.												X	T/W	(Int)o
L. densiflorum Schrad.	Common Pepper-grass							X				X	T/W	(Int)o
L. ramosissimum Nels.												X	T/W	(Int)r
Lesquerella arctica (Wormsk.) S.Wats.	Arctic Bladderepod											X	AST/X	R(NWC)r
Rorippa palustris (L.) Besser	Yellow Cress													
var. *glabrata* (Lun.) Vict. (var. fernaldiana (Butters & Abbe) Stuck.; incl. var. *williamsii* (Britt.) Hult.)		X	O	X	X	X	X	X	X	X	X	X	aST/X	Wa
var. *hispida* (Desv.) Rydb.		X		O	X		X	X			O	X	aST/X	Wa
Sisymbrium altissimum L.	Tumble Mustard							X				X	–	(Int)r
S. loeselii L.	Hedge Mustard											O	–	(Int)r
Thlaspi arvense L.	Penny Cress							X				X	–	(Int)o
SARRACENIACEAE	Pitcher Plant Family													
Sarracenia purpurea L.	Pitcher Plant		X	X	X	X	X	X	X				ST/X	R(I)ot
DROSERACEAE	Sundew Family													
Drosera anglica Huds.			X	X	X	X	X	X	X	O	X	X	ST/X	Wa
D. linearis Goldie	Linear-leaf Sundew	X		X	X	X	X	X				P	T/X	ot
D. rotundifolia L.	Round-leaf Sundew	X	X	X	X	X	X	X	X		X	X	aST/X	Wa
CRASSULACEAE	Stonecrop Family													
Sedum telephium L. (var. *purpureum* L.)	Live Forever							O					–	(Int)r
SAXIFRAGACEAE	Saxifrage Family													
Chrysosplenium tetrandum (Lund) Fries (C. *alternifolium* L. var. *t.* Lund)	Golden Saxifrage			X						X	X	X	AST/X	R(N)o
Mitella nuda L.	Mitrewort	X		X	X	X	X	X	X	O	X	X	ST/X	Wa
Parnassia kotzebuei Cham. & Schlecht.										X	X	X	aST/X	R(NC)o
P. palustris L. (var. *neogaea* Fern.)	Grass-of-Parnassus	X	X	X	X	X	X	X	X	X	X	X	ST/X	Wa
P. parviflora DC.	Small Grass-of-Parnassus	X	X	O	X	X	X	O	X	X	X	X	ST/X	Wo
Ribes glandulosum Grauer	Skunk Currant			X	X	O	O	X	X	O	X	P	ST/X	Wo
R. hirtellum Michx.	Wild Gooseberry	X		O	X	X	X	X			X		ST/X	(W)o
R. hudsonianum Rich.	Hudson Bay Currant			X	X	X	X	X	X	X	X	X	ST/X	Wa

Taxa		James Bay Coast – HB1	Akimiski Island – HB2	Severn Basin – HB3	Winisk-Ekwan Basin – HB4	Attawapiskat Basin – HB5	Albany Basin – HB6	Moose Basin – HB7	Sutton Ridges – HB8	Cape Henrietta-Maria – MT1	Ont. Maritime Tundra – MT2	Manitoba Lowland – MT3	Continental Distribution	HBL Distribution and Frequency
R. lacustre (Pers.) Poir.	Bristly Black Currant	X		X	X	X	X	X	X	X	X	X	ST/X	Wa
R. oxyacanthoides L. (incl. *R. setosum* Lindl.)	Bristly Wild Gooseberry	X	X	X	X	X	O	X	X		X	X	ST/X	Wa
R. triste Pall.	Wild Red Currant	X	X	O	X	X	X	X	X	X	X	X	ST/X	Wa
Saxifraga aizoides L.	Yellow Mountain Saxifrage			X	X				X	X	X	X	AST/X	R(N)o
S. cespitosa L. (incl. ssp. *uniflora* (R.Br.) Porsild and ssp. *exaratoides* (Simm.) Engl. & Irmsch., *sensu* Scoggan)										X	X	X	AST/X	R(NC)o
S. cernua L.	Nodding Saxifrage										X	X	AST/X	R(NC)r
S. hirculus L.	Yellow Marsh Saxifrage	X	X							X	X	X	AST/X	R(N)o
S. oppositifolia L.	Purple Saxifrage									X		X	AST/X	R(NC)o
S. paniculata Mill. (*S. aizoon* Jacq. var. *neogaea* Butters)	White Alpine Saxifrage	+							X				aST/X	R(LS)r
S. rivularis L.	Brook Saxifrage											X	AST/X	R(NWC)r
S. tricuspidata Rottb.	Prickly Saxifrage		X							X	X	X	AST/X	R(NC)o
ROSACEAE	Rose Family													
Amelanchier alnifolia Nutt.	Saskatoon Berry													
var. *alnifolia*				O	X	X	X					P	ST/X	R(I)ot
var. *compacta* (Niels.) McKay						O	X	X					ST/X	R(SI)o
A. bartramiana (Tausch) Roem.	Mountain Juneberry				X		X	X	X				ST/E	R(I)ot
A. sanguinea (Pursh) DC. (incl. *A. wiegandii* Niels. and *A. huronensis* Wieg. of reports and var. *gaspensis* Wieg.)	Shadbush	O			X	X	X	X					T/E	ot
A. spicata (Lam.) K.Koch (incl. var. *stolonifera* (Wieg.) Cinq-Mars)					X			X					T/E	ot
Aronia prunifolia (Marsh.) Rehder (*sensu* Voss 1985; *A. melanocarpa* (Michx.) Elliot)	Chokeberry						X						T/E	R(SI)r
Dryas integrifolia M.Vahl	Mountain Avens	+	X							X	X	X	AST/X	R(NC)o
Fragaria vesca L. (var. americana Port.)	Thin-leaf Strawberry						X	X					sT/X	R(S)o
F. virginiana Mill. (incl. var. *terrae-novae* (Rydb.) Fern. & Wieg. and var. *glauca* (S.Wats.) Rydb.)	Thick-leaf Strawberry	X	X	X	X	X	X	X	X	O	X	X	ST/X	Wa
Geum aleppicum Jacq. (var. *strictum* (Ait.) Fern.)	Yellow Avens	X				X	X	O			X	X	ST/X	(W)o

Taxa		James Bay Coast – HB1	Akimiski Island – HB2	Severn Basin – HB3	Winisk-Ekwan Basin – HB4	Attawapiskat Basin – HB5	Albany Basin – HB6	Moose Basin – HB7	Sutton Ridges – HB8	Cape Henrietta-Maria – MT1	Ont. Maritime Tundra – MT2	Manitoba Lowland – MT3	Continental Distribution	HBL Distribution and Frequency
G. macrophyllum Willd. (var. *perincisum* (Rydb.) Raup)	Large-leaf Avens		X	X	X	X	X	X		X	X	X	ST/X	Wa
G. rivale L.	Water Avens	X					X	X	X	X	X		ST/X	(W)o
Physocarpus opulifolius (L.) Maxim.	Ninebark					X	X	X					T/X	R(SI)o
Potentilla anserina L.	Silverweed													
var *anserina* (incl. var. *grandis* Torr. & Gray, of regional reports)		X	X	O	X	X	X	X	X	X	X	X	aST/X	Wa
var. *groenlandica* Tratt. (*P. egedii* Wormsk. var. *g.* (Tratt.) Pol.)		X	X							X	X	X	aST/X	R(C)a
P. arguta Pursh	Tall Cinquefoil											P	sT/X	R(NWI)
P. crantzii (Crantz) Beck			O							X			aST/E	R(NC)r
P. fruticosa L.	Shrubby Cinquefoil	X	X	X	X	X	X	X	X	X	X	X	aST/X	Wa
P. multifida L.		X	X								X	X	ST/X	R(C)o
P. nivea L.			X						X	X	X	X	AST/X	R(N)o
P. norvegica L.	Rough Cinquefoil	X	X	X	X	X	X	X	X		X	X	aST/X	Wa
P. palustris (L.) Scop.	Marsh Cinquefoil													
var. *palustris*		X	X	X	X	X	X	X	X			X	aST/X	Wa(t)
var. *parvifolia* (Raf.) Fern. & Long		X								X	X	X	aSs/X	R(N)o
P. pensylvanica L. (var. *littoralis* (Rydb.) Boivin; *P. pectinata* auct. *non* Raf.)			X			X		X	X	X	X	X	ST/X	(W)o
P. pulchella R.Br. (*P. subarctica* Rydb.)			X							X	X	X	ASs/X	R(NC)o
P. tridentata Sol.				X		X			X		X	X	aST/E	(W)o
Prunus pensylvanica L.f.	Firecherry					X	X	X	X			P	sT/X	R(I)ot
P. pumila L. (var. *depressa* (Pursh) Bean)	Sandcherry						X						T/E	R(SI)r
P. virginiana L.	Chokecherry					X	X	X					sT/X	R(S)o
Rosa acicularis Lindl. (var. *bourgeauiana* Crepin)	Prickly Wild Rose	X		X	X	X	X	X	X	X	X	X	ST/X	Wa
R. blanda Ait.	Smooth Wild Rose	X				X	X	X					sT/X	R(S)o
Rubus acaulis Michx. (*R. arcticus* L. var. *acaulis* (Michx.) Boivin)	Arctic Raspberry	X	X	X	X	X	X	X	X	X	X	X	ST/X	Wa
R. chamaemorus L.	Cloudberry	X	X	X	X	X	O	X	X	X	X	X	aST/X	Wa

Taxa		James Bay Coast – HB1	Akimiski Island – HB2	Severn Basin – HB3	Winisk-Ekwan Basin – HB4	Attawapiskat Basin – HB5	Albany Basin – HB6	Moose Basin – HB7	Sutton Ridges – HB8	Cape Henrietta-Maria – MT1	Ont. Maritime Tundra – MT2	Manitoba Lowland – MT3	Continental Distribution	HBL Distribution and Frequency
R. pubescens Raf.	Dwarf Raspberry													
var. *pubescens*		X		O	X	X	X	X	X		O	X	ST/X	Wo
var. *paracaulis* (Bailey) Boivin (*R.* X*paracaulis* Bailey)		X		X	X	X					X	X	sT/X	(W)o
R. strigosus Michx. (*R. idaeus* L. var. *s.* (Michx.) Maxim., var. *canadensis* Fern., and var. *aculeatissimus* Regel & Tiling)	Wild Red Raspberry	X		X		X	X	X	X		X	X	ST/X	Wa
Sorbus americana Marsh.	Mountain Ash							X					T/E	R(Sl)r
S. decora (Sarg.) Schneid. (incl. var. *groenlandica* (Schneid.) Jones)	Showy Mountain Ash					X	X	X					aST/E	R(Sl)o
Spiraea alba Duroi	Meadowsweet	X					X	X					sT/E	R(S)o
FABACEAE	Pea Family													
Astragalus adsurgens Pall. (*A. striatus* Nutt.)	Ascending Milk-vetch											X	ST/W	R(NW)r
A. agrestis Dougl. (*A. goniatus* Nutt.)	Purple Milk-vetch						X					X	ST/W	R(JB,NW)r
A. alpinus L.	Alpine Milk-vetch	X	X	X	X		X	X		X	X	X	aST/X	Wa
A. americanus (Hook.) Jones (*A. frigidus* (L.) Gray var. *americanus* (Hook.) S.Wats.)	America Milk-vetch			X								P	ST/X	R(I)ot
A. canadensis L.	Canada Milk-vetch	X					X	X					ST/X	R(S)o
A. eucosmus Robins.	Elegant Milk-vetch	X		X	X	X		X		X	X	X	aST/X	Wo
A. tenellus Pursh	Pulse Milk-vetch						X					P	sT/W	R(Sl)r
Hedysarum alpinum L. (var. *americanum* Michx.; *H. boreale* Nutt.)	Liquorice-root	X	X	X			X	X			X	P	aST/X	(W)o
H. mackenzii Rich. (*H. boreale* Nutt. var. *m.* (Rich.) Hitchc.; incl. f. *niveum* Boivin)	Mackenzie Hedysarum	X	X	X	X					X	X	X	aST/X	R(N)a
Lathyrus japonicus Willd.	Beach Pea													
var. *japonicus* (*L. maritimus* L. var. *glaber* (Ser.) Eames; *L. m.* ssp. *m.*, *sensu* Hultén)											X	X	aST/X	R(NC)r
var. *pellitus* Fern. (incl. var. *aleuticus* (Greene) Fern.; *L. m.* var. *pellitus* (Fern.) Gl., and *L. m.* ssp. *pubescens* (Hartm.) C.Regel)		X	X							X		X	aST/E	R(C)o

Taxa		James Bay Coast – HB1	Akimiski Island – HB2	Severn Basin – HB3	Winisk-Ekwan Basin – HB4	Attawapiskat Basin – HB5	Albany Basin – HB6	Moose Basin – HB7	Sutton Ridges – HB8	Cape Henrietta-Maria – MT1	Ont. Maritime Tundra – MT2	Manitoba Lowland – MT3	Continental Distribution	HBL Distribution and Frequency
L. ochroleucus Hook.	Cream Pea		X	X		X	X	X				P	ST/X	R(I)ot
L. palustris L.	Wild Pea	X	X	X	X	X	X	X		X	X	X	ST/X	Wa
(incl. var. *myrtifolius* (Muhl.) Gray, var. *linearifolius* Sm.,and var. *pilosus* (Cham.) Ledeb.)														
L. pratensis L.	Meadow Pea							X					–	(Int)r
Medicago lupulina L.	Black Medick						X						–	(Int)r
M. sativa L.	Alfalfa							X				P	–	(Int)r
Melilotus alba Medic.	White Sweet Clover						X	X				X	–	(Int)o
M. officinalis (L.) Pall.	Yellow Sweet Clover							X				X	–	(Int)o
Oxytropis arctica R.Br. (var. *bellii* (Britt.) Boivin)												X	AS/X	R(NWC)r
O. campestris (L.) DC.	Late Yellow Locoweed													
var. *johannensis* Fern. (incl. O. *terra-novae* Fern.)		X		O	X		X	X				X	aST/E	(W)o
var. *varians* (Rydb.) Barn.												X	aST/W	R(NWC)r
O. deflexa (Pall.) DC.	Reflexed Locoweed													
var. *foliolosa* (Hook.) Barn.				X						X	X	X	aST/X	R(N)o
var. *sericea* Torr. & Gray					X							P	aST/W	R(N)r
O. splendens Dougl.	Showy Locoweed			X	X						X		ST/W	R(N)o
O. viscida Nutt. (var. *hudsonica* (Greene) Barneby)	Viscid Locoweed		O		X	O				X	X	X	aST/X	R(N)o
Thermopsis rhombifolia (Nutt.) Rich. (point introduction in 1950s, since extirpated; Staniforth and Scott 1991)	Golden Bean											X	T/W	(Int)r
Trifolium hybridum L.	Alsike Clover	X					X	X				X	–	(Int)o
T. pratense L.	Red Clover	X					X	X				P	–	(Int)o
T. repens L.	White Clover	X				X	X	X				X	–	(Int)a
Vicia americana Muhl.	America Vetch	X		X		X	X	X			X	X	ST/X	Wa
V. cracca L.	Common Vetch	X				X	X	X				X	–	(Int)a
GERANIACEAE	Geranium Family													
Geranium bicknellii Britt.	Bicknell Geranium					X	X	X					ST/X	R(SI)o
OXALIDACEAE	Wood Sorrel Family													
Oxalis acetosella L. (*O. montana* Raf.)	Northern Wood Sorrel							X					T/E	R(SI)r

Taxa		James Bay Coast – HB1	Akimiski Island – HB2	Severn Basin – HB3	Winisk-Ekwan Basin – HB4	Attawapiskat Basin – HB5	Albany Basin – HB6	Moose Basin – HB7	Sutton Ridges – HB8	Cape Henrietta-Maria – MT1	Ont. Maritime Tundra – MT2	Manitoba Lowland – MT3	Continental Distribution	HBL Distribution and Frequency
LINACEAE	Flax Family													
Linum lewisii Pursh														
var. *lepagei* (Boiv.) Rogers	Lepage Wild Flax	X	X							X	X	X	aS/End	R(NC)o
(*L. perenne* L. var. *lepagei* Boivin)														
POLYGALACEAE	Milkwort Family													
Polygala paucifolia Willd.	Fringed Milkwort				X		X	X	X				T/E	R(I)o
P. senega L.	Seneca Snakeroot					X	X	X					T/X	R(I)ot
CALLITRICHACEAE	Water Starwort Family													
Callitriche hermaphroditica L.					X	X	X		X	X	X	X	aST/X	Wo
C. verna L.		X			X	X	X	X	X	X		X	aST/X	Wo
(*C. palustris* L.)														
EMEPETRACEAE	Crowberry Family													
Empetrum nigrum L.	Black Crowberry	X	X	X	X	X	X	X	X	X	X	X	AST/X	Wa
(var. *hermaphroditicum* (Lange) Sor.;														
var. *purpureum* (Raf.) DC. *sensu* Boivin)														
ANACARDIACEAE	Sumach Family													
Rhus radicans L.	Poison Ivy						X	X					T/X	R(Sl)o
(var. *rydbergii* (Sm.) Rehd.;														
Toxicodendron r. (Sm.) Greene)														
ACERACEAE	Maple Family													
Acer spicatum Lam.	Mountain Maple	X					X	X					sT/E	R(S)o
BALSAMINACEAE	Touch-me-not Family													
Impatiens capensis Meerb.	Spotted Touch-me-not	X					X	X					sT/X	R(S)o
(*I. biflora* Willd.)														
RHAMNACEAE	Buckthorn Family													
Rhamnus alnifolia L'Hér.	Alder-leaf Buckthorn	X		X	X	X	X	X	X			X	sT/X	R(I)at
HYPERICACEAE	St. John's-wort Family													
Hypericum majus (A.Gray) Britt.	Large Canada St. John's-wort						X	X					T/E	R(Sl)o
Triadenum fraseri (Spach.) Gl.	Marsh St. John's-wort						X	O					T/E	R(Sl)r
VIOLACEAE	Violet Family													
Viola adunca Sm.	Early Blue Violet	X					X	X				P	aST/X	R(S)o
V. incognita Brain.	Sweet White Violet	X						X	X				T/E	R(S)o
(part of *V. blanda* Willd. *sensu* Boivin)														

Taxa		James Bay Coast – HB1	Akimiski Island – HB2	Severn Basin – HB3	Winisk-Ekwan Basin – HB4	Attawapiskat Basin – HB5	Albany Basin – HB6	Moose Basin – HB7	Sutton Ridges – HB8	Cape Henrietta-Maria – MT1	Ont. Maritime Tundra – MT2	Manitoba Lowland – MT3	Continental Distribution	HBL Distribution and Frequency
V. labradorica Schrank (*V. adunca* Sm. var. *minor* (Hook.) Fern.)	Labrador Violet	O		X	X	X	X	X	X	X		P	aSs/X	(W)o
V. maclowskeyi Lloyd (*V. pallens* (Banks) Brain.; *V. rotundifolia* Michx. var. *pallens* Banks)	Wild White Violet	X		X	X	X	X	X		X		X	ST/X	Wa
V. nephrophylla Greene	Northern Bog Violet		X		X	X	X	X					ST/X	R(I)ot
V. palustris L. (*V. epipsila* Ledeb. ssp. *repens* (Turcz.) Becker)	Marsh Violet	X			X	X	X					X	aST/X	ot
V. pubescens Ait. (var. *eriocarpa* (Schwein.) Russ.; var. *scabriuscula* Torr. & Gray; *V. pensylvanica* Michx. var. *leiocarpa* (Fern. & Wieg.) Fern.)	Downy Yellow Violet						X	X					T/E	R(SI)r
V. renifolia A.Gray (incl. var. *brainerdii* (Greene) Fern.)	Kidney-leaf Violet	X	X	X	X	X	X	O		X	X	X	ST/X	Wa
V. selkirkii Pursh	Great-spur Violet	X					X	X					aST/X	R(S)o
ELAEAGNACEAE	Oleaster Family													
Elaeagnus commutata Bernh.	Silverberry	X		X	X	X	X	X		X	X	X	ST/X	Wa
Shepherdia canadensis (L.) Nutt.	Buffalo Berry	X	X	X	X	X	X	X	X	X	X	X	ST/X	Wa
ONAGRACEAE	Evening Primrose Family													
Circaea alpina L.	Small Enchanter's-nightshade	X					X	X					ST/X	R(S)o
Epilobium angustifolium L. (incl. ssp. *circumvagum* Mosq.)	Fireweed	X	X	X	X	X	X	X	X	X	X	X	aST/X	Wa
E. ciliatum Raf. (*E. glandulosum* Lehm. var. *adenocaulon* (Haussk.) Fern.)	Northern Willowherb	X	X	X	X	X	X	X		X	X	X	sT/X	Wa
E. davuricum Fisch.		X			X		X	X	X	X	X	X	aST/X	Wo
E. latifolium L.	River Beauty			X	X					X	X	X	AST/X	R(N)o
E. leptophyllum Raf. (*E. nesophilum* Fern.)	Narrow-leaf Willowherb	X		X	X	X	X	X		X	X		aST/X	Wo
E. palustre L.	Marsh Willowherb	X	X	X	X	X	X	X	X	X	X	X	aST/X	Wa
E. strictum Muhl. (*E. densum* Raf.)	Downy Willowherb							X					sT/E	R(S)r
Oenothera biennis L.	Yellow Evening Primrose							O				P	T/X	R(S)r
O. parviflora L. (*O. biennis* f. *muricata* (L.) Boivin; incl. *O. perangusta* Gates of regional reports)	Small-flower Evening Primrose	X					X	X					T/X	R(S)o

Taxa		James Bay Coast – HB1	Akimiski Island – HB2	Severn Basin – HB3	Winisk-Ekwan Basin – HB4	Attawapiskat Basin – HB5	Albany Basin – HB6	Moose Basin – HB7	Sutton Ridges – HB8	Cape Henrietta-Maria – MT1	Ont. Maritime Tundra – MT2	Manitoba Lowland – MT3	Continental Distribution	HBL Distribution and Frequency
HALORAGACEAE	Water Milfoil Family													
Myriophyllum alterniflorum DC.							X	X					aST/X	R(S)r
M. sibiricum Kom. (*M. spicatum* L. var. *exalbescens* (Fern.) Jeps.)	Common Water Milfoil	X	X	O	X	X	X	X	X	X	X	X	aST/X	Wa
M. verticillatum L.	Whorled Water Milfoil	X	X		X	X	X	X		X	X		aST/X	Wo
HIPPURIDACEAE	Mare's-tail Family													
Hippuris vulgaris L.	Common Mare's-tail	X	X	X	X	X	X	X	X	X	X	X	AST/X	Wa
H. tetraphylla L.f. (*H. vulgaris* f. *maritima* Hell.)	Maritime Mare's-tail	X	X		X					X	X	X	AST/X	R(C)a
ARALIACEAE	Ginseng Family													
Aralia hispida Vent.	Bristly Sarsaparilla							X					T/E	R(Sl)r
A. nudicaulis L.	Wild Sarsaparilla			X		X	X	X					sT/X	R(I)ot
APIACEAE	Carrot Family													
Angelica atropurpurea L.	Purplestem Angelica	X				X	X	X					sT/E	R(JB)o
Carum carvi L.	Caraway		O			X	X	X			X	X	–	(Int)a
Cicuta bulbifera L.	Bulbed Water-hemlock	X	X		X	X	X	X	X	X	X	X	ST/X	Wa
C. mackenzieana Raup (part of *C. virosa* L. *s.l.*)	Mackenzie Water-hemlock	X	X		X	X		X		X	X	X	ST/X	Wo
C. maculata L. (var. *angustifolia* Hook.)	Spotted Water-hemlock	X	X		X	X	X	X				X	sT/X	Wo
Conioselinum chinense (L.) BSP.	Hemlock-parsley	X						X			X		ST/X	R(JB,NC)o
Heracleum maximum Bart. (*H. lanatum* Michx.)	Cow-parsnip	X		X	X	X	X	X	X	X	X	X	ST/X	Wa
Ligusticum scothicum L.	Scotch Lovage	X	X										ST/E	R(JB)o
Osmorhiza claytoni (Michx.) Clarke	Bland Sweet Cicely						X						T/E	R(Sl)r
O. depauperata Phil. (*O. obtusa* (C. & R.) Fern.: *O. chilensis* H. & A. var. *cupressimontana* Boivin)	Blunt Sweet Cicely	X						X					sT/X	R(S)r
O. longistylis (Torr.) DC.	Long-styled Sweet Cicely							O					T/X	R(Sl)r
Pastinaca sativa L.	Parsnip						X						–	(Int)r
Sanicula marilandica L.	Black Snakeroot	X			X	X	X	X					T/X	ot
Sium suave Walt.	Water-parsnip	X	O		X	X	X	X			O	X	ST/X	Wa

Taxa		James Bay Coast – HB1	Akimiski Island – HB2	Severn Basin – HB3	Winisk-Ekwan Basin – HB4	Attawapiskat Basin – HB5	Albany Basin – HB6	Moose Basin – HB7	Sutton Ridges – HB8	Cape Henrietta-Maria – MT1	Ont. Maritime Tundra – MT2	Manitoba Lowland – MT3	Continental Distribution	HBL Distribution and Frequency
CORNACEAE	Dogwood Family													
Cornus canadensis L. (incl. var. *intermedia* Fern.)	Bunchberry	X	X	X	X	X	X	X	X		X	X	aST/X	Wa
C. stolonifera Michx.	Red Osier	X		X	X	X	X	X	X			X	ST/X	at
PYROLACEAE	Shinleaf Family													
Chimaphila umbellata (L.) Bart. (var. *cisatlantica* Blake)	Pipsissewa						X	O					ST/E	R(SI)r
Moneses uniflora (L.) A.Gray	One-flower Shinleaf	X	X	X	X	X	X	X		X	X	X	ST/X	Wa
Monotropa uniflora L.	Indian Pipe						X	X					sT/X	R(SI)o
Orthila secunda (L.) House (*Pyrola secunda* L.)	One-side Shinleaf	X	X	X	X	X	X	X	X	X	X	X	aST/X	Wa
Pyrola asarifolia Michx.	Pink Shinleaf	X	X	X	X	X	X	X	X		X	X	ST/X	Wa
P. chlorantha Sw. (*P. virens* Schweigg.)	Green Wintergreen	X	X		X	X	X	X				X	ST/X	Wo
P. elliptica Nutt.	Common Shinleaf							X					sT/X	R(SI)r
P. grandiflora Rad.	Arctic Shinleaf		O		X					X	X	X	AST/E	R(N)o
P. minor L.	Lesser Shinleaf	X		X	X	X	X			X	X	X	aST/X	Wo
ERICACEAE	Heath Family													
Andromeda glaucophylla Link (*A. polifolia* L. var. *g.* (Link) DC.)	Bog Rosemary	X	X	X	X	X	X	X	X				aST/E	at
A. polifolia L.	Dwarf Bog Rosemary		X	X	X	O			X	X	X	X	aST/X	R(N)a
Arctostaphylos alpina (L.) Spreng.	Alpine Bearberry													
var. *alpina*			X							X	X	X	aST/X	R(NC)o
var. *rubra* (Rehd. & Wils.) Bean		X	X	X	X	X			X	X	X	X	aST/X	R(N)a
A. uva-ursi (L.) Spreng.	Bearberry													
var. *uva-ursi* (incl. var. *coactilis* Fern. & Macbr.)		X	X			X	X					P	aST/X	ot
var. *adenotricha* Fern. & Macbr.		X	X	X	X	X	X		X		X	X	aST/X	Wa
Chamaedaphne calyculata (L.) Moench (*Cassandra calyculata* (L.) D.Don)	Leatherleaf		X	X	X	X	X	X	X		X	X	ST/X	Wa
Gaultheria hispidula (L.) Muhl.	Snowberry	X	X		O	X	X	X			X		sT/X	Wa
Kalmia angustifolia L.	Sheep-laurel				X	X	X	X					sT/E	R(I)ot
K. microphylla (Hook.) Heller (*K. polifolia* var. *m.* Rehder)	Dwarf Bog-laurel										X	X	ST/W	R(NC)r
K. polifolia Wang.	Bog-laurel		X	X	X	X	X	X	X	X	X	X	ST/X	Wa

Taxa	James Bay Coast – HB1	Akimiski Island – HB2	Severn Basin – HB3	Winisk-Ekwan Basin – HB4	Attawapiskat Basin – HB5	Albany Basin – HB6	Moose Basin – HB7	Sutton Ridges – HB8	Cape Henrietta-Maria – MT1	Ont. Maritime Tundra – MT2	Manitoba Lowland – MT3	Continental Distribution	HBL Distribution and Frequency
Ledum decumbens (Ait.) Lodd. Dwarf Labrador Tea (*L. palustre* L. ssp. *decumbens* (Ait.) Hult.)			X	X				X	X	X	X	aSs/X	R(N)o
L. groenlandicum Oeder Labrador Tea (*L. palustre* L. ssp. *g.* (Oeder) Hult.)		X	X	X	X	X	X	X	X	X	X	aST/X	Wa
Loiseleuria procumbens (L.) Desv. Alpine Azalea											X	aSs/X	R(NWC)r
Phyllodoce caerulea (L.) Bab. Mountain Heather									X		X	aSs/X	R(NE,NWC)r
Rhododendron lapponicum (L.) Wahl. Lapland Rose-bay				X					X	X	X	aST/X	R(N)o
Vaccinium angustifolium Ait. Common Blueberry	X		X	X	X	O	X			X		ST/X	a(W)
V. myrtilloides Michx. Velvet-leaf Blueberry (*V. angustifolium* Ait. var. *m.* (Michx.) House)			X	X	X	X	X	X			P	sT/X	R(I)at
V. oxycoccus L. Small Cranberry													
var. *oxycoccus* (*Oxycoccus microcarpus* Turcz.)	X	X	X	X		X	X	X	O	X	X	ST/X	Wa
var. *microphyllum* (Lange) Rouss. & Raym. (*Oxycoccus quadripetalus* Gilib.)				O	X	X	X					aST/X	ot
V. uliginosum L. Bog Bilberry (incl. ssp. *alpinum* (Bigel.) Hult. and ssp. *microphyllum* Lange *sensu* Hultén)	X	X	X	X	X	X		X	X	X	X	AST/X	Wa
V. vitis-idaea L. Dry-ground Cranberry (var. *minus* Lodd.)	X	X	X	X	X	X	X	X	X	X	X	aST/X	Wa
PRIMULACEAE Primrose Family													
Androsace septentrionalis L. Pygmyflower	X	X							X	X	X	AST/X	R(C)o
Glaux maritima L. Sea Milkwort (var. *obtusifolia* Fern.)	X	X										sT/X	R(JB)o
Lysimachia ciliata L. Fringed Loosestrife (*Steironema ciliata* (L.) Raf.)	X					O	X					T/X	R(S)o
L. terrestris (L.) BSP. Swamp Candles	X				X	X						T/E	R(S)o
L. thyrsiflora L. Tufted Loosestrife	X				X	X	X				X	ST/X	(W)o
Primula egaliksensis Wormsk.	X	X		X					X	X	X	aST/X	R(N)a
P. laurentiana Fern. Mealy Primrose (*P. mistassinica* Michx. var. *macropoda* (Fern.) Boivin; *P. incana* Jones of regional reports)	X			X	X				X	X	X	ST/E	R(C)o
P. mistassinica Michx. Bird's-eye Primrose (incl. var. *intercedens* (Fern.) Boivin)			O	X	X	X	X	X		X	X	ST/X	(W)o

Taxa		James Bay Coast – HB1	Akimiski Island – HB2	Severn Basin – HB3	Winisk-Ekwan Basin – HB4	Attawapiskat Basin – HB5	Albany Basin – HB6	Moose Basin – HB7	Sutton Ridges – HB8	Cape Henrietta-Maria – MT1	Ont. Maritime Tundra – MT2	Manitoba Lowland – MT3	Continental Distribution	HBL Distribution and Frequency
P. stricta Horn.	Erect Primrose	X	X			X	X			X	X	X	aST/X	(W)o
Trientalis borealis Raf.	Starflower		X		X	X	X	X					ST/X	R(I)ot
PLUMBAGINACEAE	Leadwort Family													
Armeria maritima (Mill.) Willd. (var. *labradorica* (Wallr.) Lawr.)	Sea Thrift									X	X	X	AST/X	R(NC)o
OLEACEAE	Ash Family													
Fraxinus nigra Marsh.	Black Ash						X	X					T/E	R(Sl)o
GENTIANACEAE	Gentian Family													
Gentiana affinis Griseb.	Oblong-leaf Gentian											X	T/W	(Int)r
G. linearis Froel. (var. *lanceolata* Gray; ssp. *rubricaulis* (Schwein.) Gill.)	Closed Marsh Gentian	X				X	X	X					T/E	R(S)o
Gentianella amarella (L.) Born. (ssp. *acuta* (Michx.) Gill.)	Northern Gentian	X	X	X	X	X	X	X	X	X	X	X	ST/X	Wa
G. propinqua (Rich.) Gill.	Felwort	X		O	X					X	X	X	aST/X	R(NC)a
G. tenella (Rottb.) Born.										X			aST/X	R(NE)r
Gentianopsis detonsa (Rottb.) Ma (ssp. *raupii* (Pors.) A. & D.Löve; *Gentianella d.* ssp. *r.* (Pors.) Gill.)	Northern Fringed Gentian										X		aST/W	R(C)o
G. macounii (Holm.) Iltis (*Gentianella crinita* ssp. *m.* (Holm) Gill.; incl. *Gentiana gaspensis* Vict. of regional reports)	Fringed Gentian	X				X	X	X					sT/X	R(S)o
G. nesophila (Holm) Iltis (*Gentianella detonsa* (Rottb.) G.Don ssp. *nesophila* (Holm.) Gill.)		X	X							X			aST/E	R(C)o
Halenia deflexa (Sm.) Griseb.	Spurred Gentian	X		X	X	X	X	X	X	X	X		sT/X	Wo
Lomatogonium rotatum (L.) Fries	Marsh Felwort	X	X		X	X		X		X	X	X	aST/X	R(C)o
Menyanthes trifoliata L. (var. *minor* Raf.)	Buckbean	X	X	X	X	X	X	X	X	X	X	X	aST/X	Wa
APOCYNACEAE	Dogbane Family													
Apocynum androsaemifolium L.	Speading Dogbane						X	X				P	ST/X	R(Sl)o
A. cannabinum L. (var. *hypericifolium* Gray; *A. sibiricum* Jacq.; incl var. *cordigerum* (Greene) Fern.)	Indian Hemp	X		X		X	X	X					sT/X	ot

Taxa		James Bay Coast – HB1	Akimiski Island – HB2	Severn Basin – HB3	Winisk-Ekwan Basin – HB4	Attawapiskat Basin – HB5	Albany Basin – HB6	Moose Basin – HB7	Sutton Ridges – HB8	Cape Henrietta-Maria – MT1	Ont. Maritime Tundra – MT2	Manitoba Lowland – MT3	Continental Distribution	HBL Distribution and Frequency
CONVOLVULACEAE	Morning-glory Family													
Calystegia sepia (L.) R.Br.	Hedge Bindweed						X	X					T/X	R(SI)o
(*Convolvulus sepium* L.)														
POLEMONIACEAE	Phlox Family													
Collomia linearis Nutt.	Narrow-leaf Collomia											X	T/W	R(NW)r
HYDROPHYLLACEAE	Waterleaf Family													
Phacelia franklinii (R.Br.) A.Gray	Scorpion-weed											X	ST/W	R(NW)r
BORAGINACEAE	Borage Family													
Cynoglossum boreale Fern.	Wild Comfrey	X											T/X	R(JB)r
Lappula squarrosa (Retz.) Dumort.	Bluebur							X				X	–	(Int)r
(*L. myosotis* Moench, *L. echinata* Gil.)														
Mertensia maritima (L.) S.F.Gray	Seaside Bluebell	X	X							X	X	X	aST/X	R(C)o
M. paniculata (Ait.) G.Don	Bluebell	X		X	X	X	X	X		X	X		aST/X	Wo
Myosotis laxa Lehm.	Small Forget-me-not						X						T/X	(Int)r
LAMIACEAE	Mint Family													
Dracocephalum parviflorum Nutt.	Dragonhead			O			X					P	ST/X	o
(*Moldavica parviflora* (Nutt.) Britt.)														
Galeopsis tetrahit L.	Hemp-nettle	X					X	X				P	–	(Int)o
(var. *bifida* (Boenn.) Lej. & Court.)														
Lycopus americanus Muhl.	Water-horehound	X				X	X	X				P	T/X	R(S)o
L. asper Greene Western	Water-horehound	X											T/X	R(JB)r
(*L. lucidus* Turcz. ssp. *americanus* (A.Gray) Hult.)														
L. uniflorus Michx.	Northern Water-horehound	X					X	X					T/X	R(S)o
(*L. virginicus* L. var. *pauciflorus* Benth.)														
Mentha arvensis L.	Field Mint	X		X	X	X	X	X	X	X	O	X	ST/X	Wa
(var. *glabrata* (Benth.) Fern.)														
Monarda fistulosa L.	Wild Bergamot						X	X					sT/X	R(SI)r
Physostegia virginiana (L.) Benth.												P	sT/X	R(NWI)
(var. *ledinghamii* Boivin)	False Dragonhead													
Prunella vulgaris L.	Selfheal		O	X	X	X	X	X					ST/X	at
(var. *lanceolata* (Bart.) Fern.).														
Scutellaria galericulata L.	Marsh Skullcap	X		O		X	X	X				X	ST/X	(W)o
(var. *pubescens* Benth.; *S. epilobifolia* Hamilt.)														
S. lateriflora L.	Blue Skullcap	X			X	X	X	X					sT/X	R(S)o

Taxa		James Bay Coast – HB1	Akimiski Island – HB2	Severn Basin – HB3	Winisk-Ekwan Basin – HB4	Attawapiskat Basin – HB5	Albany Basin – HB6	Moose Basin – HB7	Sutton Ridges – HB8	Cape Henrietta-Maria – MT1	Ont. Maritime Tundra – MT2	Manitoba Lowland – MT3	Continental Distribution	HBL Distribution and Frequency
Stachys pilosa Nutt. (*S. palustris* L. var. *homotricha* Fern.)	Marsh Hedge-nettle	X		O	X	X	X	X				X	ST/X	(W)o
SOLANACEAE	Potato Family													
Solanum triflorum Nutt.	Wild Tomato											X	–	(Int)r
SCROPHULARIACEAE	Figwort Family													
Bartsia alpina L.	Velvet Bells	X	X	X	X				X	X	X	X	aST/E	R(N)o
Castilleja raupii Penn.	Purple Paintbrush		X							X	X	X	aSs/X	R(N)a
C. septentrionalis Lindl. (*C. pallida* (L.) Spreng. var. *s.* (Lindl.) A.Gray)	Northeastern Paintbrush	X		X	X	X	X	X	X	X	X	X	aST/X	Wa
Chaenorrhinum minus (L.) Lange	Dwarf Snapdragon							X				P	–	(Int)r
Chelone glabra L. (incl. var. *linifolia* Coleman)	Turtlehead	X						X					T/E	R(S)o
Euphrasia arctica Lange (*E. frigida* Pugs.; incl. *E. disjuncta* Fern. & Wieg. and *E. vinacea* Sell & Yeo)	Arctic Eyebright	X	X	X	X	X	X	X	X	X	X	X	aST/X	Wa
E. hudsoniana Fern. & Wieg.	Hudson Bay Eyebright	X		X		X	X	X				X	ST/X	(W)o
Limosella aquatica L.	Mudwort	X		X								X	aST/X	R(C)o
Linaria vulgaris Mill.	Toad-flax							X				X	–	(Int)r
Melampyrum lineare Desr.	Cow-wheat						X	X					ST/X	R(Sl)r
Mimulus ringens L.	Blue Monkeyflower						X	X					T/E	R(S)o
Pedicularis flammea L.	Flaming Lousewort									X	X	X	aST/X	R(NC)a
P. groenlandica Retz. (incl. f. *pallida* Lepage)	Elephant-heads	X	X	X	X	X	X	X		X	X	X	aST/X	Wa
P. labradorica Wirsing	Labrador Lousewort									X	X	X	aST/X	R(NC)o
P. lapponica L.	Lapland Lousewort									X	X	X	aSs/X	R(NC)a
P. parviflora Sm. (*P. macrodonta* Rich.)	Purple Lousewort	X	X	X	X	X	X	X	X	X	X	X	ST/X	Wo
P. sudetica Willd.	Purple Rattle	X	X							X	X	X	ASs/X	R(NC)o
Rhinanthus crista-galli L.	Yellow Rattle	X		X	X	X	X	X	X		X	X	aST/X	Wa
R. borealis (Stern.) Chab.	Northern Yellow Rattle		X							X	X	X	aST/X	R(NC)o
Veronica americana (Raf.) Schw.	Common Speedwell						X	X					ST/X	R(S)o

Taxa		James Bay Coast – HB1	Akimiski Island – HB2	Severn Basin – HB3	Winisk-Ekwan Basin – HB4	Attawapiskat Basin – HB5	Albany Basin – HB6	Moose Basin – HB7	Sutton Ridges – HB8	Cape Henrietta-Maria – MT1	Ont. Maritime Tundra – MT2	Manitoba Lowland – MT3	Continental Distribution	HBL Distribution and Frequency
V. peregrina L. (var. *xalapensis* (HBK.) St. John & Warr.)	Hairy Speedwell											X	T/X	R(NWC)r
V. scutellata L.	Marsh Speedwell	X		X	X	X	X	O				P	ST/X	ot
LENTIBULARIACEAE	Bladderwort Family													
Pinguicula villosa L.	Small Butterwort				X				X	X	X	X	aSs/X	R(N)o
P. vulgaris L.	Butterwort	X	X	X	X	X	X	X	X	X	X	X	aST/X	Wa
Utricularia cornuta Michx.	Naked Bladderwort				X	X	X	X	X				sT/E	R(I)at
U. intermedia Hayne	Flat-leaf Bladderwort	X	X	X	X	X	X	X	X	X	X	X	aST/X	Wa
U. minor L.	Lesser Bladderwort			X	X	X	X	X	X	X	X	X	aST/X	Wo
U. vulgaris L. (var. *americana* A.Gray)	Common Bladderwrt	X	X	X	X	X	X	X	X	X	X	X	ST/X	Wa
U. X*ochroleuca* Hartm. (*U. intermedia* X *U. minor*)											X		ST/X	R(NC)r
PLANTAGINACEAE	Plantain Family													
Plantago lanceolata L.	English Plantain							X					–	(Int)r
P. major L. (incl. var. *ungavensis* Lepage)	Common Plantain	X	X	X	X	X	X	X	X			X	sT/X	Wa
P. maritima L. (*P. juncoides* Lam.)	Seaside Plantain	X	X							X	X	X	aST/X	R(C)a
RUBIACEAE	Madder Family													
Galium asprellum Michx.	Rough Bedstraw	X				X	X	X					T/E	R(S)o
G. boreale L.	Northern Bedstraw	X		X	X	X	X	X			X	X	aST/X	Wa
G. brandegei A.Gray (incl. *G. trifidum* var. *halophilum* Fern. & Wieg.)						X		X		X	X	X	aST/X	o
G. labradoricum (Wieg.) Wieg.	Labrador Bedstraw	X	X	X	X	X	X	X	X		X	X	sT/X	Wa
Galium trifidum L.	Northern Three-lobe Bedstraw	X	O		X	X	X	X	X		X	X	aST/X	Wo
G. tinctorium L.	Southern Three-lobe Bedstraw	X			X	X	X	X					aST/E	ot
G. triflorum Michx.	Sweet Bedstraw	X				X	X	X					aST/X	R(S)o
G. verum L.	Yellow Bedstraw							X					–	(Int)r
CAPRIFOLIACEAE	Honeysuckle Family													
Diervilla lonicera Mill.	Bush Honeysuckle						X	X					T/E	R(SI)o
Linnaea borealis L. (var. *americana* (Forbes) Rehd.)	Twinflower	X	X	O	X	X	X	X	X		X	X	aST/X	Wa
Lonicera dioica L. (incl. var. *glaucescens* (Rydb.) Butters)	Glaucous Honeysuckle			X		X	X	X				P	sT/X	ot

Taxa		James Bay Coast – HB1	Akimiski Island – HB2	Severn Basin – HB3	Winisk-Ekwan Basin – HB4	Attawapiskat Basin – HB5	Albany Basin – HB6	Moose Basin – HB7	Sutton Ridges – HB8	Cape Henrietta-Maria – MT1	Ont. Maritime Tundra – MT2	Manitoba Lowland – MT3	Continental Distribution	HBL Distribution and Frequency
L. hirsuta Eat.	Hairy Honeysuckle					X	X	X					sT/E	R(S)o
L. involucrata (Rich.) Banks	Northern Honeysuckle	X		X	X	X	X	X	X		X	P	sT/X	a(W)
L. oblongifolia (Goldie) Hook.	Fly Honeysuckle	X					X	X	X				sT/E	R(S)o
L. villosa (Michx.) Roem. & Schult. (*L. caerulea* L. var. *v.* (Michx.) Torr. & Gray; incl. *L. v.* var. *solonis* (Eat.) Fern.)	Waterberry	X		X	X	X	X	X	X			P	ST/X	at
Sambucus racemosa L. (var. *pubens* (Michx.) Koehne)	Red Elderberry						X	X					sT/X	R(Sl)o
Symphoricarpos albus (L.) Blake	Snowberry					X	X	X				P	sT/X	R(Sl)o
Viburnum edule (Michx.) Raf.	Low-bush Cranberry	X		X	X	X	X	X	X		X	X	ST/X	Wa
V. trilobum Marsh. (*V. opulus* L. var. *americanum* Ait.)	High-bush Cranberry						X	X					T/X	R(S)o
VALERIANACEAE	Valerian Family													
Valeriana dioica L. (var. *sylvatica* (Rich.) A.Gray; *V. septentrionalis* Rydb.)	Northern Valerian	X	X	X	X	X	X	X	X	X	X	X	ST/X	Wo
CAMPANULACEAE	Bluebell Family													
Campanula aparinoides Pursh (var. *grandiflora* Holz.; *C. uliginosa* Rydb.)	Marsh Bellflower	X				X	X	X		O		O	ST/E	o(W)
C. rotundifolia L.	Harebell	X	X		X	O	X	X	X	X		X	aST/X	Wo
C. uniflora L.	Alpine Bluebell											X	AST/X	R(NWC)r
LOBELIACEAE	Lobelia Family													
Lobelia kalmii L.	Kalm Lobelia	X	X	X	X	X	X	X	X			X	sT/X	Wa
ASTERACEAE	Aster Family													
Achillea millefolium L.	Yarrow													
var. *millefolium*		X						X			O	O	–	(Int)o
var. *nigrescens* Meyer (*A. borealis* Bong.; *A. m.* ssp. *atrotegula* Boiv.)		X	O	X	X	X		O		X	X	X	aST/X	Wa
var. *occidentalis* DC. (*A. lanulosa* Nutt.; *A. m.* ssp. *l.* (Nutt.) Piper; *A. m.* ssp. *palludotegula* Boiv.)		X	X	X	X	X	X	X	X	X	X	X	aST/X	Wo
A. ptarmica L.	Sneezeweed							X					–	(Int)r
A. sibirica Ledeb.												P	ST/W	R(NWI)
Agoseris glauca (Pursh) Raf.	False Dandelion	X		X	X	O			X		X	X	ST/W	R(N)o

Taxa		James Bay Coast – HB1	Akimiski Island – HB2	Severn Basin – HB3	Winisk-Ekwan Basin – HB4	Attawapiskat Basin – HB5	Albany Basin – HB6	Moose Basin – HB7	Sutton Ridges – HB8	Cape Henrietta-Maria – MT1	Ont. Maritime Tundra – MT2	Manitoba Lowland – MT3	Continental Distribution	HBL Distribution and Frequency
Anaphalis margaritacea (L.) Benth. & Hook.	Pearly Everlasting					X		X			O		sT/X	o
Antennaria neglecta Greene (incl. *A. howellii* Greene, *A. neodioica* Greene, and *A. campestris* Rydb.)	Common Pussytoes					X		X	X			P	sT/X	ot
A. parvifolia Nutt. (incl. *A. nitida* Greene (type location))	Small-leaf Pussytoes	X										P	sT/X	R(JB)r
A. pulcherrima (Hook.) Greene	Showy Pussytoes	X	X	X	X	X	X	X	X	X	X	X	ST/X	Wo
A. rosea Greene	Rosy Pussytoes	X	X		X							X	ST/X	R(JB,NW)r
Arctium minus (Hill.) Bernh.	Lesser Burdock							X					–	(Int)r
Arnica attenuata M.Vahl (*Arnica alpina* (L.) Olin ssp. *attenuata* (Greene) Maguire and var. *ungavensis* Boivin)	Alpine Arnica			X	X	X			X	X	X	X	AST/X	R(N)a
A. chamissonis Less. (ssp. *foliosa* (Nutt.) Maguire; var. *angustifolia* Herder; *A. wilsonii* Rydb.)	Leafy Arnica	X			O	X	X	X		X	X	P	ST/X	Wo
A. lonchophylla Greene	Spear-leaf Arnica			X		X						X	ST/X	R(I)ot
Artemisia absinthium L.	Common Wormwood							X					–	(Int)r
A. biennis Willd.	Biennial Wormwood				X	X	X	X			X	X	ST/X	(W)o
A. borealis Pall. (*A. campestris* L. var. *wormskjoldii* (Bess.) Cronq.)	Northern Wormwood	O	X		O					X	X	X	aSs/X	R(NC)a
A. canadensis Michx. (*A. campestris* L. var. *scouleriana* (Bess.) Cronq., in part)	Plains Wormwood	X		X		X	X				X	X	aST/X	(W)o
A. caudata Michx. (incl. var. *calvens* Lun.; *A. campestris* L. var. *scouleriana* (Bess.) Cronq., in part)	Sagewort Wormwood						X	X					aST/X	R(S)o
A. ludoviciana Nutt. (var. *latifolia* (Bess.) Torr. & Gray, and var. *gnaphalodes* (Nutt.) Torr. & Gray)	Prairie Sage											X	T/W	(Int)r
A. tilesii Ledeb. (var. *elatior* Torr. & Gray; incl. *A. diversifolia* Rydb.)		X	X									X	aST/X	R(JB,NWC)a
Aster alpinus L. (var. *vierhapperi* Onno; *Diplactis alpina* (L.) Semple)	Alpine Aster										X		ST/W	R(NC)r
A. borealis (Torr. & Gray) Prov. (*A. junciformis* Rydb.)	Northern Bog Aster	X			X	X	X	X	X	X	X	X	ST/X	Wa

Taxa		James Bay Coast – HB1	Akimiski Island – HB2	Severn Basin – HB3	Winisk-Ekwan Basin – HB4	Attawapiskat Basin – HB5	Albany Basin – HB6	Moose Basin – HB7	Sutton Ridges – HB8	Cape Henrietta-Maria – MT1	Ont. Maritime Tundra – MT2	Manitoba Lowland – MT3	Continental Distribution	HBL Distribution and Frequency
A. brachyactis Blake	Rayless Aster	X	X			X							sT/X	R(JB)o
A. ciliolatus Lindl.	Fringed Blue Aster	X		X	X	X	X	X	X		X	X	sT/X	Wa
A. lanceolatus Willd.	Lance-leaf Aster	X		X		X	X	X				P	ST/E	(W)o
A. lateriflorus (L.) Britt.	Calico Aster						X	X					T/E	R(S)r
A. macrophyllus L.	Large-leaf Aster						X	X					T/E	R(Sl)a
A. modestus Lindl.	Western Bog Aster	X			X	X	X	X					ST/X	ot
A. puniceus L. (incl. var. *calderi* (Boivin) Lepage)	Swamp Aster	X		X		X	X	X			X	X	sT/X	Wo
A. robbynsianus Rous. (*A. johannensis* Fern. of regional reports; incl. *A. hesperius* A.Gray (var. *gaspensis* (Vict.) Boiv.), and *A. novi-belgii* L. of regional reports)	Long-leaf Blue Aster	X	X	X		X	X	X	X	X	X	X	sT/X	Wa
A. umbellatus Mill. (incl. var. *pubens* A.Gray; *Doellingeria umbellata* (Mill.) Nees)	Flat-topped White Aster	X		O	X	X	X	X					T/X	at
Bidens cernua L.	Nodding Beggar-ticks	X						X					sT/X	R(JB)o
B. hyperborea Greene (type location)	Coastal Beggar-ticks	X						X					sT/E	R(JB)r
Chrysanthemum arcticum L. (var. *polaris* (Hult.) Boivin)	Arctic Daisy	X	O							X	X	X	aSs/X	R(C)o
C. leucanthemum L.	Ox-eye Daisy		X			X	X	X				P	–	(Int)a
Cirsium arvense (L.) Scop.	Common Thistle	X					X	X				X	–	(Int)o
C. muticum Michx.	Swamp Thistle	X			X	X	X	X					T/E	ot
Conyza canadensis (L.) Cronq.	Horseweed							X					sT/X	(Int)r
Crepis tectorum L.	Hawk's-beard				X								–	(Int)r
Erigeron acris L. (var. *kamtschaticus* (DC.) Herder; incl. var. *asteroides* (Andrz.) DC. and var. *elatus* (Greene) Cronq.)	Northern Fleabane	X	X	X	X	X	X	X	X	X	X	X	ST/X	Wa
E. annuus (L.) Pers. (ssp. *strigosus* (Muhl.) Wagen.)	Whitetop Fleabane					X	X	X					ST/X	R(S)o
E. humilis Graham (*E. unalaschkensis* (DC.) Vierh; *E. uniflorus* L. var. *u.* (DC.) Ostenf.)	Purple Fleabane									X		X	ASs/X	R(NC)r
E. hyssopifolius Michx.	Hyssop Fleabane	X		X	X	X	X	X	X	X	X	X	ST/X	Wa

Taxa		James Bay Coast – HB1	Akimiski Island – HB2	Severn Basin – HB3	Winisk-Ekwan Basin – HB4	Attawapiskat Basin – HB5	Albany Basin – HB6	Moose Basin – HB7	Sutton Ridges – HB8	Cape Henrietta-Maria – MT1	Ont. Maritime Tundra – MT2	Manitoba Lowland – MT3	Continental Distribution	HBL Distribution and Frequency
E. lonchophyllus Hook.	Hirsute Fleabane	X	X							X	X	X	ST/X	R(C)o
E. philadelphicus L.	Philadelphia Fleabane	X		X	X	X	X	X					ST/X	ot
Eupatorium maculatum L. (incl. var. *foliosum* (Fern.) Wieg.)	Spotted Joe-Pye Weed	X		O	X	X	X	X					T/X	ot
Euthamia graminifolia (L.) Nutt. (*Solidago graminifolia* var. *major* (Michx.) Fern.)	Grass-leaf Goldenrod	X		X	X	X	X	X					sT/X	at
Helianthus annuus L.	Sunflower											O	T/W	(Int)r
Hieracium aurantiacum L.	Orange Hawkweed							X					–	(Int)r
H. kalmii L. (*H. canadense* Michx. var. *canadense* and var. *hirtirameum* Fern.)	Canada Hawkweed	X				X	X	X	X				T/X	ot
H. scabriusculum Schwein. (incl. var. *columbianum* (Rydb.) Lepage)	Northern Hawkweed	X		X	X	X	X	X	X	X	X	X	ST/X	Wa
Iva xanthifolia Nutt.	Marsh-elder											O	–	(Int)r
Lactuca biennis (Moench) Fern.	Tall Blue Lettuce	X						X					sT/X	R(S)r
L. pulchella (Pursh) DC. (*L. tatarica* (L.) C.A.Meyer ssp. *p.* (Pursh) Stebb.)	Blue Lettuce	X		X	X		X	X					ST/X	(W)o
Matricaria matricarioides (Less.) Port.	Pineapple-weed	X				X	X	X	X		O	X	–	(Int)a
Megalodonta beckii (Torr.) Greene (*Bidens beckii* Torr.)	Water Beggar-ticks						X						sT/X	R(S)r
Petasites frigidus (L.) Fries (var. *palmatus* (Ait.) Cronq.)	Palm-leaf Coltsfoot			X	X	X	X	X	X	O	X	X	ST/X	Wo
P. sagittatus (Banks) A.Gray	Spear-leaf Coltsfoot	X	X			X	O	X	X	X	X	X	aST/X	Wo
P. vitifolius Greene	Vine-leaf Coltsfoot			X	X	X	X	X	X		X	X	ST/X	(W)o
Prenanthes alba L.	Rattlesnake-root	X					X	X					T/E	R(S)o
P. racemosa Michx.	Glaucous White Lettuce	X			X	X	X	X				X	sT/X	at
Senecio aureus L.	Golden Ragwort	X				X	X	X				O	sT/E	R(S,NW)o
S. congestus (R.Br.) DC. (var. *palustris* (L.) Fern.)	Marsh Ragwort	X	X							X	X	X	AST/X	R(C)a
S. indecorus Greene	Rayless Ragwort	X		X	X		X	X	X		O	X	ST/X	Wo
S. pauperculus Michx.	Northern Ragwort		X	X	X	X	X	X	X	X	X	X	ST/X	Wa
S. vulgaris L.	Common Groundsel							X					–	(Int)r

Taxa	James Bay Coast – HB1	Akimiski Island – HB2	Severn Basin – HB3	Winisk-Ekwan Basin – HB4	Attawapiskat Basin – HB5	Albany Basin – HB6	Moose Basin – HB7	Sutton Ridges – HB8	Cape Henrietta-Maria – MT1	Ont. Maritime Tundra – MT2	Manitoba Lowland – MT3	Continental Distribution	HBL Distribution and Frequency
Solidago altissima L. Tall Goldenrod (*S. canadensis* L. var. *scabra* (Muhl.) Torr. & Gray)	X				X		X					ST/E	R(S)o
S. canadensis L. Canada Goldenrod (incl. *S. lepida* DC. var. *elongata* (Nutt.) Fern., *S. c.* var. *salebrosa* (Piper) Jones, *S. c.* var. *gilvocanescens* Rydb.)	X		X	X	X	X	X		X	O	X	ST/X	Wa
S. decumbens Greene Mountain Goldenrod (var. *oreophila* (Rydb.) Fern.; *S. spathulata* DC. var. *neomexicana* (A.Gray) Cronq.)											X	ST/W	R(NWI)r
S. hispida Muhl. Hairy Goldenrod (*S. bicolor* L. var. *concolor* Torr. & Gray; incl. var. *lanata* (Hook.) Fern.)			X	X	X	X	X	X			X	sT/E	R(I)at
S. juncea Ait. Early Goldenrod	X					X	X					T/E	R(S)o
S. multiradiata Ait. Northern Goldenrod	X	X	X	X	X			X	X	X	X	aST/X	Wo
S. nemoralis Ait. Gray Goldenrod						X	O					sT/X	R(SI)r
S. ptarmicoides (Nees) Boivin Sneezeweed Goldenrod (*Aster ptarmicoides* (Nees) Torr. & Gray)			X	X	X	X	X					T/X	R(I)ot
S. rugosa Mill. Rough-leaf Goldenrod							X					T/E	R(SI)r
S. serotina Ait. Late Goldenrod (*S. gigantea* Ait. var. *serotina* (Kuntze) Cronq.)	X						X					ST/X	R(S)o
S. uliginosa Nutt. Bog Goldenrod (incl. var. *peracuta* (Fern.) Friesn.; *S purshii* Porter)	X	X		X	X	X	X	X				sT/X	at
Sonchus arvensis L. Glandular Sow-thistle	X				X	X	X					–	(Int)o
S. uliginosus Bieb. Field Sow-thistle (*S. arvensis* var. *glabrescens* G., G. & W.)	X					X	X				X	–	(Int)o
Tanacetum huronense Nutt. Huron Tansy (*Chrysanthemum bipinnatum* L. ssp. *h.* (Nutt.) Hult.)	O		X	X		X	X		X	X	X	ST/X	Wo
Taraxacum ceratophorum (Ledeb.) DC. Northern Dandelion (subsection Ceratophora Hand.-Mazz. other than *T. lacerum*; incl. *T. lapponicum* Kihlm. and *T. dumentorum* Greene)		X		X	X			X	X	X	X	AST/X	(W)o
T. lacerum Greene	X	X	X	X						X	X	aST/X	R(N)o

Taxa		James Bay Coast – HB1	Akimiski Island – HB2	Severn Basin – HB3	Winisk-Ekwan Basin – HB4	Attawapiskat Basin – HB5	Albany Basin – HB6	Moose Basin – HB7	Sutton Ridges – HB8	Cape Henrietta-Maria – MT1	Ont. Maritime Tundra – MT2	Manitoba Lowland – MT3	Continental Distribution	HBL Distribution and Frequency
T. officinale Weber	Common Dandelion	X	X		X			X			O	X	–	(Int)o
Tripleurospermum phaeocephalum (Rupr.) Pobed. (*Matricaria ambigua* (Ledeb.) Kryl.; *M. maritima* L. ssp. *phaeocephala* (Rupr.) Rausch)		X	X							X	X	X	aS/X	R(NC)o

APPENDIX C

Excluded Records

Collection vouchers are cited in the checklist for particular data collection areas (X). Other literature reports were also considered. These reports, however, are not cited if the record deserves exclusion from the flora based on the redetermination of originally cited materials or the absence of any relevant voucher collections at CAN, DAO, TRT, WIN, MMMN, SSMF, TRTE, QFA, or the personal herbaria of P.F. Maycock, J.C. Ritchie, or others. Some of the most critical of these unsubstantiated records are listed below. Some unverified literature reports are, however, included in the checklist, on the basis of their nearby distribution range and habitat preferences. These are noted in the checklist with "O"; their eventual documentation seems highly probable.

Agropyron dasystachyum (Hook.) Scribn.

Reports from the Ontario Lowland in the Severn River basin were based on a voucher specimen (*Moir 747*, Fawn River at Mink Creek, CAN), which is X*Agroelymus hirtiflorus*.

Alisma triviale Pursh

Reported by Scoggan (1978) from near York Factory, but no specimen from the Manitoba Lowland was at CAN, DAO, TRT, WIN, MMMN, QFA, and no record of the particular specimen appears in Scoggan's Flora Canada files at CAN.

Andromeda glaucophylla Link

The report from the Manitoba Lowland (or vicinity) at Gillam by Scoggan (1979) was based on a specimen (*Schofield 981*, Gillam, 19 June 1950, WIN, DAO), which is *A. polifolia*.

Arnica rhizomata A.Nels.

The report from the Ontario Lowland at Renison, Moose River, by Fagerstrom (1948), was based on a specimen (*Hustich*, 11 Aug. 1946, HEL), which is *A. chamissonis* (ann. Riley, xerox TRT).

Aster cordifolius L.

The report from the Ontario Lowland at Renison, Moose River by Fagerstrom (1948) was based on a specimen (*Hustich 17*, 28 Aug. 1947, HEL), which is *A. ciliolatus* (ann. Riley, xerox TRT).

A. lanceolatus Willd.

Reports from Ontario maritime tundra coast (Moir 1958) were based on two collections; the only one found (*Moir 1406*, vicinity of mouth of Severn R., 14 Aug. 1952, MINN) was *A. borealis*.

Botrychium ascendens W.H.Wagner

Reports from the Ontario Lowland (e.g., Argus et al. 1982–1987) from Severn River mouth were based on a specimen (*Moir 1444,* CAN) originally determined as *B. simplex* var. *tenebrosum*, but redetermined as *B. lunaria* by Wagner in 1986.

B. simplex Hitch.

Reported from the Ontario Lowland from "1 mi up Severn" on the basis of the same voucher specimen (*Moir 1444*, CAN), which was also annotated to *B. lunaria* by W.H. Wagner (with whom I agree).

Bromus pumpellianus Scribn.

The report from the Manitoba Lowland by Scoggan (1957) (and presumably the mapping by Porsild and Cody 1980) was based on a specimen (*Buller 114*, Churchill, 15 August 1939, DAO), which is *B. inermis*.

Caltha natans Pall.

Moir (1958) reported this species in the northwestern Ontario Lowland on the basis of a specimen (*Scott*, Severn River, TRT) since annotated to *C. palustris*, and another specimen (*Moir 2116A*, Black Duck R., July 1953, MINN), which is also *C. palustris*.

Carex angustior Mack.

Reports from the Ontario Lowland by Dutilly and Lepage (1963) and Lepage (1966) are probably referable to *C. interior* or *C. sterilis*. No relevant vouchers of *C. angustior* were found at CAN, DAO, or TRT. This taxon is now largely referable to *C. echinata* (Reznicek 1978), which is not reported from the Lowland.

C. haydenii Dewey

Reported from the Ontario Lowland from the Renison area, Moose River, by Fagerstrom (1948; cited by Scoggan 1978) and from the Abitibi River by Dutilly and Lepage (1963), but no voucher specimens were at TRT, CAN, DAO, QFA, or TRTE.

C. lachenalii Schkuhr

Reported from the Manitoba Lowland at Churchill by Scoggan (1978) and Johnson (1987), but no voucher seen at CAN, DAO, TRT, WIN, MMMN, QFA, or herbarium of J.C. Ritchie.

C. langeana Fern.

Reported from the Manitoba Lowland at Churchill by Scoggan (1978) and Johnson (1987), but no voucher seen at CAN, DAO, TRT, WIN, MMMN, QFA, or herbarium of J.C. Ritchie.

C. membranacea Hook.

Reported from the Manitoba Lowland at Churchill by Scoggan (1978) and Porsild and Cody (1980; mapping) based on specimen (*Schofield and Crum 6776*, Churchill, 17 July 1956, CAN), which is *C. rotundata*.

C. misandra R. Br.

Reported from the Manitoba Lowland at Churchill by Porsild (1957) and Scoggan (1978), but no specimen seen at CAN, DAO, TRT, WIN, MMMN, or QFA, and no clarification in Scoggan's Flora Canada files at CAN. Gardner (1946) indicates his collection No. *109B* is the relevant collection, but *Gardner 109* (QFA) is *C. atrofusca*.

C. nigra (L.) Reich.

The report by Dutilly and Lepage (1963) for the Ontario Lowland at Albany is improbable and, so far, unsubstantiated.

C. oligosperma Michx.

Reported from the Manitoba Lowland at Churchill by Scoggan (1957), but the only voucher seen (Beckett, TYPE of var. *churchilliana* Raymond, WIN) has been annotated to *C. livida*.

C. rossii F. Boott

Cited by Scoggan (1957, 1978) from the Manitoba Lowland at Gillam, based on a Schofield collection. The only Schofield collection found was *Schofield 1284*, Gillam, 29 July 1950 (DAO), which is *C. deflexa*.

C. rotundata Wahl.

Previously identified specimens from the Ontario Lowland (*Sims 2737A*, *Riley 7271*, TRT; *Lumsden*, CAN) are all *C. saxatilis* (B. Ford, personal communication).

C. supina Willd.

Reported from the Manitoba Lowland at Churchill by Porsild (1957) and Scoggan (1978), but no voucher at CAN, DAO, TRT, WIN, or MMMN.

C. tenera Dewey

Mapped from the Manitoba Lowland in Gillam area by Porsild and Cody (1980), but no voucher material is available from either Gillam or the Manitoba Lowland at CAN, DAO, TRT, WIN, or MMMN.

C. tribuloides Wahl.

The report from the Ontario Lowland at Renison, Moose River (Fagerstrom 1948), was based on a specimen (*Hustich 49*, 28 August 1947, HEL), which is *C. bebbii* (ann. Riley, xerox TRT).

Cerastium beeringianum Cham. & Schlecht.

The mapping of this species from the Manitoba Lowland at Churchill by Porsild and Cody (1980) is based on no voucher specimens at CAN, DAO, WIN, MMMN, or TRT.

Coptis trifolia (L.) Salisb.

Mapped immediately upstream of York Factory by Porsild and Cody (1980), but there is no voucher material from the Manitoba Lowland *s.s.* at CAN, DAO, TRT, WIN, or MMMN; it is certainly expected to occur in the Manitoba Lowland.

Cypripedium acaule Ait.

Reportedly collected by Wilson from the Ontario Lowland along the Kapiskau River (Macoun 1906), but no voucher specimens are extant at CAN, DAO, TRT, or QFA.

Deschampsia flexuosa (L.) Trin.

The report from the Ontario Lowland at Renison, Moose River, by Fagerstrom (1948) was based on a specimen (Hustich, 1 August 1948, HEL), which is *D. cespitosa* (ann. Riley, xerox TRT).

Draba cinerea Adams

Reported from Churchill by Scoggan (1957) and from the Manitoba Lowland by Ritchie (1962), based on a specimen (*Beckett 184a*, Churchill, 1 August 1953, WIN) annotated to *D. glabella* by Riley (1981). The CAN duplicate of this specimen was annotated by Boivin (1968–1969) to *D. lanceolata* (*D. cana*), but doesn't have the stellate silicles of that taxon.

D. lanceolata Royle (*D. cana* Rydb.)

Reported from Churchill by White and Johnson (1980), based on specimen (*Johnson 73-374*, Churchill area, August 1973, WIN, MMMN), which is *D. incana;* and reported from northern Manitoba (Boivin 1968–1969) and the Churchill area (Porsild and Cody 1980, mapped) presumably on the basis of the specimen (*Beckett 184a*) discussed under *D. cinerea*.

Drosera intermedia Hayne

All the Lowland specimens seen (*Hustich 1436*, *Baldwin 7451*; CAN) are referable to *D. anglica*.

Galium aparine L.

Reports from the Lowland by Dutilly et al. (1962) and Scoggan (1957) require verification; no specimens are at CAN, DAO, TRT, QFA, WIN, MMMN, or the personal herbarium of J.C. Ritchie. The Manitoba Lowland report from Gillam was based on a specimen (*Schofield 1165,* CAN, WIN), which is *G. labradoricum.*

Geum macrophyllum Willd. var. *macrophyllum*

The report from the Ontario Lowland from Moose Factory by Dutilly and Lepage (1963) is probably referable to var. *perincisum*; no Lowland specimens of the typical variety are at CAN, DAO, TRT, or QFA.

Goodyera tesselata Lodd.

The report from the Ontario Lowland at Renison, Moose River, by Fagerstrom (1948) was based on a specimen (*Hustich*, 11 August 1946, HEL), which is *G. repens* (ann. Riley, xerox TRT).

Hieracium canadense Michx.

Reported from the Manitoba Lowland by Ritchie (1962), presumably based on a specimen (*Schofield 1417*, Gillam, 11 August 1950, WIN) annotated by E. Lepage to *H. scabriusculum.*

Iris versicolor L.

Reported from the Manitoba Lowland by Ritchie (1962) but no voucher material seen at CAN, DAO, TRT, WIN, MMMN, or the personal herbarium of J.C. Ritchie.

Juncus biglumis L.

The report from the Ontario Lowland at Cape Henrietta Maria (Dutilly et al. 1954) is based on a specimen (*Dutilly and Lepage 31193*, QFA), which is *J. alpinus* (ann. Riley 1981).

Loiseleuria procumbens (L.) Desv.

The report from the Ontario Lowland at Hawley Lake (Kershaw and Rouse 1971) is not based on any collected voucher specimens (K. Kershaw, personal communication), nor has it been collected or reported by other workers.

Lonicera canadensis Bartr. ex Marsh.

The report from the Ontario Lowland at Renison, Moose River, by Hustich (1955; repeated by Dutilly and Lepage 1963, Baldwin et al. 1959, and Scoggan 1979, but not reported by Fagerstrom 1948) is not based on any voucher specimens at HEL, CAN, DAO, TRT, or QFA.

Luzula hyperborea R. Br. (*L. nivalis* (Laest.) Beurl., *L. arctica* Blytt)

Reported from the Manitoba Lowland at Churchill by Boivin (1979) on the basis of a specimen (*Gardner 48*, Churchill, 3 July 1930; QFA), which has the narrower leaves, frequent leaf hairs, and longer inflorescence bract of *L. multiflora* var. *frigida.*

Minuartia rossii (R. Br.) Graebn.

Reported from the Manitoba Lowland by Ritchie (1962) and from Churchill by Scoggan (1978; Scoggan's Flora Canada files at CAN cite a literature reference, Schofield 1958), but no voucher material at CAN, DAO, TRT, WIN, MMMN, or the personal herbarium of J.C. Ritchie.

Parnassia glauca Raf. (*P. caroliniana* Michx.)

The reports from the Ontario Lowland at Renison, Moose River, by Hustich (1955) and Scoggan (1978) are not based on vouchers at HEL, CAN, DAO, QFA, or TRT, and were not reported by Dawson (1973).

Potentilla hyparctica Malte

The only Lowland specimen seen at TRT, CAN or DAO was *Brown 18*, Ft. Churchill, 1950, CAN, which was correctly annotated to *P. nivea* by Boivin. The mapping of the species by Hultén (1968) from Akimiski Island is also unsubstantiated and was probably based on a Bear Island collection (*Coates*, 1947, CAN), which is not in the Lowland as considered here.

Potamogeton zosteriformis Fern.

Reported from the Manitoba Lowland at York Factory by Scoggan (1978) on the basis of a citation by Hooker (1839), but there is no voucher specimen at CAN, DAO, QFA, TRT, WIN, or MMMN.

Puccinellia angustata (Krecz.) Kitagawa

Dutilly & Lepage 16915, Winisk (CAN), was cited by Lepage (1966) as *P. hauptiana* (Krecz.) Kitagawa and later redetermined to *P.a.* by E. Lepage (personal communication). The specimen appears to be a variant of *P. vaginata s.l.*

P. coarctata Fern.

Dutilly & Lepage 36741, Pointe Mesakonan (DAO), is better referred to *P. lucida*; it does not correspond with *P. coarctata*, a paratype of which is at CAN.

P. deschampsioides Sorens.

Gillett 2196, Churchill (DAO), was tentatively annotated to this species by Sorensen, but the specimen closely resembles *P. lucida*. *Gillett 2223*, Fort Churchill (WIN) was also re-determined by J.R. Swallen as *P. lucida.*

P. distans (L.) Parl.

A specimen from the Ontario Lowland (*Scott*, "near Ft. Severn") was cited by Moir (1958), but is not at TRT, CAN or DAO, and is almost certainly referable to *P. lucida*, the native species most resembling it. *P. distans* is a western halophyte sparingly introduced eastward in disturbed saline sites, the closest of which is at Cochrane, Ontario, in the railroad yards (TRT).

P. laurentiana Fern. & Weath.

The report from the Ontario Lowland from western James Bay (Scoggan 1978) is probably also referable to *P. lucida*, and no voucher specimen is at CAN, DAO, or TRT.

P. macra Fern. & Weath.

The report from the Ontario Lowland from the western James Bay coast (Dutilly et al. 1954) is based on specimens, for example at CAN, which are referable to *P. lucida*.

P. nuttalliana (Schultes) Hitchc.

Mapped by Hultén (1968) from the southern Lowland, but the only Lowland specimen found (CAN) was *P. lucida*.

Ribes americanum Mill.

The report from the Ontario Lowland at Renison, Moose River, by Fagerstrom (1948) was based on a specimen (*Hustich*, 11 August 1946, HEL), which is *R. hudsonianum* (ann. Riley, xerox TRT).

Rumex arcticus Trautv.

Mapped as disjunct to Churchill by Porsild and Cody (1980) and cited as such by Scoggan (1957, 1978), this species was not reported from the Manitoba Lowland by Dawson (1979) or Boivin (1968–1969); the citations were probably based on a single immature specimen at CAN, indeterminable.

Salix athabascensis Raup

Mapped from the Ontario Lowland in the Attawapiskat area by Porsild and Cody (1980) on the basis of a specimen (*Dutilly & Lepage 15466*, DAO) which Ball in 1946 and Argus in 1981 considered a possible hybrid with *S. pedicellaris*; the specimen is the isotype of *S. jamesensis* Lepage.

S. brachycarpa Nutt. ssp. *niphoclada* (Rydb.) Argus

Mapped by Porsild (1957) and Porsild and Cody (1980) (*sub S. niphoclada*) for Churchill, but excluded from the Lowland flora by G. Argus (personal communication).

S. glauca L. ssp. *acutifolia* (Hook.) Hult.

Mapped by Porsild and Cody (1980) for the Lowland based on specimens, for examples *Moir 1321* (CAN) and *Schofield and Crum 6874* (CAN), which deserve annotation (G. Argus, personal communication) to *S. athabascensis* or *S. glauca* var. *callicarpaea*.

S. herbacea L.

Mapped by Porsild (1957) and cited by Scoggan (1978) from Churchill, but without Lowland voucher material at CAN, DAO, TRT, or WIN. The species is largely restricted to Precambrian sites in the southern part of its range and may eventually be found on the Sutton Ridges or other Precambrian outliers in the Lowland.

Salix richardsonii Hook.

Manitoba Lowland reports from Churchill are based on specimens dating from the early 1900s (see Raup 1943), and others such as *Beckett 50*, Churchill, 28 July 1956 (DAO), which are better considered the intergrading northeastern *S. calcicola*. Map records from Cape Henrietta Maria (Porsild and Cody 1980) are based on specimens (e.g., *Lumsden 32*, Site 415, CAN) annotated to *S. calcicola* by G. Argus.

Scirpus validus Vahl

Reported from the Manitoba Lowland by Ritchie (1962) but no voucher material seen at CAN, DAO, TRT, WIN, MMMN, or the personal herbarium of J.C. Ritchie.

Sibbaldia procumbens L.

The report from the Ontario Lowland at Hawley Lake (Kershaw and Rouse 1971) was not based on any collected voucher material (K. Kershaw, personal communication) nor has it been collected or reported in the Lowland by other workers; it may have been a field misidentification of *Potentilla tridentata*.

Silene acaulis L.

The reports from the Ontario Lowland at East Pen Island (Kershaw and Rouse 1973) and near Cape Henrietta Maria (Neal and Kershaw 1973) were not based on any collected voucher material (K. Kershaw, personal communication) nor has the species been collected or reported in the Lowland by other workers.

Sparganium angustifolium Michx.

Reported from the Manitoba Lowland at York Factory by Scoggan (1957), based on a specimen (*Scoggan 6012a*, York Factory, 22–26 July 1949, WIN), which is *S. hyperboreum*; the only other relevant specimen from CAN, DAO, TRT, WIN, and MMMN is from Gillam (DAO).

Thalictrum polygamum Muhl.

The report from the Ontario Lowland at Renison, Moose River, by Hustich (1955) was also the basis of the citation by Dutilly and Lepage (1963); it is not substantiated by collected specimens. The species was not listed among Hustich's voucher materials by Fagerstrom (1948), and Baldwin's Mammamattawa collection (*Baldwin 6474*, 20 July 1956, CAN, TRT) is *T. dasycarpum*.

Trillium undulatum Willd.

The report from the Ontario Lowland at Renison, Moose River, by Fagerstrom (1948) was based on a specimen (*Hustich*, 21 June 1947, HEL), which is *T. cernuum* (ann. Riley, xerox at TRT).

Vaccinium oxycoccus L. var. *microphyllum* (Lange) Rouss. & Raym.

Mapped from the Manitoba Lowland in the Churchill area by Porsild and Cody (1980; *sub Oxycoccus quadripetalus* Gil.), but without voucher specimens at CAN, DAO, TRT, WIN, or MMMN.